T0310590

'An understanding of the remarkable properties of the Poisson process is essential for anyone interested in the mathematical theory of probability or in its many fields of application. This book is a lucid and thorough account, rigorous but not pedantic, and accessible to any reader familiar with modern mathematics at first-degree level. Its publication is most welcome.'

— J. F. C. Kingman, *University of Bristol*

'I have always considered the Poisson process to be a cornerstone of applied probability. This excellent book demonstrates that it is a whole world in and of itself. The text is exciting and indispensable to anyone who works in this field.'

— Dietrich Stoyan, *TU Bergakademie Freiberg*

'Last and Penrose's *Lectures on the Poisson Process* constitutes a splendid addition to the monograph literature on point processes. While emphasising the Poisson and related processes, their mathematical approach also covers the basic theory of random measures and various applications, especially to stochastic geometry. They assume a sound grounding in measure-theoretic probability, which is well summarised in two appendices (on measure and probability theory). Abundant exercises conclude each of the twenty-two "lectures" which include examples illustrating their "course" material. It is a first-class complement to John Kingman's essay on the Poisson process.'

— Daryl Daley, *University of Melbourne*

'Pick n points uniformly and independently in a cube of volume n in Euclidean space. The limit of these random configurations as $n \to \infty$ is the Poisson process. This book, written by two of the foremost experts on point processes, gives a masterful overview of the Poisson process and some of its relatives. Classical tenets of the theory, like thinning properties and Campbell's formula, are followed by modern developments, such as Liggett's extra heads theorem, Fock space, permanental processes and the Boolean model. Numerous exercises throughout the book challenge readers and bring them to the edge of current theory.'

— Yuval Peres, *Principal Researcher, Microsoft Research, and Foreign Associate, National Academy of Sciences*

Lectures on the Poisson Process

The Poisson process, a core object in modern probability, enjoys a richer theory than is sometimes appreciated. This volume develops the theory in the setting of a general abstract measure space, establishing basic results and properties as well as certain advanced topics in the stochastic analysis of the Poisson process. Also discussed are applications and related topics in stochastic geometry, including stationary point processes, the Boolean model, the Gilbert graph, stable allocations and hyperplane processes. Comprehensive, rigorous, and self-contained, this text is ideal for graduate courses or for self-study, with a substantial number of exercises for each chapter. Mathematical prerequisites, mainly a sound knowledge of measure-theoretic probability, are kept in the background, but are reviewed comprehensively in an appendix. The authors are well-known researchers in probability theory, especially stochastic geometry. Their approach is informed both by their research and by their extensive experience in teaching at undergraduate and graduate levels.

GÜNTER LAST is Professor of Stochastics at the Karlsruhe Institute of Technology. He is a distinguished probabilist with particular expertise in stochastic geometry, point processes and random measures. He has coauthored a research monograph on marked point processes on the line as well as two textbooks on general mathematics. He has given many invited talks on his research worldwide.

MATHEW PENROSE is Professor of Probability at the University of Bath. He is an internationally leading researcher in stochastic geometry and applied probability and is the author of the influential monograph *Random Geometric Graphs*. He received the Friedrich Wilhelm Bessel Research Award from the Humboldt Foundation in 2008, and has held visiting positions as guest lecturer in New Delhi, Karlsruhe, San Diego, Birmingham and Lille.

INSTITUTE OF MATHEMATICAL STATISTICS TEXTBOOKS

IMS Textbooks give introductory accounts of topics of current concern suitable for advanced courses at master's level, for doctoral students and for individual study. They are typically shorter than a fully developed textbook, often arising from material created for a topical course. Lengths of 100–290 pages are envisaged. The books typically contain exercises.

Other Books in the Series

1. *Probability on Graphs*, by Geoffrey Grimmett
2. *Stochastic Networks*, by Frank Kelly and Elena Yudovina
3. *Bayesian Filtering and Smoothing*, by Simo Särkkä
4. *The Surprising Mathematics of Longest Increasing Subsequences*, by Dan Romik
5. *Noise Sensitivity of Boolean Functions and Percolation*, by Christophe Garban and Jeffrey E. Steif
6. *Core Statistics*, by Simon N. Wood
7. *Lectures on the Poisson Process*, by Günter Last and Mathew Penrose

Lectures on the Poisson Process

GÜNTER LAST
Karlsruhe Institute of Technology

MATHEW PENROSE
University of Bath

Shaftesbury Road, Cambridge CB2 8EA, United Kingdom

One Liberty Plaza, 20th Floor, New York, NY 10006, USA

477 Williamstown Road, Port Melbourne, VIC 3207, Australia

314–321, 3rd Floor, Plot 3, Splendor Forum, Jasola District Centre, New Delhi – 110025, India

103 Penang Road, #05–06/07, Visioncrest Commercial, Singapore 238467

Cambridge University Press is part of Cambridge University Press & Assessment, a department of the University of Cambridge.

We share the University's mission to contribute to society through the pursuit of education, learning and research at the highest international levels of excellence.

www.cambridge.org
Information on this title: www.cambridge.org/9781107088016

DOI: 10.1017/9781316104477

First published 2018

A catalogue record for this publication is available from the British Library

Library of Congress Cataloging-in-Publication data
Names: Last, Günter, author. | Penrose, Mathew, author.
Title: Lectures on the Poisson process / Günter Last, Karlsruhe Institute of Technology, Mathew Penrose, University of Bath.
Description: Cambridge : Cambridge University Press, 2018. | Series: Institute of Mathematical Statistics textbooks | Includes bibliographical references and index.
Identifiers: LCCN 2017027687 | ISBN 9781107088016
Subjects: LCSH: Poisson processes. | Stochastic processes. | Probabilities.
Classification: LCC QA274.42 .L36 2018 | DDC 519.2/4–dc23
LC record available at https://lccn.loc.gov/2017027687

ISBN 978-1-107-08801-6 Hardback
ISBN 978-1-107-45843-7 Paperback

To Our Families

Contents

Preface

The Poisson process generates point patterns in a purely random manner. It plays a fundamental role in probability theory and its applications, and enjoys a rich and beautiful theory. While many of the applications involve point processes on the line, or more generally in Euclidean space, many others do not. Fortunately, one can develop much of the theory in the abstract setting of a general measurable space.

We have prepared the present volume so as to provide a modern textbook on the general Poisson process. Despite its importance, there are not many monographs or graduate texts with the Poisson process as their main point of focus, for example by comparison with the topic of Brownian motion. This is probably due to a viewpoint that the theory of Poisson processes on its own is too insubstantial to merit such a treatment. Such a viewpoint now seems out of date, especially in view of recent developments in the stochastic analysis of the Poisson process. We also extend our remit to topics in stochastic geometry, which is concerned with mathematical models for random geometric structures [4, 5, 23, 45, 123, 126, 147]. The Poisson process is fundamental to stochastic geometry, and the applications areas discussed in this book lie largely in this direction, reflecting the taste and expertise of the authors. In particular, we discuss Voronoi tessellations, stable allocations, hyperplane processes, the Boolean model and the Gilbert graph.

Besides stochastic geometry, there are many other fields of application of the Poisson process. These include Lévy processes [10, 83], Brownian excursion theory [140], queueing networks [6, 149], and Poisson limits in extreme value theory [139]. Although we do not cover these topics here, we hope nevertheless that this book will be a useful resource for people working in these and related areas.

This book is intended to be a basis for graduate courses or seminars on the Poisson process. It might also serve as an introduction to point process theory. Each chapter is supposed to cover material that can be presented (at least in principle) in a single lecture. In practice, it may not always be possible to get through an entire chapter in one lecture; however, in most chapters the most essential material is presented in the early part of the chapter, and the later part could feasibly be left as background reading if necessary. While it is recommended to read the earlier chapters in a linear order at least up to Chapter 5, there is some scope for the reader to pick and choose from the later chapters. For example, a reader more interested in stochastic geometry could look at Chapters 8–11 and 16–17. A reader wishing to focus on the general abstract theory of Poisson processes could look at Chapters 6, 7, 12, 13 and 18–21. A reader wishing initially to take on slightly easier material could look at Chapters 7–9, 13 and 15–17.

The book divides loosely into three parts. In the first part we develop basic results on the Poisson process in the general setting. In the second part we introduce models and results of stochastic geometry, most but not all of which are based on the Poisson process, and which are most naturally developed in the Euclidean setting. Chapters 8, 9, 10, 16, 17 and 22 are devoted exclusively to stochastic geometry while other chapters use stochastic geometry models for illustrating the theory. In the third part we return to the general setting and describe more advanced results on the stochastic analysis of the Poisson process.

Our treatment requires a sound knowledge of measure-theoretic probability theory. However, specific knowledge of stochastic processes is not assumed. Since the focus is always on the probabilistic structure, technical issues of measure theory are kept in the background, whenever possible. Some basic facts from measure and probability theory are collected in the appendices.

When treating a classical and central subject of probability theory, a certain overlap with other books is inevitable. Much of the material of the earlier chapters, for instance, can also be found (in a slightly more restricted form) in the highly recommended book [75] by J.F.C. Kingman. Further results on Poisson processes, as well as on general random measures and point processes, are presented in the monographs [6, 23, 27, 53, 62, 63, 69, 88, 107, 134, 139]. The recent monograph Kallenberg [65] provides an excellent systematic account of the modern theory of random measures. Comments on the early history of the Poisson process, on the history of the main results presented in this book and on the literature are given in Appendix C.

In preparing this manuscript we have benefited from comments on earlier versions from Daryl Daley, Fabian Gieringer, Christian Hirsch, Daniel Hug, Olav Kallenberg, Paul Keeler, Martin Möhle, Franz Nestmann, Jim

Pitman, Matthias Schulte, Tomasz Rolski, Dietrich Stoyan, Christoph Thäle, Hermann Thorisson and Hans Zessin, for which we are most grateful. Thanks are due to Franz Nestmann for producing the figures. We also wish to thank Olav Kallenberg for making available to us an early version of his monograph [65].

Günter Last
Mathew Penrose

Symbols

$\mathbb{Z} = \{0, 1, -1, 2, -2, \dots\}$	set of integers
$\mathbb{N} = \{1, 2, 3, 4, \dots\}$	set of positive integers
$\mathbb{N}_0 = \{0, 1, 2, \dots\}$	set of non-negative integers
$\overline{\mathbb{N}} = \mathbb{N} \cup \{\infty\}$	extended set of positive integers
$\overline{\mathbb{N}}_0 = \mathbb{N}_0 \cup \{\infty\}$	extended set of non-negative integers
$\mathbb{R} = (-\infty, \infty), \mathbb{R}_+ = [0, \infty)$	real line (resp. non-negative real half-line)
$\overline{\mathbb{R}} = \mathbb{R} \cup \{-\infty, \infty\}$	extended real line
$\overline{\mathbb{R}}_+ = \mathbb{R}_+ \cup \{\infty\} = [0, \infty]$	extended half-line
$\mathbb{R}(\mathbb{X}), \mathbb{R}_+(\mathbb{X})$	$\mathbb{R}$-valued (resp. $\mathbb{R}_+$-valued) measurable functions on $\mathbb{X}$
$\overline{\mathbb{R}}(\mathbb{X}), \overline{\mathbb{R}}_+(\mathbb{X})$	$\overline{\mathbb{R}}$-valued (resp. $\overline{\mathbb{R}}_+$-valued) measurable functions on $\mathbb{X}$
u^+, u^-	positive and negative part of an $\overline{\mathbb{R}}$-valued function u
$a \wedge b, a \vee b$	minimum (resp. maximum) of $a, b \in \overline{\mathbb{R}}$
$\mathbf{1}\{\cdot\}$	indicator function
$a^{\oplus} := \mathbf{1}\{a \neq 0\}a^{-1}$	generalised inverse of $a \in \mathbb{R}$
$\operatorname{card} A = \lvert A \rvert$	number of elements of a set A
$[n]$	$\{1, \dots, n\}$
Σ_n	group of permutations of $[n]$
Π_n, Π_n^*	set of all partitions (resp. subpartitions) of $[n]$
$(n)_k = n \cdots (n - k + 1)$	descending factorial
δ_x	Dirac measure at the point x
$\mathbf{N}_{<\infty}(\mathbb{X}) \equiv \mathbf{N}_{<\infty}$	set of all finite counting measures on $\mathbb{X}$
$\mathbf{N}(\mathbb{X}) \equiv \mathbf{N}$	set of all countable sums of measures from $\mathbf{N}_{<\infty}$
$\mathbf{N}_l(\mathbb{X}), \mathbf{N}_s(\mathbb{X})$	set of all locally finite (resp. simple) measures in $\mathbf{N}(\mathbb{X})$
$\mathbf{N}_{ls}(\mathbb{X}) := \mathbf{N}_l(\mathbb{X}) \cap \mathbf{N}_s(\mathbb{X})$	set of all locally finite and simple measures in $\mathbf{N}(\mathbb{X})$
$x \in \mu$	short for $\mu\{x\} = \mu(\{x\}) > 0, \mu \in \mathbf{N}$
ν_B	restriction of a measure ν to a measurable set B

$\mathcal{B}(\mathbb{X})$	Borel σ-field on a metric space $\mathbb{X}$
$\mathcal{X}_b$	bounded Borel subsets of a metric space $\mathbb{X}$
$\mathbb{R}^d$	Euclidean space of dimension $d \in \mathbb{N}$
$\mathcal{B}^d := \mathcal{B}(\mathbb{R}^d)$	Borel σ-field on $\mathbb{R}^d$
λ_d	Lebesgue measure on $(\mathbb{R}^d, \mathcal{B}^d)$
$\|\cdot\|$	Euclidean norm on $\mathbb{R}^d$
$\langle \cdot, \cdot \rangle$	Euclidean scalar product on $\mathbb{R}^d$
$\mathcal{C}^d, \mathcal{C}^{(d)}$	compact (resp. non-empty compact) subsets of $\mathbb{R}^d$
$\mathcal{K}^d, \mathcal{K}^{(d)}$	compact (resp. non-empty compact) convex subsets of $\mathbb{R}^d$
$\mathcal{R}^d$	convex ring in $\mathbb{R}^d$ (finite unions of convex sets)
$K + x, K - x$	translation of $K \subset \mathbb{R}^d$ by x (resp. $-x$)
$K \oplus L$	Minkowski sum of $K, L \subset \mathbb{R}^d$
$V_0, \dots, V_d$	intrinsic volumes
$\phi_i = \int V_i(K)\, \mathbb{Q}(dK)$	i-th mean intrinsic volume of a typical grain
$B(x, r)$	closed ball with centre x and radius $r \geq 0$
$\kappa_d = \lambda_d(B^d)$	volume of the unit ball in $\mathbb{R}^d$
$<$	strict lexicographical order on $\mathbb{R}^d$
$l(B)$	lexicographic minimum of a non-empty finite set $B \subset \mathbb{R}^d$
$(\Omega, \mathcal{F}, \mathbb{P})$	probability space
$\mathbb{E}[X]$	expectation of a random variable X
$\mathbb{V}\mathrm{ar}[X]$	variance of a random variable X
$\mathbb{C}\mathrm{ov}[X, Y]$	covariance between random variables X and Y
L_η	Laplace functional of a random measure η
$\overset{d}{=}, \overset{d}{\to}$	equality (resp. convergence) in distribution

1

Poisson and Other Discrete Distributions

The Poisson distribution arises as a limit of the binomial distribution. This chapter contains a brief discussion of some of its fundamental properties as well as the Poisson limit theorem for null arrays of integer-valued random variables. The chapter also discusses the binomial and negative binomial distributions.

1.1 The Poisson Distribution

A random variable X is said to have a *binomial distribution* $\mathrm{Bi}(n, p)$ with parameters $n \in \mathbb{N}_0 := \{0, 1, 2, \ldots\}$ and $p \in [0, 1]$ if

$$\mathbb{P}(X = k) = \mathrm{Bi}(n, p; k) := \binom{n}{k} p^k (1-p)^{n-k}, \quad k = 0, \ldots, n, \tag{1.1}$$

where $0^0 := 1$. In the case $n = 1$ this is the *Bernoulli distribution* with parameter p. If $X_1, \ldots, X_n$ are independent random variables with such a Bernoulli distribution, then their sum has a binomial distribution, that is

$$X_1 + \cdots + X_n \overset{d}{=} X, \tag{1.2}$$

where X has the distribution $\mathrm{Bi}(n, p)$ and where $\overset{d}{=}$ denotes equality in distribution. It follows that the expectation and variance of X are given by

$$\mathbb{E}[X] = np, \qquad \mathbb{V}\mathrm{ar}[X] = np(1-p). \tag{1.3}$$

A random variable X is said to have a *Poisson distribution* $\mathrm{Po}(\gamma)$ with parameter $\gamma \geq 0$ if

$$\mathbb{P}(X = k) = \mathrm{Po}(\gamma; k) := \frac{\gamma^k}{k!} e^{-\gamma}, \quad k \in \mathbb{N}_0. \tag{1.4}$$

If $\gamma = 0$, then $\mathbb{P}(X = 0) = 1$, since we take $0^0 = 1$. Also we allow $\gamma = \infty$; in this case we put $\mathbb{P}(X = \infty) = 1$ so $\mathrm{Po}(\infty; k) = 0$ for $k \in \mathbb{N}_0$.

The Poisson distribution arises as a limit of binomial distributions as follows. Let $p_n \in [0, 1]$, $n \in \mathbb{N}$, be a sequence satisfying $np_n \to \gamma$ as $n \to \infty$, with $\gamma \in (0, \infty)$. Then, for $k \in \{0, \dots, n\}$,

$$\binom{n}{k} p_n^k (1 - p_n)^{n-k} = \frac{(np_n)^k}{k!} \cdot \frac{(n)_k}{n^k} \cdot (1 - p_n)^{-k} \cdot \left(1 - \frac{np_n}{n}\right)^n \to \frac{\gamma^k}{k!} e^{-\gamma}, \quad (1.5)$$

as $n \to \infty$, where

$$(n)_k := n(n - 1) \cdots (n - k + 1) \qquad (1.6)$$

is the k-th *descending factorial* (of n) with $(n)_0$ interpreted as 1.

Suppose X is a Poisson random variable with finite parameter γ. Then its expectation is given by

$$\mathbb{E}[X] = e^{-\gamma} \sum_{k=0}^{\infty} k \frac{\gamma^k}{k!} = e^{-\gamma} \gamma \sum_{k=1}^{\infty} \frac{\gamma^{k-1}}{(k-1)!} = \gamma. \qquad (1.7)$$

The *probability generating function* of X (or of $\text{Po}(\gamma)$) is given by

$$\mathbb{E}[s^X] = e^{-\gamma} \sum_{k=0}^{\infty} \frac{\gamma^k}{k!} s^k = e^{-\gamma} \sum_{k=0}^{\infty} \frac{(\gamma s)^k}{k!} = e^{\gamma(s-1)}, \quad s \in [0, 1]. \qquad (1.8)$$

It follows that the *Laplace transform* of X (or of $\text{Po}(\gamma)$) is given by

$$\mathbb{E}[e^{-tX}] = \exp[-\gamma(1 - e^{-t})], \quad t \geq 0. \qquad (1.9)$$

Formula (1.8) is valid for each $s \in \mathbb{R}$ and (1.9) is valid for each $t \in \mathbb{R}$. A calculation similar to (1.8) shows that the *factorial moments* of X are given by

$$\mathbb{E}[(X)_k] = \gamma^k, \quad k \in \mathbb{N}_0, \qquad (1.10)$$

where $(0)_0 := 1$ and $(0)_k := 0$ for $k \geq 1$. Equation (1.10) implies that

$$\mathbb{V}\text{ar}[X] = \mathbb{E}[X^2] - \mathbb{E}[X]^2 = \mathbb{E}[(X)_2] + \mathbb{E}[X] - \mathbb{E}[X]^2 = \gamma. \qquad (1.11)$$

We continue with a characterisation of the Poisson distribution.

Proposition 1.1 *An $\mathbb{N}_0$-valued random variable X has distribution* $\text{Po}(\gamma)$ *if and only if, for every function $f: \mathbb{N}_0 \to \mathbb{R}_+$, we have*

$$\mathbb{E}[Xf(X)] = \gamma \mathbb{E}[f(X + 1)]. \qquad (1.12)$$

Proof By a similar calculation to (1.7) and (1.8) we obtain for any function $f: \mathbb{N}_0 \to \mathbb{R}_+$ that (1.12) holds. Conversely, if (1.12) holds for all such functions f, then we can make the particular choice $f := \mathbf{1}_{\{k\}}$ for $k \in \mathbb{N}$, to obtain the recursion

$$k \mathbb{P}(X = k) = \gamma \mathbb{P}(X = k - 1).$$

This recursion has (1.4) as its only (probability) solution, so the result follows. □

1.2 Relationships Between Poisson and Binomial Distributions

The next result says that if X and Y are independent Poisson random variables, then $X+Y$ is also Poisson and the conditional distribution of X given $X+Y$ is binomial:

Proposition 1.2 *Let X and Y be independent with distributions* $\mathrm{Po}(\gamma)$ *and* $\mathrm{Po}(\delta)$, *respectively, with* $0 < \gamma+\delta < \infty$. *Then $X+Y$ has distribution* $\mathrm{Po}(\gamma+\delta)$ *and*

$$\mathbb{P}(X = k \mid X + Y = n) = \mathrm{Bi}(n, \gamma/(\gamma + \delta); k), \qquad n \in \mathbb{N}_0,\ k = 0, \ldots, n.$$

Proof For $n \in \mathbb{N}_0$ and $k \in \{0, \ldots, n\}$,

$$\begin{aligned}\mathbb{P}(X = k, X + Y = n) = \mathbb{P}(X = k, Y = n - k) &= \frac{\gamma^k}{k!}e^{-\gamma}\frac{\delta^{n-k}}{(n-k)!}e^{-\delta}\\ &= e^{-(\gamma+\delta)}\left(\frac{(\gamma+\delta)^n}{n!}\right)\binom{n}{k}\left(\frac{\gamma}{\gamma+\delta}\right)^k\left(\frac{\delta}{\gamma+\delta}\right)^{n-k}\\ &= \mathrm{Po}(\gamma+\delta; n)\,\mathrm{Bi}(n, \gamma/(\gamma+\delta); k),\end{aligned}$$

and the assertions follow. □

Let Z be an $\mathbb{N}_0$-valued random variable and let $Z_1, Z_2, \ldots$ be a sequence of independent random variables that have a Bernoulli distribution with parameter $p \in [0, 1]$. If Z and $(Z_n)_{n\geq 1}$ are independent, then the random variable

$$X := \sum_{j=1}^{Z} Z_j \tag{1.13}$$

is called a *p-thinning* of Z, where we set $X := 0$ if $Z = 0$. This means that the conditional distribution of X given $Z = n$ is binomial with parameters n and p.

The following partial converse of Proposition 1.2 is a noteworthy property of the Poisson distribution.

Proposition 1.3 *Let $p \in [0, 1]$. Let Z have a Poisson distribution with parameter $\gamma \geq 0$ and let X be a p-thinning of Z. Then X and $Z - X$ are independent and Poisson distributed with parameters $p\gamma$ and $(1 - p)\gamma$, respectively.*

Proof We may assume that $\gamma > 0$. The result follows once we have shown that

$$\mathbb{P}(X = m, Z - X = n) = \mathrm{Po}(p\gamma; m)\,\mathrm{Po}((1-p)\gamma; n), \quad m, n \in \mathbb{N}_0. \tag{1.14}$$

Since the conditional distribution of X given $Z = m + n$ is binomial with parameters $m + n$ and p, we have

$$\begin{aligned}\mathbb{P}(X = m, Z - X = n) &= \mathbb{P}(Z = m + n)\,\mathbb{P}(X = m \mid Z = m + n)\\ &= \left(\frac{e^{-\gamma}\gamma^{m+n}}{(m+n)!}\right)\binom{m+n}{m} p^m (1-p)^n\\ &= \left(\frac{p^m\gamma^m}{m!}\right)e^{-p\gamma}\left(\frac{(1-p)^n\gamma^n}{n!}\right)e^{-(1-p)\gamma},\end{aligned}$$

and (1.14) follows. □

1.3 The Poisson Limit Theorem

The next result generalises (1.5) to sums of Bernoulli variables with unequal parameters, among other things.

Proposition 1.4 *Suppose for $n \in \mathbb{N}$ that $m_n \in \mathbb{N}$ and $X_{n,1}, \dots, X_{n,m_n}$ are independent random variables taking values in $\mathbb{N}_0$. Let $p_{n,i} := \mathbb{P}(X_{n,i} \geq 1)$ and assume that*

$$\lim_{n\to\infty} \max_{1\leq i\leq m_n} p_{n,i} = 0. \tag{1.15}$$

Assume further that $\lambda_n := \sum_{i=1}^{m_n} p_{n,i} \to \gamma$ as $n \to \infty$, where $\gamma > 0$, and that

$$\lim_{n\to\infty} \sum_{i=1}^{m_n} \mathbb{P}(X_{n,i} \geq 2) = 0. \tag{1.16}$$

Let $X_n := \sum_{i=1}^{m_n} X_{n,i}$. Then for $k \in \mathbb{N}_0$ we have

$$\lim_{n\to\infty} \mathbb{P}(X_n = k) = \mathrm{Po}(\gamma; k). \tag{1.17}$$

Proof Let $X'_{n,i} := \mathbf{1}\{X_{n,i} \geq 1\} = \min\{X_{n,i}, 1\}$ and $X'_n := \sum_{i=1}^{m_n} X'_{n,i}$. Since $X'_{n,i} \neq X_{n,i}$ if and only if $X_{n,i} \geq 2$, we have

$$\mathbb{P}(X'_n \neq X_n) \leq \sum_{i=1}^{m_n} \mathbb{P}(X_{n,i} \geq 2).$$

By assumption (1.16) we can assume without restriction of generality that

$X'_{n,i} = X_{n,i}$ for all $n \in \mathbb{N}$ and $i \in \{1, \ldots, m_n\}$. Moreover it is no loss of generality to assume for each (n, i) that $p_{n,i} < 1$. We then have

$$\mathbb{P}(X_n = k) = \sum_{1 \le i_1 < i_2 < \cdots < i_k \le m_n} p_{n,i_1} p_{n,i_2} \cdots p_{n,i_k} \frac{\prod_{j=1}^{m_n}(1 - p_{n,j})}{(1 - p_{n,i_1}) \cdots (1 - p_{n,i_k})}. \quad (1.18)$$

Let $\mu_n := \max_{1 \le i \le m_n} p_{n,i}$. Since $\sum_{j=1}^{m_n} p_{n,j}^2 \le \lambda_n \mu_n \to 0$ as $n \to \infty$, we have

$$\log\left(\prod_{j=1}^{m_n}(1 - p_{n,j})\right) = \sum_{j=1}^{m_n}(-p_{n,j} + O(p_{n,j}^2)) \to -\gamma \text{ as } n \to \infty, \quad (1.19)$$

where the function $O(\cdot)$ satisfies $\limsup_{r \to 0} |r|^{-1}|O(r)| < \infty$. Also,

$$\inf_{1 \le i_1 < i_2 < \cdots < i_k \le m_n} (1 - p_{n,i_1}) \cdots (1 - p_{n,i_k}) \ge (1 - \mu_n)^k \to 1 \text{ as } n \to \infty. \quad (1.20)$$

Finally, with $\sum^{\neq}_{i_1,\ldots,i_k \in \{1,2,\ldots,m_n\}}$ denoting summation over all ordered k-tuples of distinct elements of $\{1, 2, \ldots, m_n\}$, we have

$$k! \sum_{1 \le i_1 < i_2 < \cdots < i_k \le m_n} p_{n,i_1} p_{n,i_2} \cdots p_{n,i_k} = \sum^{\neq}_{i_1,\ldots,i_k \in \{1,2,\ldots,m_n\}} p_{n,i_1} p_{n,i_2} \cdots p_{n,i_k},$$

and

$$0 \le \left(\sum_{i=1}^{m_n} p_{n,i}\right)^k - \sum^{\neq}_{i_1,\ldots,i_k \in \{1,2,\ldots,m_n\}} p_{n,i_1} p_{n,i_2} \cdots p_{n,i_k}$$
$$\le \binom{k}{2} \sum_{i=1}^{m_n} p_{n,i}^2 \left(\sum_{j=1}^{m_n} p_{n,j}\right)^{k-2},$$

which tends to zero as $n \to \infty$. Therefore

$$k! \sum_{1 \le i_1 < i_2 < \cdots < i_k \le m_n} p_{n,i_1} p_{n,i_2} \cdots p_{n,i_k} \to \gamma^k \text{ as } n \to \infty. \quad (1.21)$$

The result follows from (1.18) by using (1.19), (1.20) and (1.21). □

1.4 The Negative Binomial Distribution

A random element Z of $\mathbb{N}_0$ is said to have a *negative binomial distribution* with parameters $r > 0$ and $p \in (0, 1]$ if

$$\mathbb{P}(Z = n) = \frac{\Gamma(n + r)}{\Gamma(n + 1)\Gamma(r)}(1 - p)^n p^r, \quad n \in \mathbb{N}_0, \quad (1.22)$$

where the *Gamma function* $\Gamma\colon (0,\infty) \to (0,\infty)$ is defined by

$$\Gamma(a) := \int_0^\infty t^{a-1} e^{-t}\, dt, \quad a > 0. \tag{1.23}$$

(In particular $\Gamma(a) = (a-1)!$ for $a \in \mathbb{N}$.) This can be seen to be a probability distribution by Taylor expansion of $(1-x)^{-r}$ evaluated at $x = 1-p$. The probability generating function of Z is given by

$$\mathbb{E}[s^Z] = p^r(1 - s + sp)^{-r}, \quad s \in [0,1]. \tag{1.24}$$

For $r \in \mathbb{N}$, such a Z may be interpreted as the number of failures before the rth success in a sequence of independent Bernoulli trials. In the special case $r = 1$ we get the *geometric distribution*

$$\mathbb{P}(Z = n) = (1-p)^n p, \quad n \in \mathbb{N}_0. \tag{1.25}$$

Another interesting special case is $r = 1/2$. In this case

$$\mathbb{P}(Z = n) = \frac{(2n-1)!!}{2^n n!}(1-p)^n p^{1/2}, \quad n \in \mathbb{N}_0, \tag{1.26}$$

where we recall the definition (B.6) for $(2n-1)!!$. This follows from the fact that $\Gamma(n + 1/2) = (2n-1)!!\, 2^{-n}\sqrt{\pi}$, $n \in \mathbb{N}_0$.

The negative binomial distribution arises as a mixture of Poisson distributions. To explain this, we need to introduce the *Gamma distribution* with *shape parameter* $a > 0$ and *scale parameter* $b > 0$. This is a probability measure on $\mathbb{R}_+$ with Lebesgue density

$$x \mapsto b^a \Gamma(a)^{-1} x^{a-1} e^{-bx} \tag{1.27}$$

on $\mathbb{R}_+$. If a random variable Y has this distribution, then one says that Y is Gamma distributed with shape parameter a and scale parameter b. In this case Y has Laplace transform

$$\mathbb{E}[e^{-tY}] = \left(\frac{b}{b+t}\right)^a, \quad t \geq 0. \tag{1.28}$$

In the case $a = 1$ we obtain the *exponential distribution* with parameter b. Exercise 1.11 asks the reader to prove the following result.

Proposition 1.5 *Suppose that the random variable $Y \geq 0$ is Gamma distributed with shape parameter $a > 0$ and scale parameter $b > 0$. Let Z be an $\mathbb{N}_0$-valued random variable such that the conditional distribution of Z given Y is* Po(Y). *Then Z has a negative binomial distribution with parameters a and $b/(b+1)$.*

1.5 Exercises

Exercise 1.1 Prove equation (1.10).

Exercise 1.2 Let X be a random variable taking values in $\mathbb{N}_0$. Assume that there is a $\gamma \geq 0$ such that $\mathbb{E}[(X)_k] = \gamma^k$ for all $k \in \mathbb{N}_0$. Show that X has a Poisson distribution. (Hint: Derive the Taylor series for $g(s) := \mathbb{E}[s^X]$ at $s_0 = 1$.)

Exercise 1.3 Confirm Proposition 1.3 by showing that

$$\mathbb{E}[s^X t^{Z-X}] = e^{p\gamma(s-1)} e^{(1-p)\gamma(t-1)}, \quad s, t \in [0, 1],$$

using a direct computation and Proposition B.4.

Exercise 1.4 (Generalisation of Proposition 1.2) Let $m \in \mathbb{N}$ and suppose that $X_1, \dots, X_m$ are independent random variables with Poisson distributions $\text{Po}(\gamma_1), \dots, \text{Po}(\gamma_m)$, respectively. Show that $X := X_1 + \cdots + X_m$ is Poisson distributed with parameter $\gamma := \gamma_1 + \cdots + \gamma_m$. Assuming $\gamma > 0$, show moreover for any $k \in \mathbb{N}$ that

$$\mathbb{P}(X_1 = k_1, \dots, X_m = k_m \mid X = k) = \frac{k!}{k_1! \cdots k_m!} \left(\frac{\gamma_1}{\gamma}\right)^{k_1} \cdots \left(\frac{\gamma_m}{\gamma}\right)^{k_m} \tag{1.29}$$

for $k_1 + \cdots + k_m = k$. This is a *multinomial distribution* with parameters k and $\gamma_1/\gamma, \dots, \gamma_m/\gamma$.

Exercise 1.5 (Generalisation of Proposition 1.3) Let $m \in \mathbb{N}$ and suppose that Z_n, $n \in \mathbb{N}$, is a sequence of independent random vectors in $\mathbb{R}^m$ with common distribution $\mathbb{P}(Z_1 = e_i) = p_i$, $i \in \{1, \dots, m\}$, where e_i is the i-th unit vector in $\mathbb{R}^m$ and $p_1 + \cdots + p_m = 1$. Let Z have a Poisson distribution with parameter γ, independent of $(Z_1, Z_2, \dots)$. Show that the components of the random vector $X := \sum_{j=1}^{Z} Z_j$ are independent and Poisson distributed with parameters $p_1\gamma, \dots, p_m\gamma$.

Exercise 1.6 (Bivariate extension of Proposition 1.4) Let $\gamma > 0, \delta \geq 0$. Suppose for $n \in \mathbb{N}$ that $m_n \in \mathbb{N}$ and for $1 \leq i \leq m_n$ that $p_{n,i}, q_{n,i} \in [0, 1)$ with $\sum_{i=1}^{m_n} p_{n,i} \to \gamma$ and $\sum_{i=1}^{m_n} q_{n,i} \to \delta$, and $\max_{1 \leq i \leq m_n} \max\{p_{n,i}, q_{n,i}\} \to 0$ as $n \to \infty$. Suppose for $n \in \mathbb{N}$ that $(X_n, Y_n) = \sum_{i=1}^{m_n} (X_{n,i}, Y_{n,i})$, where each $(X_{n,i}, Y_{n,i})$ is a random 2-vector whose components are Bernoulli distributed with parameters $p_{n,i}, q_{n,i}$, respectively, and satisfy $X_{n,i} Y_{n,i} = 0$ almost surely. Assume the random vectors $(X_{n,i}, Y_{n,i})$, $1 \leq i \leq m_n$, are independent. Prove that X_n, Y_n are asymptotically (as $n \to \infty$) distributed as a pair of independent Poisson variables with parameters γ, δ, i.e. for $k, \ell \in \mathbb{N}_0$,

$$\lim_{n\to\infty} \mathbb{P}(X_n = k, Y_n = \ell) = e^{-(\gamma+\delta)} \frac{\gamma^k \delta^\ell}{k!\ell!}.$$

Exercise 1.7 (Probability of a Poisson variable being even) Suppose X is Poisson distributed with parameter $\gamma > 0$. Using the fact that the probability generating function (1.8) extends to $s = -1$, verify the identity $\mathbb{P}(X/2 \in \mathbb{Z}) = (1 + e^{-2\gamma})/2$. For $k \in \mathbb{N}$ with $k \geq 3$, using the fact that the probability generating function (1.8) extends to a k-th complex root of unity, find a closed-form formula for $\mathbb{P}(X/k \in \mathbb{Z})$.

Exercise 1.8 Let $\gamma > 0$, and suppose X is Poisson distributed with parameter γ. Suppose $f : \mathbb{N} \to \mathbb{R}_+$ is such that $\mathbb{E}[f(X)^{1+\varepsilon}] < \infty$ for some $\varepsilon > 0$. Show that $\mathbb{E}[f(X + k)] < \infty$ for any $k \in \mathbb{N}$.

Exercise 1.9 Let $0 < \gamma < \gamma'$. Give an example of a random vector (X, Y) with X Poisson distributed with parameter γ and Y Poisson distributed with parameter γ', such that $Y-X$ *is not* Poisson distributed. (Hint: First consider a pair X', Y' such that $Y'-X'$ *is* Poisson distributed, and then modify finitely many of the values of their joint probability mass function.)

Exercise 1.10 Suppose $n \in \mathbb{N}$ and set $[n] := \{1, \dots, n\}$. Suppose that Z is a uniform random permutation of $[n]$, that is a random element of the space Σ_n of all bijective mappings from $[n]$ to $[n]$ such that $\mathbb{P}(Z = \pi) = 1/n!$ for each $\pi \in \Sigma_n$. For $a \in \mathbb{R}$ let $\lceil a \rceil := \min\{k \in \mathbb{Z} : k \geq a\}$. Let $\gamma \in [0, 1]$ and let $X_n := \mathrm{card}\{i \in [\lceil \gamma n \rceil] : Z(i) = i\}$ be the number of fixed points of Z among the first $\lceil \gamma n \rceil$ integers. Show that the distribution of X_n converges to $\mathrm{Po}(\gamma)$, that is

$$\lim_{n\to\infty} \mathbb{P}(X_n = k) = \frac{\gamma^k}{k!} e^{-\gamma}, \quad k \in \mathbb{N}_0.$$

(Hint: Establish an explicit formula for $\mathbb{P}(X_n = k)$, starting with the case $k = 0$.)

Exercise 1.11 Prove Proposition 1.5.

Exercise 1.12 Let $\gamma > 0$ and $\delta > 0$. Find a random vector (X, Y) such that X, Y and $X + Y$ are Poisson distributed with parameter γ, δ and $\gamma + \delta$, respectively, but X and Y are not independent.

2

Point Processes

A point process is a random collection of at most countably many points, possibly with multiplicities. This chapter defines this concept for an arbitrary measurable space and provides several criteria for equality in distribution.

2.1 Fundamentals

The idea of a point process is that of a random, at most countable, collection Z of points in some space $\mathbb{X}$. A good example to think of is the d-dimensional Euclidean space $\mathbb{R}^d$. Ignoring measurability issues for the moment, we might think of Z as a mapping $\omega \mapsto Z(\omega)$ from Ω into the system of countable subsets of $\mathbb{X}$, where $(\Omega, \mathcal{F}, \mathbb{P})$ is an underlying probability space. Then Z can be identified with the family of mappings

$$\omega \mapsto \eta(\omega, B) := \mathrm{card}(Z(\omega) \cap B), \quad B \subset \mathbb{X},$$

counting the number of points that Z has in B. (We write $\mathrm{card}\, A$ for the number of elements of a set A.) Clearly, for any fixed $\omega \in \Omega$ the mapping $\eta(\omega, \cdot)$ is a measure, namely the *counting measure* supported by $Z(\omega)$. It turns out to be a mathematically fruitful idea to define point processes as random counting measures.

To give the general definition of a point process let $(\mathbb{X}, \mathcal{X})$ be a measurable space. Let $\mathbf{N}_{<\infty}(\mathbb{X}) \equiv \mathbf{N}_{<\infty}$ denote the space of all measures μ on $\mathbb{X}$ such that $\mu(B) \in \mathbb{N}_0 := \mathbb{N} \cup \{0\}$ for all $B \in \mathcal{X}$, and let $\mathbf{N}(\mathbb{X}) \equiv \mathbf{N}$ be the space of all measures that can be written as a countable sum of measures from $\mathbf{N}_{<\infty}$. A trivial example of an element of $\mathbf{N}$ is the *zero measure* 0 that is identically zero on $\mathcal{X}$. A less trivial example is the *Dirac measure* δ_x at a point $x \in \mathbb{X}$ given by $\delta_x(B) := \mathbf{1}_B(x)$. More generally, any (finite or infinite) sequence $(x_n)_{n=1}^k$ of elements of $\mathbb{X}$, where $k \in \overline{\mathbb{N}} := \mathbb{N} \cup \{\infty\}$ is the number of terms in the sequence, can be used to define a measure

$$\mu = \sum_{n=1}^{k} \delta_{x_n}. \tag{2.1}$$

Then $\mu \in \mathbf{N}$ and

$$\mu(B) = \sum_{n=1}^{k} \mathbf{1}_B(x_n), \quad B \in \mathcal{X}.$$

More generally we have, for any measurable $f : \mathbb{X} \to [0, \infty]$, that

$$\int f \, d\mu = \sum_{n=1}^{k} f(x_n). \tag{2.2}$$

We can allow for $k = 0$ in (2.1). In this case μ is the zero measure. The points $x_1, x_2, \dots$ are not assumed to be pairwise distinct. If $x_i = x_j$ for some $i, j \leq k$ with $i \neq j$, then μ is said to have *multiplicities*. In fact, the multiplicity of x_i is the number $\mathrm{card}\{j \leq k : x_j = x_i\}$. Any μ of the form (2.1) is interpreted as a counting measure with possible multiplicities.

In general one cannot guarantee that any $\mu \in \mathbf{N}$ can be written in the form (2.1); see Exercise 2.5. Fortunately, only weak assumptions on $(\mathbb{X}, \mathcal{X})$ and μ are required to achieve this; see e.g. Corollary 6.5. Moreover, large parts of the theory can be developed without imposing further assumptions on $(\mathbb{X}, \mathcal{X})$, other than to be a measurable space.

A measure ν on $\mathbb{X}$ is said to be *s-finite* if ν is a countable sum of finite measures. By definition, each element of $\mathbf{N}$ is s-finite. We recall that a measure ν on $\mathbb{X}$ is said to be *σ-finite* if there is a sequence $B_m \in \mathcal{X}$, $m \in \mathbb{N}$, such that $\cup_m B_m = \mathbb{X}$ and $\nu(B_m) < \infty$ for all $m \in \mathbb{N}$. Clearly every σ-finite measure is s-finite. Any $\overline{\mathbb{N}}_0$-valued σ-finite measure is in $\mathbf{N}$. In contrast to σ-finite measures, any countable sum of s-finite measures is again s-finite. If the points x_n in (2.1) are all the same, then this measure μ is not σ-finite. The counting measure on $\mathbb{R}$ (supported by $\mathbb{R}$) is an example of a measure with values in $\overline{\mathbb{N}}_0 := \overline{\mathbb{N}} \cup \{0\}$, that is not s-finite. Exercise 6.10 gives an example of an s-finite $\overline{\mathbb{N}}_0$-valued measure that is not in $\mathbf{N}$.

Let $\mathcal{N}(\mathbb{X}) \equiv \mathcal{N}$ denote the σ-field generated by the collection of all subsets of $\mathbf{N}$ of the form

$$\{\mu \in \mathbf{N} : \mu(B) = k\}, \quad B \in \mathcal{X}, \, k \in \mathbb{N}_0.$$

This means that $\mathcal{N}$ is the smallest σ-field on $\mathbf{N}$ such that $\mu \mapsto \mu(B)$ is measurable for all $B \in \mathcal{X}$.

Definition 2.1 A *point process* on $\mathbb{X}$ is a random element η of $(\mathbf{N}, \mathcal{N})$, that is a measurable mapping $\eta\colon \Omega \to \mathbf{N}$.

If η is a point process on $\mathbb{X}$ and $B \in \mathcal{X}$, then we denote by $\eta(B)$ the mapping $\omega \mapsto \eta(\omega, B) := \eta(\omega)(B)$. By the definitions of η and the σ-field $\mathcal{N}$ these are random variables taking values in $\overline{\mathbb{N}}_0$, that is

$$\{\eta(B) = k\} \equiv \{\omega \in \Omega : \eta(\omega, B) = k\} \in \mathcal{F}, \quad B \in \mathcal{X},\ k \in \overline{\mathbb{N}}_0. \qquad (2.3)$$

Conversely, a mapping $\eta\colon \Omega \to \mathbf{N}$ is a point process if (2.3) holds. In this case we call $\eta(B)$ the *number of points* of η in B. Note that the mapping $(\omega, B) \mapsto \eta(\omega, B)$ is a *kernel* from Ω to $\mathbb{X}$ (see Section A.1) with the additional property that $\eta(\omega, \cdot) \in \mathbf{N}$ for each $\omega \in \Omega$.

Example 2.2 Let X be a random element in $\mathbb{X}$. Then

$$\eta := \delta_X$$

is a point process. Indeed, the required measurability property follows from

$$\{\eta(B) = k\} = \begin{cases} \{X \in B\}, & \text{if } k = 1, \\ \{X \notin B\}, & \text{if } k = 0, \\ \emptyset, & \text{otherwise.} \end{cases}$$

The above one-point process can be generalised as follows.

Example 2.3 Let $\mathbb{Q}$ be a probability measure on $\mathbb{X}$ and suppose that $X_1, \dots, X_m$ are independent random elements in $\mathbb{X}$ with distribution $\mathbb{Q}$. Then

$$\eta := \delta_{X_1} + \cdots + \delta_{X_m}$$

is a point process on $\mathbb{X}$. Because

$$\mathbb{P}(\eta(B) = k) = \binom{m}{k} \mathbb{Q}(B)^k (1 - \mathbb{Q}(B))^{m-k}, \quad k = 0, \dots, m,$$

η is referred to as a *binomial process* with *sample size* m and *sampling distribution* $\mathbb{Q}$.

In this example, the random measure η can be written as a sum of Dirac measures, and we formalise the class of point processes having this property in the following definition. Here and later we say that two point processes η and η' are *almost surely equal* if there is an $A \in \mathcal{F}$ with $\mathbb{P}(A) = 1$ such that $\eta(\omega) = \eta'(\omega)$ for each $\omega \in A$.

Definition 2.4 We shall refer to a point process η on $\mathbb{X}$ as a *proper point process* if there exist random elements $X_1, X_2, \dots$ in $\mathbb{X}$ and an $\overline{\mathbb{N}}_0$-valued random variable κ such that almost surely

$$\eta = \sum_{n=1}^{\kappa} \delta_{X_n}. \tag{2.4}$$

In the case $\kappa = 0$ this is interpreted as the zero measure on $\mathbb{X}$.

The motivation for this terminology is that the intuitive notion of a point process is that of a (random) set of points, rather than an integer-valued measure. A proper point process is one which can be interpreted as a countable (random) set of points in $\mathbb{X}$ (possibly with repetitions), thereby better fitting this intuition.

The class of proper point processes is very large. Indeed, we shall see later that if $\mathbb{X}$ is a Borel subspace of a complete separable metric space, then *any* locally finite point process on $\mathbb{X}$ (see Definition 2.13) is proper, and that, for general $(\mathbb{X}, \mathcal{X})$, if η is a *Poisson* point process on $\mathbb{X}$ there is a proper point process on $\mathbb{X}$ having the same distribution as η (these concepts will be defined in due course); see Corollary 6.5 and Corollary 3.7. Exercise 2.5 shows, however, that not all point processes are proper.

2.2 Campbell's Formula

A first characteristic of a point process is the mean number of points lying in an arbitrary measurable set:

Definition 2.5 The *intensity measure* of a point process η on $\mathbb{X}$ is the measure λ defined by

$$\lambda(B) := \mathbb{E}[\eta(B)], \quad B \in \mathcal{X}. \tag{2.5}$$

It follows from basic properties of expectation that the intensity measure of a point process is indeed a measure.

Example 2.6 The intensity measure of a binomial process with sample size m and sampling distribution $\mathbb{Q}$ is given by

$$\lambda(B) = \mathbb{E}\left[\sum_{k=1}^{m} \mathbf{1}\{X_k \in B\}\right] = \sum_{k=1}^{m} \mathbb{P}(X_k \in B) = m\,\mathbb{Q}(B).$$

Independence of the random variables $X_1, \dots, X_m$ is not required for this calculation.

Let $\overline{\mathbb{R}} := [-\infty, \infty]$ and $\overline{\mathbb{R}}_+ := [0, \infty]$. Let us denote by $\mathbb{R}(\mathbb{X})$ (resp. $\overline{\mathbb{R}}(\mathbb{X})$) the set of all measurable functions $u\colon \mathbb{X} \to \mathbb{R}$ (resp. $u\colon \mathbb{X} \to \overline{\mathbb{R}}$). Let $\mathbb{R}_+(\mathbb{X})$ (resp. $\overline{\mathbb{R}}_+(\mathbb{X})$) be the set of all those $u \in \mathbb{R}(\mathbb{X})$ (resp. $u \in \overline{\mathbb{R}}(\mathbb{X})$) with $u \geq 0$. Given $u \in \overline{\mathbb{R}}(\mathbb{X})$, define the functions $u^+, u^- \in \overline{\mathbb{R}}_+(X)$ by $u^+(x) := \max\{u(x), 0\}$ and $u^-(x) := \max\{-u(x), 0\}$, $x \in \mathbb{X}$. Then $u(x) = u^+(x) - u^-(x)$. We recall from measure theory (see Section A.1) that, for any measure ν on $\mathbb{X}$, the integral $\int u\, d\nu \equiv \int u(x)\, \nu(dx)$ of $u \in \overline{\mathbb{R}}(\mathbb{X})$ with respect to ν is defined as

$$\int u(x)\, \nu(dx) \equiv \int u\, d\nu := \int u^+\, d\nu - \int u^-\, d\nu$$

whenever this expression is not of the form $\infty - \infty$. Otherwise we use here the convention $\int u(x)\, \nu(dx) := 0$. We often write

$$\nu(u) := \int u(x)\, \nu(dx),$$

so that $\nu(B) = \nu(\mathbf{1}_B)$ for any $B \in \mathcal{X}$. If η is a point process, then $\eta(u) \equiv \int u\, d\eta$ denotes the mapping $\omega \mapsto \int u(x)\, \eta(\omega, dx)$.

Proposition 2.7 (Campbell's formula) *Let η be a point process on $(\mathbb{X}, \mathcal{X})$ with intensity measure λ. Let $u \in \overline{\mathbb{R}}(\mathbb{X})$. Then $\int u(x)\, \eta(dx)$ is a random variable. Moreover,*

$$\mathbb{E}\Big[\int u(x)\, \eta(dx)\Big] = \int u(x)\, \lambda(dx) \tag{2.6}$$

whenever $u \geq 0$ or $\int |u(x)|\, \lambda(dx) < \infty$.

Proof If $u(x) = \mathbf{1}_B(x)$ for some $B \in \mathcal{X}$ then $\int u(x)\, \eta(dx) = \eta(B)$ and both assertions are true by definition. By standard techniques of measure theory (linearity and monotone convergence) this can be extended, first to measurable simple functions and then to arbitrary $u \in \overline{\mathbb{R}}_+(\mathbb{X})$.

Let $u \in \overline{\mathbb{R}}(\mathbb{X})$. We have just seen that $\eta(u^+)$ and $\eta(u^-)$ are random variables, so that $\eta(u)$ is a random variable too. Assume that $\int |u(x)|\, \lambda(dx) < \infty$. Then the first part of the proof shows that $\eta(u^+)$ and $\eta(u^-)$ both have a finite expectation and that

$$\mathbb{E}[\eta(u)] = \mathbb{E}[\eta(u^+)] - \mathbb{E}[\eta(u^-)] = \lambda(u^+) - \lambda(u^-) = \lambda(u).$$

This concludes the proof. □

2.3 Distribution of a Point Process

In accordance with the terminology of probability theory (see Section B.1), the *distribution* of a point process η on $\mathbb{X}$ is the probability measure $\mathbb{P}_\eta$ on $(\mathbf{N}, \mathcal{N})$, given by $A \mapsto \mathbb{P}(\eta \in A)$. If η' is another point process with the same distribution, we write $\eta \overset{d}{=} \eta'$.

The following device is a powerful tool for analysing point processes. We use the convention $e^{-\infty} := 0$.

Definition 2.8 The *Laplace* (or *characteristic*) *functional* of a point process η on $\mathbb{X}$ is the mapping $L_\eta \colon \mathbb{R}_+(\mathbb{X}) \to [0, 1]$ defined by

$$L_\eta(u) := \mathbb{E}\Big[\exp\Big(-\int u(x)\,\eta(dx)\Big)\Big], \quad u \in \mathbb{R}_+(\mathbb{X}).$$

Example 2.9 Let η be the binomial process of Example 2.3. Then, for $u \in \mathbb{R}_+(\mathbb{X})$,

$$L_\eta(u) = \mathbb{E}\Big[\exp\Big(-\sum_{k=1}^{m} u(X_k)\Big)\Big] = \mathbb{E}\Big[\prod_{k=1}^{m} \exp[-u(X_k)]\Big]$$
$$= \prod_{k=1}^{m} \mathbb{E}[\exp[-u(X_k)]] = \Big[\int \exp[-u(x)]\,\mathbb{Q}(dx)\Big]^m.$$

The following proposition characterises equality in distribution for point processes. It shows, in particular, that the Laplace functional of a point process determines its distribution.

Proposition 2.10 *For point processes η and η' on $\mathbb{X}$ the following assertions are equivalent:*

(i) $\eta \overset{d}{=} \eta'$;
(ii) $(\eta(B_1), \dots, \eta(B_m)) \overset{d}{=} (\eta'(B_1), \dots, \eta'(B_m))$ *for all* $m \in \mathbb{N}$ *and all pairwise disjoint* $B_1, \dots, B_m \in \mathcal{X}$;
(iii) $L_\eta(u) = L_{\eta'}(u)$ *for all* $u \in \mathbb{R}_+(\mathbb{X})$;
(iv) *for all* $u \in \mathbb{R}_+(\mathbb{X})$, $\eta(u) \overset{d}{=} \eta'(u)$ *as random variables in* $\overline{\mathbb{R}}_+$.

Proof First we prove that (i) implies (iv). Given $u \in \mathbb{R}_+(\mathbb{X})$, define the function $g_u \colon \mathbf{N} \to \overline{\mathbb{R}}_+$ by $\mu \mapsto \int u\,d\mu$. By Proposition 2.7 (or a direct check based on first principles), g_u is a measurable function. Also,

$$\mathbb{P}_{\eta(u)}(\cdot) = \mathbb{P}(\eta(u) \in \cdot) = \mathbb{P}(\eta \in g_u^{-1}(\cdot)),$$

and likewise for η'. So if $\eta \overset{d}{=} \eta'$ then also $\eta(u) \overset{d}{=} \eta'(u)$.

Next we show that (iv) implies (iii). For any $\overline{\mathbb{R}}_+$-valued random variable

Y we have $\mathbb{E}[\exp(-Y)] = \int e^{-y}\,\mathbb{P}_Y(dy)$, which is determined by the distribution $\mathbb{P}_Y$. Hence, if (iv) holds,

$$L_\eta(u) = \mathbb{E}[\exp(-\eta(u))] = \mathbb{E}[\exp(-\eta'(u))] = L_{\eta'}(u)$$

for all $u \in \mathbb{R}_+(\mathbb{X})$, so (iii) holds.

Assume now that (iii) holds and consider a *simple function* of the form $u = c_1\mathbf{1}_{B_1} + \cdots + c_m\mathbf{1}_{B_m}$, where $m \in \mathbb{N}$, $B_1,\dots,B_m \in \mathcal{X}$ and $c_1,\dots,c_m \in (0,\infty)$. Then

$$L_\eta(u) = \mathbb{E}\Big[\exp\Big(-\sum_{j=1}^m c_j\eta(B_j)\Big)\Big] = \hat{\mathbb{P}}_{(\eta(B_1),\dots,\eta(B_m))}(c_1,\dots,c_m) \tag{2.7}$$

where for any measure μ on $[0,\infty]^m$ we write $\hat{\mu}$ for its multivariate Laplace transform. Since a finite measure on $\mathbb{R}_+^m$ is determined by its Laplace transform (this follows from Proposition B.4), we can conclude that the restriction of $\mathbb{P}_{(\eta(B_1),\dots,\eta(B_m))}$ (a measure on $[0,\infty]^m$) to $(0,\infty)^m$ is the same as the restriction of $\mathbb{P}_{(\eta'(B_1),\dots,\eta'(B_m))}$ to $(0,\infty)^m$. Then, using the fact that $\mathbb{P}_{(\eta(B_1),\dots,\eta(B_m))}$ and $\mathbb{P}_{(\eta'(B_1),\dots,\eta'(B_m))}$ are probability measures on $[0,\infty]^m$, by forming suitable complements we obtain $\mathbb{P}_{(\eta(B_1),\dots,\eta(B_m))} = \mathbb{P}_{(\eta'(B_1),\dots,\eta'(B_m))}$ (these details are left to the reader). In other words, (iii) implies (ii).

Finally we assume (ii) and prove (i). Let $m \in \mathbb{N}$ and $B_1,\dots,B_m \in \mathcal{X}$, not necessarily pairwise disjoint. Let $C_1,\dots,C_n$ be the atoms of the field generated by $B_1,\dots,B_m$; see Section A.1. For each $i \in \{1,\dots,m\}$ there exists $J_i \subset \{1,\dots,n\}$ such that $B_i = \cup_{j\in J_i} C_j$. (Note that $J_i = \emptyset$ if $B_i = \emptyset$.) Let $D_1,\dots,D_m \subset \mathbb{N}_0$. Then

$$\begin{aligned}&\mathbb{P}(\eta(B_1) \in D_1,\dots,\eta(B_m) \in D_m)\\ &\quad = \int \mathbf{1}\Big\{\sum_{j\in J_1} k_j \in D_1,\dots,\sum_{j\in J_m} k_j \in D_m\Big\}\,\mathbb{P}_{(\eta(C_1),\dots,\eta(C_n))}(d(k_1,\dots,k_n)).\end{aligned}$$

Therefore $\mathbb{P}_\eta$ and $\mathbb{P}_{\eta'}$ coincide on the system $\mathcal{H}$ consisting of all sets of the form

$$\{\mu \in \mathbf{N} : \mu(B_1) \in D_1,\dots,\mu(B_m) \in D_m\},$$

where $m \in \mathbb{N}$, $B_1,\dots,B_m \in \mathcal{X}$ and $D_1,\dots,D_m \subset \mathbb{N}_0$. Clearly $\mathcal{H}$ is a π-system; that is, closed under pairwise intersections. Moreover, the smallest σ-field $\sigma(\mathcal{H})$ containing $\mathcal{H}$ is the full σ-field $\mathcal{N}$. Hence (i) follows from the fact that a probability measure is determined by its values on a generating π-system; see Theorem A.5. □

2.4 Point Processes on Metric Spaces

Let us now assume that $\mathbb{X}$ is a metric space with metric ρ; see Section A.2. Then it is always to be understood that $\mathcal{X}$ is the Borel σ-field $\mathcal{B}(\mathbb{X})$ of $\mathbb{X}$. In particular, the singleton $\{x\}$ is in $\mathcal{X}$ for all $x \in \mathbb{X}$. If ν is a measure on $\mathbb{X}$ then we often write $\nu\{x\} := \nu(\{x\})$. If $\nu\{x\} = 0$ for all $x \in \mathbb{X}$, then ν is said to be *diffuse*. Moreover, if $\mu \in \mathbf{N}(\mathbb{X})$ then we write $x \in \mu$ if $\mu(\{x\}) > 0$.

A set $B \subset \mathbb{X}$ is said to be *bounded* if it is empty or its *diameter*

$$d(B) := \sup\{\rho(x, y) : x, y \in B\}$$

is finite.

Definition 2.11 Suppose that $\mathbb{X}$ is a metric space. The system of bounded measurable subsets of $\mathbb{X}$ is denoted by $\mathcal{X}_b$. A measure ν on $\mathbb{X}$ is said to be *locally finite* if $\nu(B) < \infty$ for every $B \in \mathcal{X}_b$. Let $\mathbf{N}_l(\mathbb{X})$ denote the set of all locally finite elements of $\mathbf{N}(\mathbb{X})$ and let $\mathcal{N}_l(\mathbb{X}) := \{A \cap \mathbf{N}_l(\mathbb{X}) : A \in \mathcal{N}(\mathbb{X})\}$.

Fix some $x_0 \in \mathbb{X}$. Then any bounded set B is contained in the closed *ball* $B(x_0, r) = \{x \in \mathbb{X} : \rho(x, x_0) \leq r\}$ for sufficiently large $r > 0$. In fact, if $B \neq \emptyset$, then we can take, for instance, $r := d(B) + \rho(x_1, x_0)$ for some $x_1 \in B$. Note that $B(x_0, n) \uparrow \mathbb{X}$ as $n \to \infty$. Hence a measure ν on $\mathbb{X}$ is locally finite if and only if $\nu(B(x_0, n)) < \infty$ for each $n \in \mathbb{N}$. In particular, the set $\mathbf{N}_l(\mathbb{X})$ is measurable, that is $\mathbf{N}_l(\mathbb{X}) \in \mathcal{N}(\mathbb{X})$. Moreover, any locally finite measure is σ-finite.

Proposition 2.12 *Let η and η' be point processes on a metric space $\mathbb{X}$. Suppose $\eta(u) \overset{d}{=} \eta'(u)$ for all $u \in \mathbb{R}_+(\mathbb{X})$ such that $\{u > 0\}$ is bounded. Then $\eta \overset{d}{=} \eta'$.*

Proof Suppose that

$$\eta(u) \overset{d}{=} \eta'(u), \quad u \in \mathbb{R}_+(\mathbb{X}),\ \{u > 0\} \text{ bounded.} \tag{2.8}$$

Then $L_\eta(u) = L_{\eta'}(u)$ for any $u \in \mathbb{R}_+(\mathbb{X})$ such that $\{u > 0\}$ is bounded. Given any $v \in \mathbb{R}_+(\mathbb{X})$, we can choose a sequence u_n, $n \in \mathbb{N}$, of functions in $\mathbb{R}_+(\mathbb{X})$ such that $\{u_n > 0\}$ is bounded for each n, and $u_n \uparrow v$ pointwise. Then, by dominated convergence and (2.8),

$$L_\eta(v) = \lim_{n\to\infty} L_\eta(u_n) = \lim_{n\to\infty} L_{\eta'}(u_n) = L_{\eta'}(v),$$

so $\eta \overset{d}{=} \eta'$ by Proposition 2.10. □

Definition 2.13 A point process η on a metric space $\mathbb{X}$ is said to be *locally finite* if $\mathbb{P}(\eta(B) < \infty) = 1$ for every bounded $B \in \mathcal{X}$.

If required, we could interpret a locally finite point process η as a random element of the space $(\mathbf{N}_l(\mathbb{X}), \mathcal{N}_l(\mathbb{X}))$, introduced in Definition 2.11. Indeed, we can define another point process $\tilde{\eta}$ by $\tilde{\eta}(\omega, \cdot) := \eta(\omega, \cdot)$ if the latter is locally finite and by $\tilde{\eta}(\omega, \cdot) := 0$ (the zero measure) otherwise. Then $\tilde{\eta}$ is a random element of $(\mathbf{N}_l(\mathbb{X}), \mathcal{N}_l(\mathbb{X}))$ that coincides $\mathbb{P}$-almost surely ($\mathbb{P}$-a.s.) with η.

The reader might have noticed that the proof of Proposition 2.12 has not really used the metric on $\mathbb{X}$. The proof of the next refinement of this result (not used later in the book) exploits the metric in an essential way.

Proposition 2.14 *Let η and η' be locally finite point processes on a metric space $\mathbb{X}$. Suppose $\eta(u) \overset{d}{=} \eta'(u)$ for all continuous $u\colon \mathbb{X} \to \mathbb{R}_+$ such that $\{u > 0\}$ is bounded. Then $\eta \overset{d}{=} \eta'$.*

Proof Let $\mathbf{G}$ be the space of continuous functions $u\colon \mathbb{X} \to \mathbb{R}_+$ such that $\{u > 0\}$ is bounded. Assume that $\eta(u) \overset{d}{=} \eta'(u)$ for all $u \in \mathbf{G}$. Since $\mathbf{G}$ is closed under non-negative linear combinations, it follows, as in the proof that (iii) implies (ii) in Proposition 2.10, that

$$(\eta(u_1), \eta(u_2), \dots) \overset{d}{=} (\eta'(u_1), \eta'(u_2), \dots),$$

first for any finite sequence and then (by Theorem A.5 in Section A.1) for any infinite sequence $u_n \in \mathbf{G}$, $n \in \mathbb{N}$. Take a bounded closed set $C \subset \mathbb{X}$ and, for $n \in \mathbb{N}$, define

$$u_n(x) := \max\{1 - nd(x, C), 0\}, \quad x \in \mathbb{X},$$

where $d(x, C) := \inf\{\rho(x, y) : y \in C\}$ and $\inf \emptyset := \infty$. By Exercise 2.8, $u_n \in \mathbf{G}$. Moreover, $u_n \downarrow \mathbf{1}_C$ as $n \to \infty$, and since η is locally finite we obtain $\eta(u_n) \to \eta(C)$ $\mathbb{P}$-a.s. The same relation holds for η'. It follows that statement (ii) of Proposition 2.10 holds whenever $B_1, \dots, B_m$ are closed and bounded, but not necessarily disjoint. Hence, fixing a closed ball $C \subset \mathbb{X}$, $\mathbb{P}_\eta$ and $\mathbb{P}_{\eta'}$ coincide on the π-system $\mathcal{H}_C$ consisting of all sets of the form

$$\{\mu \in \mathbf{N}_l : \mu(B_1 \cap C) \le k_1, \dots, \mu(B_m \cap C) \le k_m\}, \tag{2.9}$$

where $m \in \mathbb{N}$, $B_1, \dots, B_m \subset \mathbb{X}$ are closed and $k_1, \dots, k_m \in \mathbb{N}_0$. Another application of Theorem A.5 shows that $\mathbb{P}_\eta$ and $\mathbb{P}_{\eta'}$ coincide on $\sigma(\mathcal{H}_C)$ and then also on $\mathcal{N}_l' := \sigma(\cup_{i=1}^{\infty} \sigma(\mathcal{H}_{B_i}))$, where $B_i := B(x_0, i)$ and $x_0 \in \mathbb{X}$ is fixed.

It remains to show that $\mathcal{N}_l' = \mathcal{N}_l$. Let $i \in \mathbb{N}$ and let $\mathcal{N}_i$ denote the smallest σ-field on $\mathbf{N}_l$ containing the sets $\{\mu \in \mathbf{N}_l : \mu(B \cap B_i) \le k\}$ for all closed sets $B \subset \mathbb{X}$ and each $k \in \mathbb{N}_0$. Let $\mathcal{D}$ be the system of all Borel sets $B \subset \mathbb{X}$ such that $\mu \mapsto \mu(B \cap B_i)$ is $\mathcal{N}_i$-measurable. Then $\mathcal{D}$ is a Dynkin system containing the π-system of all closed sets, so that the monotone class theorem shows $\mathcal{D} = \mathcal{X}$. Therefore $\sigma(\mathcal{H}_{B_i})$ contains $\{\mu \in \mathbf{N}_l : \mu(B \cap B_i) \le k\}$ for all $B \in \mathcal{X}$ and all $k \in \mathbb{N}_0$. Letting $i \to \infty$ we see that $\mathcal{N}'_l$ contains $\{\mu \in \mathbf{N}_l : \mu(B) \le k\}$ and therefore every set from $\mathcal{N}_l$. □

2.5 Exercises

Exercise 2.1 Give an example of a point process η on a measurable space $(\mathbb{X}, \mathcal{X})$ with intensity measure λ and $u \in \mathbb{R}(\mathbb{X})$ (violating the condition that $u \ge 0$ or $\int |u(x)|\lambda(dx) < \infty$), such that Campbell's formula (2.6) fails.

Exercise 2.2 Let $\mathcal{X}^* \subset \mathcal{X}$ be a π-system generating $\mathcal{X}$. Let η be a point process on $\mathbb{X}$ that is σ-*finite* on $\mathcal{X}^*$, meaning that there is a sequence $C_n \in \mathcal{X}^*$, $n \in \mathbb{N}$, such that $\cup_{n=1}^{\infty} C_n = \mathbb{X}$ and $\mathbb{P}(\eta(C_n) < \infty) = 1$ for all $n \in \mathbb{N}$. Let η' be another point process on $\mathbb{X}$ and suppose that the equality in Proposition 2.10(ii) holds for all $B_1, \dots, B_m \in \mathcal{X}^*$ and $m \in \mathbb{N}$. Show that $\eta \overset{d}{=} \eta'$.

Exercise 2.3 Let $\eta_1, \eta_2, \dots$ be a sequence of point processes and define $\eta := \eta_1 + \eta_2 + \cdots$, that is $\eta(\omega, B) := \eta_1(\omega, B) + \eta_2(\omega, B) + \cdots$ for all $\omega \in \Omega$ and $B \in \mathcal{X}$. Show that η is a point process. (Hint: Prove first that $\mathbf{N}(\mathbb{X})$ is closed under countable summation.)

Exercise 2.4 Let $\eta_1, \eta_2, \dots$ be a sequence of proper point processes. Show that $\eta := \eta_1 + \eta_2 + \cdots$ is a proper point process.

Exercise 2.5 Suppose that $\mathbb{X} = [0, 1]$. Find a σ-field $\mathcal{X}$ and a measure μ on $(\mathbb{X}, \mathcal{X})$ such that $\mu(\mathbb{X}) = 1$ and $\mu(B) \in \{0, 1\}$ for all $B \in \mathcal{X}$, which is not of the form $\mu = \delta_x$ for some $x \in \mathbb{X}$. (Hint: Take the system of all finite subsets of $\mathbb{X}$ as a generator of $\mathcal{X}$.)

Exercise 2.6 Let η be a point process on $\mathbb{X}$ with intensity measure λ and let $B \in \mathcal{X}$ such that $\lambda(B) < \infty$. Show that

$$\lambda(B) = -\frac{d}{dt} L_\eta(t\mathbf{1}_B)\Big|_{t=0}.$$

Exercise 2.7 Let η be a point process on $\mathbb{X}$. Show for each $B \in \mathcal{X}$ that

$$\mathbb{P}(\eta(B) = 0) = \lim_{t\to\infty} L_\eta(t\mathbf{1}_B).$$

Exercise 2.8 Let $(\mathbb{X}, \rho)$ be a metric space. Let $C \subset \mathbb{X}$, $C \neq \emptyset$. For $x \in \mathbb{X}$ let $d(x, C) := \inf\{\rho(x, z) : z \in C\}$. Show that $d(\cdot, C)$ has the Lipschitz property

$$|d(x, C) - d(y, C)| \le \rho(x, y), \quad x, y \in \mathbb{X}.$$

(Hint: Take $z \in C$ and bound $\rho(x, z)$ by the triangle inequality.)

3

Poisson Processes

For a Poisson point process the number of points in a given set has a Poisson distribution. Moreover, the numbers of points in disjoint sets are stochastically independent. A Poisson process exists on a general s-finite measure space. Its distribution is characterised by a specific exponential form of the Laplace functional.

3.1 Definition of the Poisson Process

In this chapter we fix an arbitrary measurable space $(\mathbb{X}, \mathcal{X})$. We are now ready for the definition of the main subject of this volume. Recall that for $\gamma \in [0, \infty]$, the Poisson distribution $\mathrm{Po}(\gamma)$ was defined at (1.4).

Definition 3.1 Let λ be an s-finite measure on $\mathbb{X}$. A *Poisson process* with intensity measure λ is a point process η on $\mathbb{X}$ with the following two properties:

(i) For every $B \in \mathcal{X}$ the distribution of $\eta(B)$ is Poisson with parameter $\lambda(B)$, that is to say $\mathbb{P}(\eta(B) = k) = \mathrm{Po}(\lambda(B); k)$ for all $k \in \mathbb{N}_0$.
(ii) For every $m \in \mathbb{N}$ and all pairwise disjoint sets $B_1, \dots, B_m \in \mathcal{X}$ the random variables $\eta(B_1), \dots, \eta(B_m)$ are independent.

Property (i) of Definition 3.1 is responsible for the name of the Poisson process. A point process with property (ii) is said to be *completely independent*. (One also says that η has *independent increments* or is *completely random*.) For a (locally finite) point process without multiplicities and a diffuse intensity measure (on a complete separable metric space) we shall see in Chapter 6 that the two defining properties of a Poisson process are equivalent.

If η is a Poisson process with intensity measure λ then $\mathbb{E}[\eta(B)] = \lambda(B)$, so that Definition 3.1 is consistent with Definition 2.5. In particular, if $\lambda = 0$ is the zero measure, then $\mathbb{P}(\eta(\mathbb{X}) = 0) = 1$.

Let us first record that for each s-finite λ there is at most one Poisson process with intensity measure λ, up to equality in distribution.

Proposition 3.2 *Let η and η' be two Poisson processes on $\mathbb{X}$ with the same s-finite intensity measure. Then $\eta \overset{d}{=} \eta'$.*

Proof The result follows from Proposition 2.10. □

3.2 Existence of Poisson Processes

In this section we show by means of an explicit construction that Poisson processes exist. Before we can do this, we need to deal with the *superposition* of independent Poisson processes.

Theorem 3.3 (Superposition theorem) *Let η_i, $i \in \mathbb{N}$, be a sequence of independent Poisson processes on $\mathbb{X}$ with intensity measures λ_i. Then*

$$\eta := \sum_{i=1}^{\infty} \eta_i \tag{3.1}$$

is a Poisson process with intensity measure $\lambda := \lambda_1 + \lambda_2 + \cdots$.

Proof Exercise 2.3 shows that η is a point process.

For $n \in \mathbb{N}$ and $B \in \mathcal{X}$, we have by Exercise 1.4 that $\xi_n(B) := \sum_{i=1}^{n} \eta_i(B)$ has a Poisson distribution with parameter $\sum_{i=1}^{n} \lambda_i(B)$. Also $\xi_n(B)$ converges monotonically to $\eta(B)$ so by continuity of probability, and the fact that $\mathrm{Po}(\gamma; j)$ is continuous in γ for $j \in \mathbb{N}_0$, for all $k \in \mathbb{N}_0$ we have

$$\begin{aligned}\mathbb{P}(\eta(B) \le k) &= \lim_{n\to\infty} \mathbb{P}(\xi_n(B) \le k) \\ &= \lim_{n\to\infty} \sum_{j=0}^{k} \mathrm{Po}\left(\sum_{i=1}^{n} \lambda_i(B); j\right) = \sum_{j=0}^{k} \mathrm{Po}\left(\sum_{i=1}^{\infty} \lambda_i(B); j\right)\end{aligned}$$

so that $\eta(B)$ has the $\mathrm{Po}(\lambda(B))$ distribution.

Let $B_1, \dots, B_m \in \mathcal{X}$ be pairwise disjoint. Then $(\eta_i(B_j), 1 \le j \le m, i \in \mathbb{N})$ is a family of independent random variables, so that by the grouping property of independence the random variables $\sum_i \eta_i(B_1), \dots, \sum_i \eta_i(B_m)$ are independent. Thus η is completely independent. □

Now we construct a Poisson process on $(\mathbb{X}, \mathcal{X})$ with arbitrary s-finite intensity measure. We start by generalising Example 2.3.

Definition 3.4 Let $\mathbb{V}$ and $\mathbb{Q}$ be probability measures on $\mathbb{N}_0$ and $\mathbb{X}$, respectively. Suppose that $X_1, X_2, \dots$ are independent random elements in $\mathbb{X}$

with distribution $\mathbb{Q}$, and let κ be a random variable with distribution $\mathbb{V}$, independent of (X_n). Then

$$\eta := \sum_{k=1}^{\kappa} \delta_{X_k} \tag{3.2}$$

is called a *mixed binomial process* with *mixing distribution* $\mathbb{V}$ and *sampling distribution* $\mathbb{Q}$.

The following result provides the key for the construction of Poisson processes.

Proposition 3.5 *Let $\mathbb{Q}$ be a probability measure on $\mathbb{X}$ and let $\gamma \geq 0$. Suppose that η is a mixed binomial process with mixing distribution* $\mathrm{Po}(\gamma)$ *and sampling distribution $\mathbb{Q}$. Then η is a Poisson process with intensity measure $\gamma\,\mathbb{Q}$.*

Proof Let κ and (X_n) be given as in Definition 3.4. To prove property (ii) of Definition 3.1 it is no loss of generality to assume that $B_1, \dots, B_m$ are pairwise disjoint measurable subsets of $\mathbb{X}$ satisfying $\cup_{i=1}^m B_i = \mathbb{X}$. (Otherwise we can add the complement of this union.) Let $k_1, \dots, k_m \in \mathbb{N}_0$ and set $k := k_1 + \cdots + k_m$. Then

$$\begin{aligned}
&\mathbb{P}(\eta(B_1) = k_1, \dots, \eta(B_m) = k_m) \\
&\quad = \mathbb{P}(\kappa = k)\,\mathbb{P}\Big(\sum_{j=1}^{k} \mathbf{1}\{X_j \in B_1\} = k_1, \dots, \sum_{j=1}^{k} \mathbf{1}\{X_j \in B_m\} = k_m\Big).
\end{aligned}$$

Since the second probability on the right is multinomial, this gives

$$\begin{aligned}
\mathbb{P}(\eta(B_1) = k_1, \dots, \eta(B_m) = k_m) &= \frac{\gamma^k}{k!} e^{-\gamma} \frac{k!}{k_1! \cdots k_m!} \mathbb{Q}(B_1)^{k_1} \cdots \mathbb{Q}(B_m)^{k_m} \\
&= \prod_{j=1}^{m} \frac{(\gamma \mathbb{Q}(B_j))^{k_j}}{k_j!} e^{-\gamma \mathbb{Q}(B_j)}.
\end{aligned}$$

Summing over $k_2, \dots, k_m$ shows that $\eta(B_1)$ is Poisson distributed with parameter $\gamma\,\mathbb{Q}(B_1)$. A similar statement applies to $\eta(B_2), \dots, \eta(B_m)$. Therefore $\eta(B_1), \dots, \eta(B_m)$ are independent. □

Theorem 3.6 (Existence theorem) *Let λ be an s-finite measure on $\mathbb{X}$. Then there exists a Poisson process on $\mathbb{X}$ with intensity measure λ.*

Proof The result is trivial if $\lambda(\mathbb{X}) = 0$.

Suppose for now that $0 < \lambda(\mathbb{X}) < \infty$. On a suitable probability space, assume that $\kappa, X_1, X_2, \dots$ are independent random elements, with κ taking values in $\mathbb{N}_0$ and each X_i taking values in $\mathbb{X}$, with κ having the $\mathrm{Po}(\lambda(\mathbb{X}))$ distribution and each X_i having $\lambda(\cdot)/\lambda(\mathbb{X})$ as its distribution. Here the probability space can be taken to be a suitable product space; see the proof of Corollary 3.7 below. Let η be the mixed binomial process given by (3.2). Then, by Proposition 3.5, η is a Poisson process with intensity measure λ, as required.

Now suppose that $\lambda(\mathbb{X}) = \infty$. There is a sequence λ_i, $i \in \mathbb{N}$, of measures on $(\mathbb{X}, \mathcal{X})$ with strictly positive and finite total measure, such that $\lambda = \sum_{i=1}^{\infty} \lambda_i$. On a suitable (product) probability space, let η_i, $i \in \mathbb{N}$, be a sequence of independent Poisson processes with η_i having intensity measure λ_i. This is possible by the preceding part of the proof. Set $\eta = \sum_{i=1}^{\infty} \eta_i$. By the superposition theorem (Theorem 3.3), η is a Poisson process with intensity measure λ, and the proof is complete. □

A corollary of the preceding proof is that on arbitrary $(\mathbb{X}, \mathcal{X})$ *every Poisson point process is proper* (see Definition 2.4), up to equality in distribution.

Corollary 3.7 *Let λ be an s-finite measure on $\mathbb{X}$. Then there is a probability space $(\Omega, \mathcal{F}, \mathbb{P})$ supporting random elements $X_1, X_2, \dots$ in $\mathbb{X}$ and κ in $\overline{\mathbb{N}}_0$, such that*

$$\eta := \sum_{n=1}^{\kappa} \delta_{X_n} \tag{3.3}$$

is a Poisson process with intensity measure λ.

Proof We consider only the case $\lambda(\mathbb{X}) = \infty$ (the other case is covered by Proposition 3.5). Take the measures λ_i, $i \in \mathbb{N}$, as in the last part of the proof of Theorem 3.6. Let $\gamma_i := \lambda_i(\mathbb{X})$ and $\mathbb{Q}_i := \gamma_i^{-1}\lambda_i$. We shall take $(\Omega, \mathcal{F}, \mathbb{P})$ to be the product of spaces $(\Omega_i, \mathcal{F}_i, \mathbb{P}_i)$, $i \in \mathbb{N}$, where each $(\Omega_i, \mathcal{F}_i, \mathbb{P}_i)$ is again an infinite product of probability spaces $(\Omega_{ij}, \mathcal{F}_{ij}, \mathbb{P}_{ij})$, $j \in \mathbb{N}_0$, with $\Omega_{i0} := \mathbb{N}_0$, $\mathbb{P}_{i0} := \mathrm{Po}(\gamma_i)$ and $(\Omega_{ij}, \mathcal{F}_{ij}, \mathbb{P}_{ij}) := (\mathbb{X}, \mathcal{X}, \mathbb{Q}_i)$ for $j \geq 1$. On this space we can define independent random elements κ_i, $i \in \mathbb{N}$, and X_{ij}, $i, j \in \mathbb{N}$, such that κ_i has distribution $\mathrm{Po}(\gamma_i)$ and X_{ij} has distribution $\mathbb{Q}_i$; see Theorem B.2. The proof of Theorem 3.6 shows how to define $\kappa, X_1, X_2, \dots$ in terms of these random variables in a measurable (algorithmic) way. The details are left to the reader. □

As a consequence of Corollary 3.7, when checking a statement involving only the distribution of a Poisson process η, it is no restriction of generality to assume that η is proper. Exercise 3.9 shows that there are Poisson processes which are not proper. On the other hand, Corollary 6.5 will show that any suitably regular point process on a Borel subset of a complete separable metric space is proper.

The next result is a converse to Proposition 3.5.

Proposition 3.8 *Let η be a Poisson process on $\mathbb{X}$ with intensity measure λ satisfying $0 < \lambda(\mathbb{X}) < \infty$. Then η has the distribution of a mixed binomial process with mixing distribution* $\text{Po}(\lambda(\mathbb{X}))$ *and sampling distribution $\mathbb{Q} := \lambda(\mathbb{X})^{-1}\lambda$. The conditional distribution $\mathbb{P}(\eta \in \cdot \mid \eta(\mathbb{X}) = m)$, $m \in \mathbb{N}$, is that of a binomial process with sample size m and sampling distribution $\mathbb{Q}$.*

Proof Let η' be a mixed binomial process that has mixing distribution $\text{Po}(\lambda(\mathbb{X}))$ and sampling distribution $\mathbb{Q}$. Then $\eta' \stackrel{d}{=} \eta$ by Propositions 3.5 and 3.2. This is our first assertion. Also, by definition, $\mathbb{P}(\eta' \in \cdot \mid \eta'(X) = m)$ has the distribution of a binomial process with sample size m and sampling distribution $\mathbb{Q}$, and by the first assertion so does $\mathbb{P}(\eta \in \cdot \mid \eta(X) = m)$, yielding the second assertion. □

3.3 Laplace Functional of the Poisson Process

The following characterisation of Poisson processes is of great value for both theory and applications.

Theorem 3.9 *Let λ be an s-finite measure on $\mathbb{X}$ and let η be a point process on $\mathbb{X}$. Then η is a Poisson process with intensity measure λ if and only if*

$$L_\eta(u) = \exp\left[-\int (1 - e^{-u(x)})\,\lambda(dx)\right], \quad u \in \mathbb{R}_+(\mathbb{X}). \tag{3.4}$$

Proof Assume first that η is a Poisson process with intensity measure λ. Consider first the simple function $u := c_1\mathbf{1}_{B_1} + \cdots + c_m\mathbf{1}_{B_m}$, where $m \in \mathbb{N}$, $c_1, \dots, c_m \in (0, \infty)$ and $B_1, \dots, B_m \in \mathcal{X}$ are pairwise disjoint. Then

$$\mathbb{E}[\exp[-\eta(u)]] = \mathbb{E}\left[\exp\left(-\sum_{i=1}^m c_i\eta(B_i)\right)\right] = \mathbb{E}\left[\prod_{i=1}^m \exp[-c_i\eta(B_i)]\right].$$

The complete independence and the formula (1.9) for the Laplace transform of the Poisson distribution (this also holds for $\text{Po}(\infty)$) yield

$$L_\eta(u) = \prod_{i=1}^m \mathbb{E}\left[\exp[-c_i\eta(B_i)]\right] = \prod_{i=1}^m \exp[-\lambda(B_i)(1 - e^{-c_i})]$$

$$= \exp\left[-\sum_{i=1}^m \lambda(B_i)(1 - e^{-c_i})\right] = \exp\left[-\sum_{i=1}^m \int_{B_i} (1 - e^{-u})\,d\lambda\right].$$

Since $1 - e^{-u(x)} = 0$ for $x \notin B_1 \cup \cdots \cup B_m$, this is the right-hand side of (3.4). For general $u \in \mathbb{R}_+(\mathbb{X})$, choose simple functions u_n with $u_n \uparrow u$ as $n \to \infty$. Then, by monotone convergence (Theorem A.6), $\eta(u_n) \uparrow \eta(u)$ as $n \to \infty$, and by dominated convergence for expectations the left-hand side of

$$\mathbb{E}[\exp[-\eta(u_n)]] = \exp\left[-\int (1 - e^{-u_n(x)})\, \lambda(dx)\right]$$

tends to $L_\eta(u)$. By monotone convergence again (this time for the integral with respect to λ), the right-hand side tends to the right-hand side of (3.4).

Assume now that (3.4) holds. Let η' be a Poisson process with intensity measure λ. (By Theorem 3.6, such an η' exists.) By the preceding argument, $L_{\eta'}(u) = L_\eta(u)$ for all $u \in \mathbb{R}_+(\mathbb{X})$. Therefore, by Proposition 2.10, $\eta \overset{d}{=} \eta'$; that is, η is a Poisson process with intensity measure λ. □

3.4 Exercises

Exercise 3.1 Use Exercise 1.12 to deduce that there exist a measure space $(\mathbb{X}, \mathcal{X}, \lambda)$ and a point process on $\mathbb{X}$ satisfying part (i) but not part (ii) of the definition of a Poisson process (Definition 3.1).

Exercise 3.2 Show that there exist a measure space $(\mathbb{X}, \mathcal{X}, \lambda)$ and a point process η on $\mathbb{X}$ satisfying part (i) of Definition 3.1 and part (ii) of that definition with 'independent' replaced by 'pairwise independent', such that η is not a Poisson point process. In other words, show that we can have $\eta(B)$ Poisson distributed for all $B \in \mathcal{X}$, and $\eta(A)$ independent of $\eta(B)$ for all disjoint pairs $A, B \in \mathcal{X}$, but $\eta(A_1), \dots, \eta(A_k)$ not mutually independent for all disjoint $A_1, \dots, A_k \in \mathcal{X}$.

Exercise 3.3 Let η be a Poisson process on $\mathbb{X}$ with intensity measure λ and let $B \in \mathcal{X}$ with $0 < \lambda(B) < \infty$. Suppose $B_1, \dots, B_n$ are sets in $\mathcal{X}$ forming a partition of B. Show for all $k_1, \dots, k_n \in \mathbb{N}_0$ and $m := \sum_i k_i$ that

$$\mathbb{P}(\cap_{i=1}^n \{\eta(B_i) = k_i\} \mid \eta(B) = m) = \left(\frac{m!}{k_1!k_2!\cdots k_n!}\right) \prod_{i=1}^n \left(\frac{\lambda(B_i)}{\lambda(B)}\right)^{k_i}.$$

Exercise 3.4 Let η be a Poisson process on $\mathbb{X}$ with s-finite intensity measure λ and let $u \in \mathbb{R}_+(\mathbb{X})$. Use the proof of Theorem 3.9 to show that

$$\mathbb{E}\left[\exp\left(\int u(x)\, \eta(dx)\right)\right] = \exp\left[\int (e^{u(x)} - 1)\, \lambda(dx)\right].$$

Exercise 3.5 Let $\mathbb{V}$ be a probability measure on $\mathbb{N}_0$ with generating function $G_{\mathbb{V}}(s) := \sum_{n=0}^{\infty} \mathbb{V}(\{n\})s^n$, $s \in [0,1]$. Let η be a mixed binomial process with mixing distribution $\mathbb{V}$ and sampling distribution $\mathbb{Q}$. Show that

$$L_\eta(u) = G_{\mathbb{V}}\Big(\int e^{-u}\, d\mathbb{Q}\Big), \quad u \in \mathbb{R}_+(\mathbb{X}).$$

Assume now that $\mathbb{V}$ is a Poisson distribution; show that the preceding formula is consistent with Theorem 3.9.

Exercise 3.6 Let η be a point process on $\mathbb{X}$. Using the convention $e^{-\infty} := 0$, the Laplace functional $L_\eta(u)$ can be defined for any $u \in \overline{\mathbb{R}}_+(\mathbb{X})$. Assume now that η is a Poisson process with intensity measure λ. Use Theorem 3.9 to show that

$$\mathbb{E}\Big[\prod_{n=1}^{\kappa} u(X_n)\Big] = \exp\Big[-\int (1-u(x))\,\lambda(dx)\Big], \tag{3.5}$$

for any measurable $u\colon \mathbb{X} \to [0,1]$, where η is assumed to be given by (3.3).

The left-hand side of (3.5) is called the *probability generating functional* of η. It can be defined for any point process (proper or not) by taking the expectation of $\exp\big[\int \ln u(x)\,\eta(dx)\big]$.

Exercise 3.7 Let η be a Poisson process with finite intensity measure λ. Show for all $f \in \mathbb{R}_+(\mathbf{N})$ that

$$\mathbb{E}[f(\eta)] = e^{-\lambda(\mathbb{X})} f(0) + e^{-\lambda(\mathbb{X})} \sum_{n=1}^{\infty} \frac{1}{n!} \int f(\delta_{x_1} + \cdots + \delta_{x_n})\, \lambda^n(d(x_1,\ldots,x_n)).$$

Exercise 3.8 Let η be a Poisson process with s-finite intensity measure λ and let $f \in \mathbb{R}_+(\mathbf{N})$ be such that $\mathbb{E}[f(\eta)] < \infty$. Suppose that η' is a Poisson process with intensity measure λ' such that $\lambda = \lambda' + \nu$ for some finite measure ν. Show that $\mathbb{E}[f(\eta')] < \infty$. (Hint: Use the superposition theorem.)

Exercise 3.9 In the setting of Exercise 2.5, show that there is a probability measure λ on $(\mathbb{X}, \mathcal{X})$ and a Poisson process η with intensity measure λ such that η is not proper. (Hint: Use Exercise 2.5.)

Exercise 3.10 Let $0 < \gamma < \gamma'$. Give an example of two Poisson processes η, η' on $(0,1)$ with intensity measures $\gamma\lambda_1$ and $\gamma'\lambda_1$, respectively (λ_1 denoting Lebesgue measure), such that $\eta \le \eta'$ but $\eta' - \eta$ is *not* a Poisson process. (Hint: Use Exercise 1.9.)

Exercise 3.11 Let η be a Poisson process with intensity measure λ and let $B_1, B_2 \in \mathcal{X}$ satisfy $\lambda(B_1) < \infty$ and $\lambda(B_2) < \infty$. Show that the covariance between $\eta(B_1)$ and $\eta(B_2)$ is given by $\mathbb{C}\mathrm{ov}[\eta(B_1), \eta(B_2)] = \lambda(B_1 \cap B_2)$.

4

The Mecke Equation and Factorial Measures

The Mecke equation provides a way to compute the expectation of integrals, i.e. sums, with respect to a Poisson process, where the integrand can depend on both the point process and the point in the state space. This functional equation characterises a Poisson process. The Mecke identity can be extended to integration with respect to factorial measures, i.e. to multiple sums. Factorial measures can also be used to define the Janossy measures, thus providing a local description of a general point process. The factorial moment measures of a point process are defined as the expected factorial measures. They describe the probability of the occurrence of points in a finite number of infinitesimally small sets.

4.1 The Mecke Equation

In this chapter we take $(\mathbb{X}, \mathcal{X})$ to be arbitrary measurable space and use the abbreviation $(\mathbf{N}, \mathcal{N}) := (\mathbf{N}(\mathbb{X}), \mathcal{N}(\mathbb{X}))$. Let η be a Poisson process on $\mathbb{X}$ with s-finite intensity measure λ and let $f \in \mathbb{R}_+(\mathbb{X} \times \mathbf{N})$. The complete independence of η implies for each $x \in \mathbb{X}$ that, heuristically speaking, $\eta(dx)$ and the restriction $\eta_{\{x\}^c}$ of η to $\mathbb{X} \setminus \{x\}$ are independent. Therefore

$$\mathbb{E}\left[\int f(x, \eta_{\{x\}^c})\, \eta(dx)\right] = \int \mathbb{E}[f(x, \eta_{\{x\}^c})]\, \lambda(dx), \tag{4.1}$$

where we ignore measurability issues. If $\mathbb{P}(\eta(\{x\}) = 0) = 1$ for each $x \in \mathbb{X}$ (which is the case if λ is a diffuse measure on a Borel space), then the right-hand side of (4.1) equals $\int \mathbb{E}[f(x, \eta)]\, \lambda(dx)$. (Exercise 6.11 shows a way to extend this to an arbitrary intensity measure.) We show that a proper version of the resulting integral identity holds in general and characterises the Poisson process. This equation is a fundamental tool for analysing the Poisson process and can be used in many specific calculations. In the special case where $\mathbb{X}$ has just a single element, Theorem 4.1 essentially reduces to an earlier result about the Poisson distribution, namely Proposition 1.1.

Theorem 4.1 (Mecke equation) *Let λ be an s-finite measure on $\mathbb{X}$ and η a point process on $\mathbb{X}$. Then η is a Poisson process with intensity measure λ if and only if*

$$\mathbb{E}\Big[\int f(x,\eta)\,\eta(dx)\Big] = \int \mathbb{E}[f(x,\eta+\delta_x)]\,\lambda(dx) \tag{4.2}$$

for all $f \in \mathbb{R}_+(\mathbb{X}\times\mathbf{N})$.

Proof Let us start by noting that the mapping $(x,\mu) \mapsto \mu + \delta_x$ (adding a point x to the counting measure μ) from $\mathbb{X}\times\mathbf{N}$ to $\mathbf{N}$ is measurable. Indeed, the mapping $(x,\mu) \mapsto \mu(B) + \mathbf{1}_B(x)$ is measurable for all $B \in \mathcal{X}$.

If η is a Poisson process, then (4.2) is a special case of (4.11) to be proved in Section 4.2.

Assume now that (4.2) holds for all measurable $f \geq 0$. Let $B_1,\dots,B_m$ be disjoint sets in $\mathcal{X}$ with $\lambda(B_i) < \infty$ for each i. For $k_1,\dots,k_m \in \overline{\mathbb{N}}_0$ with $k_1 \geq 1$ we define

$$f(x,\mu) = \mathbf{1}_{B_1}(x)\prod_{i=1}^{m}\mathbf{1}\{\mu(B_i)=k_i\}, \quad (x,\mu)\in\mathbb{X}\times\mathbf{N}.$$

Then

$$\mathbb{E}\Big[\int f(x,\eta)\,\eta(dx)\Big] = \mathbb{E}\Big[\eta(B_1)\prod_{i=1}^{m}\mathbf{1}\{\eta(B_i)=k_i\}\Big] = k_1\mathbb{P}\Big(\cap_{i=1}^{m}\{\eta(B_i)=k_i\}\Big),$$

with the (measure theory) convention $\infty\cdot 0 := 0$. On the other hand, we have for each $x\in\mathbb{X}$ that

$$\mathbb{E}[f(x,\eta+\delta_x)] = \mathbf{1}_{B_1}(x)\,\mathbb{P}(\eta(B_1)=k_1-1,\eta(B_2)=k_2,\dots,\eta(B_m)=k_m)$$

(with $\infty - 1 := \infty$) so that, by (4.2),

$$k_1\,\mathbb{P}(\cap_{i=1}^{m}\{\eta(B_i)=k_i\}) = \lambda(B_1)\,\mathbb{P}(\{\eta(B_1)=k_1-1\}\cap\cap_{i=2}^{m}\{\eta(B_i)=k_i\}).$$

Assume that $\mathbb{P}(\cap_{i=2}^{m}\{\eta(B_i)=k_i\}) > 0$ and note that otherwise $\eta(B_1)$ and the event $\cap_{i=2}^{m}\{\eta(B_i)=k_i\}$ are independent. Putting

$$\pi_k = \mathbb{P}(\eta(B_1)=k \mid \cap_{i=2}^{m}\{\eta(B_i)=k_i\}), \quad k\in\overline{\mathbb{N}}_0,$$

we have

$$k\pi_k = \lambda(B_1)\pi_{k-1}, \quad k\in\overline{\mathbb{N}}.$$

Since $\lambda(B_1) < \infty$ this implies $\pi_\infty = 0$. The only distribution satisfying this recursion is given by $\pi_k = \mathrm{Po}(\lambda(B_1);k)$, regardless of $k_2,\dots,k_m$; hence $\eta(B_1)$ is $\mathrm{Po}(\lambda(B_1))$ distributed, and independent of $\cap_{i=2}^{m}\{\eta(B_i)=k_i\}$. Hence, by an induction on m, the variables $\eta(B_1),\dots,\eta(B_m)$ are independent.

For general $B \in \mathcal{X}$ we still get for all $k \in \mathbb{N}$ that

$$k\,\mathbb{P}(\eta(B) = k) = \lambda(B)\,\mathbb{P}(\eta(B) = k - 1).$$

If $\lambda(B) = \infty$ we obtain $\mathbb{P}(\eta(B) = k - 1) = 0$ and hence $\mathbb{P}(\eta(B) = \infty) = 1$.

It follows that η has the defining properties of the Poisson process. □

4.2 Factorial Measures and the Multivariate Mecke Equation

Equation (4.2) admits a useful generalisation involving multiple integration. To formulate this version we consider, for $m \in \mathbb{N}$, the m-th power $(\mathbb{X}^m, \mathcal{X}^m)$ of $(\mathbb{X}, \mathcal{X})$; see Section A.1. Suppose $\mu \in \mathbf{N}$ is given by

$$\mu = \sum_{j=1}^{k} \delta_{x_j} \tag{4.3}$$

for some $k \in \overline{\mathbb{N}}_0$ and some $x_1, x_2, \ldots \in \mathbb{X}$ (not necessarily distinct) as in (2.1). Then we define another measure $\mu^{(m)} \in \mathbf{N}(\mathbb{X}^m)$ by

$$\mu^{(m)}(C) = \sum_{i_1,\ldots,i_m \le k}^{\ne} \mathbf{1}\{(x_{i_1}, \ldots, x_{i_m}) \in C\}, \quad C \in \mathcal{X}^m, \tag{4.4}$$

where the superscript $\ne$ indicates summation over m-tuples with pairwise different entries and where an empty sum is defined as zero. (In the case $k = \infty$ this involves only integer-valued indices.) In other words this means that

$$\mu^{(m)} = \sum_{i_1,\ldots,i_m \le k}^{\ne} \delta_{(x_{i_1},\ldots,x_{i_m})}. \tag{4.5}$$

To aid understanding, it is helpful to consider in (4.4) a set C of the special product form $B_1 \times \cdots \times B_m$. If these sets are pairwise disjoint, then the right-hand side of (4.4) factorises, yielding

$$\mu^{(m)}(B_1 \times \cdots \times B_m) = \prod_{j=1}^{m} \mu(B_j). \tag{4.6}$$

If, on the other hand, $B_j = B$ for all $j \in \{1, \ldots, m\}$ then, clearly,

$$\mu^{(m)}(B^m) = \mu(B)(\mu(B) - 1) \cdots (\mu(B) - m + 1) = (\mu(B))_m. \tag{4.7}$$

Therefore $\mu^{(m)}$ is called the m-th *factorial measure* of μ. For $m = 2$ and arbitrary $B_1, B_2 \in \mathcal{X}$ we obtain from (4.4) that

$$\mu^{(2)}(B_1 \times B_2) = \mu(B_1)\mu(B_2) - \mu(B_1 \cap B_2), \tag{4.8}$$

provided that $\mu(B_1 \cap B_2) < \infty$. Otherwise $\mu^{(2)}(B_1 \times B_2) = \infty$.

Factorial measures satisfy the following useful recursion:

Lemma 4.2 *Let $\mu \in \mathbf{N}$ be given by (4.3) and define $\mu^{(1)} := \mu$. Then, for all $m \in \mathbb{N}$,*

$$\mu^{(m+1)} = \int \bigg[\int \mathbf{1}\{(x_1, \dots, x_{m+1}) \in \cdot\} \mu(dx_{m+1}) \\ - \sum_{j=1}^{m} \mathbf{1}\{(x_1, \dots, x_m, x_j) \in \cdot\} \bigg] \mu^{(m)}(d(x_1, \dots, x_m)). \tag{4.9}$$

Proof Let $m \in \mathbb{N}$ and $C \in \mathcal{X}^{m+1}$. Then

$$\mu^{(m+1)}(C) = \sum_{i_1,\dots,i_m \le k}^{\neq} \sum_{\substack{j=1 \\ j \notin \{i_1,\dots,i_m\}}}^{k} \mathbf{1}\{(x_{i_1}, \dots, x_{i_m}, x_j) \in C\}.$$

Here the inner sum equals

$$\sum_{j=1}^{k} \mathbf{1}\{(x_{i_1}, \dots, x_{i_m}, x_j) \in C\} - \sum_{l=1}^{m} \mathbf{1}\{(x_{i_1}, \dots, x_{i_m}, x_{i_l}) \in C\},$$

where the latter difference is either a non-negative integer (if the first sum is finite) or ∞ (if the first sum is infinite). This proves the result. □

For a general space $(\mathbb{X}, \mathcal{X})$ there is no guarantee that a measure $\mu \in \mathbf{N}$ can be represented as in (4.3); see Exercise 2.5. Equation (4.9) suggests a recursive definition of the factorial measures of a general $\mu \in \mathbf{N}$, without using a representation as a sum of Dirac measures. The next proposition confirms this idea.

Proposition 4.3 *For any $\mu \in \mathbf{N}$ there is a unique sequence $\mu^{(m)} \in \mathbf{N}(\mathbb{X}^m)$, $m \in \mathbb{N}$, satisfying $\mu^{(1)} := \mu$ and the recursion (4.9). The mappings $\mu \mapsto \mu^{(m)}$ are measurable.*

The proof of Proposition 4.3 is given in Section A.1 (see Proposition A.18) and can be skipped without too much loss. It is enough to remember that $\mu^{(m)}$ can be defined by (4.4), whenever μ is given by (4.3). This follows from Lemma 4.2 and the fact that the solution of (4.9) must be unique. It follows by induction that (4.6) and (4.7) remain valid for general $\mu \in \mathbf{N}$; see Exercise 4.4.

Let η be a point process on $\mathbb{X}$ and let $m \in \mathbb{N}$. Proposition 4.3 shows that

$\eta^{(m)}$ is a point process on $\mathbb{X}^m$. If η is proper and given as at (2.4), then

$$\eta^{(m)} = \sum_{i_1,\dots,i_m\in\{1,\dots,\kappa\}}^{\neq} \delta_{(X_{i_1},\dots,X_{i_m})}. \tag{4.10}$$

We continue with the multivariate version of the Mecke equation (4.2).

Theorem 4.4 (Multivariate Mecke equation) *Let η be a Poisson process on $\mathbb{X}$ with s-finite intensity measure λ. Then, for every $m \in \mathbb{N}$ and for every $f \in \mathbb{R}_+(\mathbb{X}^m \times \mathbf{N})$,*

$$\mathbb{E}\Big[\int f(x_1,\dots,x_m,\eta)\,\eta^{(m)}(d(x_1,\dots,x_m))\Big] \\ = \int \mathbb{E}[f(x_1,\dots,x_m,\eta+\delta_{x_1}+\cdots+\delta_{x_m})]\,\lambda^m(d(x_1,\dots,x_m)). \tag{4.11}$$

Proof By Proposition 4.3, the map $\mu \mapsto \mu^{(m)}$ is measurable, so that (4.11) involves only the distribution of η. By Corollary 3.7 we can hence assume that η is proper and given by (2.4). Let us first assume that $\lambda(\mathbb{X}) < \infty$. Then $\lambda = \gamma\,\mathbb{Q}$ for some $\gamma \ge 0$ and some probability measure $\mathbb{Q}$ on $\mathbb{X}$. By Proposition 3.5, we can then assume that η is a mixed binomial process as in Definition 3.4, with κ having the $\mathrm{Po}(\gamma)$ distribution. Let $f \in \overline{\mathbb{R}}_+(\mathbb{X}^m \times \mathbf{N})$. Then we obtain from (4.10) and (2.2) that the left-hand side of (4.11) equals

$$e^{-\gamma}\sum_{k=m}^{\infty}\frac{\gamma^k}{k!}\mathbb{E}\Big[\sum_{i_1,\dots,i_m\in\{1,\dots,k\}}^{\neq} f(X_{i_1},\dots,X_{i_m},\delta_{X_1}+\cdots+\delta_{X_k})\Big] \\ = e^{-\gamma}\sum_{k=m}^{\infty}\frac{\gamma^k}{k!}\sum_{i_1,\dots,i_m\in\{1,\dots,k\}}^{\neq}\mathbb{E}[f(X_{i_1},\dots,X_{i_m},\delta_{X_1}+\cdots+\delta_{X_k})], \tag{4.12}$$

where we have used first independence of κ and (X_n) and then the fact that we can perform integration and summation in any order we want (since $f \ge 0$). Let us denote by $\mathbf{y} = (y_1,\dots,y_m)$ a generic element of $\mathbb{X}^m$. Since the X_i are independent with distribution $\mathbb{Q}$, the expression (4.12) equals

$$e^{-\gamma}\sum_{k=m}^{\infty}\frac{\gamma^k(k)_m}{k!}\mathbb{E}\Big[\int f\Big(\mathbf{y},\sum_{i=1}^{k-m}\delta_{X_i}+\sum_{j=1}^{m}\delta_{y_j}\Big)\mathbb{Q}^m(d\mathbf{y})\Big] \\ = e^{-\gamma}\gamma^m\sum_{k=m}^{\infty}\frac{\gamma^{k-m}}{(k-m)!}\int\mathbb{E}\Big[f\Big(\mathbf{y},\sum_{i=1}^{k-m}\delta_{X_i}+\sum_{j=1}^{m}\delta_{y_j}\Big)\Big]\mathbb{Q}^m(d\mathbf{y}) \\ = \int\mathbb{E}[f(y_1,\dots,y_m,\eta+\delta_{y_1}+\cdots+\delta_{y_m})]\,\lambda^m(d(y_1,\dots,y_m)),$$

where we have again used the mixed binomial representation. This proves (4.11) for finite λ.

Now suppose $\lambda(\mathbb{X}) = \infty$. As in the proof of Theorem 3.6 we can then assume that $\eta = \sum_i \eta_i$, where η_i are independent proper Poisson processes with intensity measures λ_i each having finite total measure. By the grouping property of independence, the point processes

$$\xi_i := \sum_{j \le i} \eta_j, \quad \chi_i := \sum_{j \ge i+1} \eta_j$$

are independent for each $i \in \mathbb{N}$. By (4.10) we have $\xi_i^{(m)} \uparrow \eta^{(m)}$ as $i \to \infty$. Hence we can apply monotone convergence (Theorem A.12) to see that the left-hand side of (4.11) is given by

$$\lim_{i\to\infty} \mathbb{E}\Big[\int f(x_1, \dots, x_m, \xi_i + \chi_i)\, \xi_i^{(m)}(d(x_1, \dots, x_m))\Big]$$
$$= \lim_{i\to\infty} \mathbb{E}\Big[\int f_i(x_1, \dots, x_m, \xi_i)\, \xi_i^{(m)}(d(x_1, \dots, x_m))\Big], \qquad (4.13)$$

where $f_i(x_1, \dots, x_m, \mu) := \mathbb{E}[f(x_1, \dots, x_m, \mu + \chi_i)]$, $(x_1, \dots, x_m, \mu) \in \mathbb{X}^m \times \mathbf{N}$. Setting $\lambda_i' := \sum_{j=1}^i \lambda_j$, we can now apply the previous result to obtain from Fubini's theorem (Theorem A.13) that the expression (4.13) equals

$$\lim_{i\to\infty} \int \mathbb{E}[f_i(x_1, \dots, x_m, \xi_i + \delta_{x_1} + \cdots + \delta_{x_m})]\, (\lambda_i')^m(d(x_1, \dots, x_m))$$
$$= \lim_{i\to\infty} \int \mathbb{E}[f(x_1, \dots, x_m, \eta + \delta_{x_1} + \cdots + \delta_{x_m})]\, (\lambda_i')^m(d(x_1, \dots, x_m)).$$

By (A.7) this is the right-hand side of (4.11). □

Next we formulate another useful version of the multivariate Mecke equation. For $\mu \in \mathbf{N}$ and $x \in \mathbb{X}$ we define the measure $\mu \setminus \delta_x \in \mathbf{N}$ by

$$\mu \setminus \delta_x := \begin{cases} \mu - \delta_x, & \text{if } \mu \ge \delta_x, \\ \mu, & \text{otherwise.} \end{cases} \qquad (4.14)$$

For $x_1, \dots, x_m \in \mathbb{X}$, the measure $\mu \setminus \delta_{x_1} \setminus \cdots \setminus \delta_{x_m} \in \mathbf{N}$ is defined inductively.

Theorem 4.5 *Let η be a proper Poisson process on $\mathbb{X}$ with s-finite intensity measure λ and let $m \in \mathbb{N}$. Then, for any $f \in \mathbb{R}_+(\mathbb{X}^m \times \mathbf{N})$,*

$$\mathbb{E}\Big[\int f(x_1, \dots, x_m, \eta \setminus \delta_{x_1} \setminus \cdots \setminus \delta_{x_m})\, \eta^{(m)}(d(x_1, \dots, x_m))\Big]$$
$$= \int \mathbb{E}[f(x_1, \dots, x_m, \eta)]\, \lambda^m(d(x_1, \dots, x_m)). \qquad (4.15)$$

Proof If $\mathbb{X}$ is a subspace of a complete separable metric space as in Proposition 6.2, then it is easy to show that $(x_1, \dots, x_m, \mu) \mapsto \mu \setminus \delta_{x_1} \setminus \cdots \setminus \delta_{x_m}$ is a measurable mapping from $\mathbb{X}^m \times \mathbf{N}_l(\mathbb{X})$ to $\mathbf{N}_l(\mathbb{X})$. In that case, and if λ is locally finite, (4.15) follows upon applying (4.11) to the function $(x_1, \dots, x_m, \mu) \mapsto f(x_1, \dots, x_m, \mu \setminus \delta_{x_1} \setminus \cdots \setminus \delta_{x_m})$. In the general case we use that η is proper. Therefore the mapping $(\omega, x_1, \dots, x_m) \mapsto \eta(\omega) \setminus \delta_{x_1} \setminus \cdots \setminus \delta_{x_m}$ is measurable, which is enough to make (4.15) a meaningful statement. The proof can proceed in exactly the same way as the proof of Theorem 4.4. □

4.3 Janossy Measures

The *restriction* ν_B of a measure ν on $\mathbb{X}$ to a set $B \in \mathcal{X}$ is a measure on $\mathbb{X}$ defined by

$$\nu_B(B') := \nu(B \cap B'), \quad B' \in \mathcal{X}. \tag{4.16}$$

If η is a point process on $\mathbb{X}$, then so is its restriction η_B. For $B \in \mathcal{X}$, $m \in \mathbb{N}$ and a measure ν on $\mathbb{X}$ we write $\nu_B^m := (\nu_B)^m$. For a point process η on $\mathbb{X}$ we write $\eta_B^{(m)} := (\eta_B)^{(m)}$.

Factorial measures can be used to describe the restriction of point processes as follows.

Definition 4.6 *Let η be a point process on $\mathbb{X}$, let $B \in \mathcal{X}$ and $m \in \mathbb{N}$. The* Janossy measure of order *m of η restricted to B is the measure on $\mathbb{X}^m$ defined by*

$$J_{\eta,B,m} := \frac{1}{m!}\mathbb{E}[\mathbf{1}\{\eta(B) = m\}\eta_B^{(m)}(\cdot)]. \tag{4.17}$$

The number $J_{\eta,B,0} := \mathbb{P}(\eta(B) = 0)$ is called the Janossy measure of order 0.

Note that the Janossy measures $J_{\eta,B,m}$ are symmetric (see (A.17))) and

$$J_{\eta,B,m}(\mathbb{X}^m) = \mathbb{P}(\eta(B) = m), \quad m \in \mathbb{N}. \tag{4.18}$$

If $\mathbb{P}(\eta(B) < \infty) = 1$, then the Janossy measures determine the distribution of the restriction η_B of η to B:

Theorem 4.7 *Let η and η' be point processes on $\mathbb{X}$. Let $B \in \mathcal{X}$ and assume that $J_{\eta,B,m} = J_{\eta',B,m}$ for each $m \in \mathbb{N}_0$. Then*

$$\mathbb{P}(\eta(B) < \infty, \eta_B \in \cdot) = \mathbb{P}(\eta'(B) < \infty, \eta'_B \in \cdot).$$

Proof For notational convenience we assume that $B = \mathbb{X}$. Let $m \in \mathbb{N}$ and suppose that $\mu \in \mathbf{N}$ satisfies $\mu(\mathbb{X}) = m$. We assert for each $A \in \mathcal{N}$ that

$$\mathbf{1}\{\mu \in A\} = \frac{1}{m!} \int \mathbf{1}\{\delta_{x_1} + \cdots + \delta_{x_m} \in A\} \mu^{(m)}(d(x_1, \ldots, x_m)). \tag{4.19}$$

Since both sides of (4.19) are finite measures in A, it suffices to prove this identity for each set A of the form

$$A = \{\nu \in \mathbf{N} : \nu(B_1) = i_1, \ldots, \nu(B_n) = i_n\},$$

where $n \in \mathbb{N}$, $B_1, \ldots, B_n \in \mathcal{X}$ and $i_1, \ldots, i_n \in \mathbb{N}_0$. Given such a set, let μ' be defined as in Lemma A.15. Then $\mu \in A$ if and only if $\mu' \in A$ and the right-hand side of (4.19) does not change upon replacing μ by μ'. Hence it suffices to check (4.19) for finite sums of Dirac measures. This is obvious from (4.4).

It follows from (4.17) that for all $m \in \mathbb{N}$ and $f \in \mathbb{R}_+(\mathbb{X})$ we have

$$\int f \, dJ_{\eta,\mathbb{X},m} = \frac{1}{m!} \mathbb{E}\Big[\mathbf{1}\{\eta(B) = m\} \int f \, d\eta^{(m)}\Big]. \tag{4.20}$$

From (4.19) and (4.20) we obtain for each $A \in \mathcal{N}$ that

$$\begin{aligned}
&\mathbb{P}(\eta(\mathbb{X}) < \infty, \eta \in A) \\
&\quad = \mathbf{1}\{0 \in A\} J_{\eta,\mathbb{X},0} + \sum_{m=1}^{\infty} \int \mathbf{1}\{\delta_{x_1} + \cdots + \delta_{x_m} \in A\} \, J_{\eta,\mathbb{X},m}(d(x_1, \ldots, x_m))
\end{aligned}$$

and hence the assertion. □

Example 4.8 Let η be a Poisson process on $\mathbb{X}$ with s-finite intensity measure λ. Let $m \in \mathbb{N}$ and $B \in \mathcal{X}$. By the multivariate Mecke equation (Theorem 4.4) we have for each $C \in \mathcal{X}^m$ that

$$\begin{aligned}
J_{\eta,B,m}(C) &= \frac{1}{m!} \mathbb{E}[\mathbf{1}\{\eta(B) = m\} \eta^{(m)}(B^m \cap C)] \\
&= \frac{1}{m!} \mathbb{E}\Big[\int_C \mathbf{1}\{(\eta + \delta_{x_1} + \cdots + \delta_{x_m})(B) = m\} \, \lambda_B^m(d(x_1, \ldots, x_m))\Big].
\end{aligned}$$

For $x_1, \ldots, x_m \in B$ we have $(\eta + \delta_{x_1} + \cdots + \delta_{x_m})(B) = m$ if and only if $\eta(B) = 0$. Therefore we obtain

$$J_{\eta,B,m} = \frac{e^{-\lambda(B)}}{m!} \lambda_B^m, \quad m \in \mathbb{N}. \tag{4.21}$$

4.4 Factorial Moment Measures

Definition 4.9 For $m \in \mathbb{N}$ the m-th *factorial moment measure* of a point process η is the measure α_m on $\mathbb{X}^m$ defined by

$$\alpha_m(C) := \mathbb{E}[\eta^{(m)}(C)], \quad C \in \mathcal{X}^m. \tag{4.22}$$

If the point process η is proper, i.e. given by (2.4), then

$$\alpha_m(C) = \mathbb{E}\Big[\sum\nolimits^{\neq}_{i_1,\dots,i_m \le \kappa} \mathbf{1}\{(X_{i_1},\dots,X_{i_m}) \in C\}\Big], \tag{4.23}$$

and hence for $f \in \mathbb{R}_+(\mathbb{X}^m)$ we have that

$$\int_{\mathbb{X}^m} f(x_1,\dots,x_m)\,\alpha_m(d(x_1,\dots,x_m)) = \mathbb{E}\Big[\sum\nolimits^{\neq}_{i_1,\dots,i_m \le \kappa} f(X_{i_1},\dots,X_{i_m})\Big].$$

The first factorial moment measure of a point process η is just the intensity measure of Definition 2.5, while the second describes the second order properties of η. For instance, it follows from (4.8) (and Exercise 4.4 if η is not proper) that

$$\alpha_2(B_1 \times B_2) = \mathbb{E}[\eta(B_1)\eta(B_2)] - \mathbb{E}[\eta(B_1 \cap B_2)], \tag{4.24}$$

provided that $\mathbb{E}[\eta(B_1 \cap B_2)] < \infty$.

Theorem 4.4 has the following immediate consequence:

Corollary 4.10 *Given $m \in \mathbb{N}$ the m-th factorial moment measure of a Poisson process with s-finite intensity measure λ is λ^m.*

Proof Apply (4.11) to the function $f(x_1,\dots,x_m,\eta) = \mathbf{1}\{(x_1,\dots,x_m) \in C\}$ for $C \in \mathcal{X}^m$. □

Let η be a point process on $\mathbb{X}$ with intensity measure λ and let $f, g \in L^1(\lambda) \cap L^2(\lambda)$. By the Cauchy–Schwarz inequality ((A.2) for $p = q = 2$) we have $fg \in L^1(\lambda)$ so that Campbell's formula (Proposition 2.7) shows that $\eta(|f|) < \infty$ and $\eta(|fg|) < \infty$ hold almost surely. Therefore it follows from the case $m = 1$ of (4.9) that

$$\int f(x)f(y)\,\eta^{(2)}(d(x,y)) = \eta(f)\eta(g) - \eta(fg), \quad \mathbb{P}\text{-a.s.}$$

Reordering terms and taking expectations gives

$$\mathbb{E}[\eta(f)\eta(g)] = \lambda(fg) + \int f(x)g(y)\,\alpha_2(d(x,y)), \tag{4.25}$$

provided that $\int |f(x)g(y)|\,\alpha_2(d(x,y)) < \infty$ or $f, g \ge 0$. If η is a Poisson process with s-finite intensity measure λ, then (4.25) and Corollary 4.10 imply the following useful generalisation of Exercise 3.11:

$$\mathbb{E}[\eta(f)\eta(g)] = \lambda(fg) + \lambda(f)\lambda(g), \quad f, g \in L^1(\lambda) \cap L^2(\lambda). \tag{4.26}$$

Under certain assumptions the factorial moment measures of a point process determine its distribution. To derive this result we need the following lemma. We use the conventions $e^{-\infty} := 0$ and $\log 0 := -\infty$.

Lemma 4.11 *Let η be a point process on $\mathbb{X}$. Let $B \in \mathcal{X}$ and assume that there exists $c > 1$ such that the factorial moment measures α_n of η satisfy*

$$\alpha_n(B^n) \le n!c^n, \quad n \ge 1. \tag{4.27}$$

Let $u \in \mathbb{R}_+(\mathbb{X})$ and $a < c^{-1}$ be such that $u(x) < a$ for $x \in B$ and $u(x) = 0$ for $x \notin B$. Then

$$\mathbb{E}\left[\exp\left(\int \log(1 - u(x))\,\eta(dx)\right)\right] = 1 + \sum_{n=1}^{\infty} \frac{(-1)^n}{n!} \int u(x_1)\cdots u(x_n)\,\alpha_n(d(x_1, \dots, x_n)). \tag{4.28}$$

Proof Since u vanishes outside B, we have

$$P := \exp\left(\int \log(1 - u(x))\,\eta(dx)\right) = \exp\left(\int \log(1 - u(x))\,\eta_B(dx)\right).$$

Hence we can assume that $\eta(\mathbb{X} \setminus B) = 0$. Since $\alpha_1(B) = \mathbb{E}[\eta(B)] < \infty$, we can also assume that $\eta(B) < \infty$. But then we obtain from Exercise 4.6 that

$$P = \sum_{n=0}^{\infty} (-1)^n P_n,$$

where $P_0 := 1$ and

$$P_n := \frac{1}{n!} \int u(x_1)\cdots u(x_n)\,\eta^{(n)}(d(x_1, \dots, x_n)),$$

and where we note that $\eta^{(n)} = 0$ if $n > \eta(\mathbb{X})$; see (4.7). Exercise 4.9 asks the reader to prove that

$$\sum_{n=0}^{2m-1} (-1)^n P_n \le P \le \sum_{n=0}^{2m} (-1)^n P_n, \quad m \ge 1. \tag{4.29}$$

These inequalities show that

$$\left| P - \sum_{n=0}^{k} (-1)^n P_n \right| \le P_k, \quad k \ge 1.$$

It follows that

$$\left|\mathbb{E}[P] - \mathbb{E}\left[\sum_{n=0}^{k}(-1)^n P_n\right]\right| \le \mathbb{E}[P_k] = \frac{1}{k!}\int u(x_1)\cdots u(x_k)\,\alpha_k(d(x_1,\dots,x_k)),$$

where we have used the definition of the factorial moment measures. The last term can be bounded by

$$\frac{a^k}{k!}\alpha_k(B^k) \le a^k c^k,$$

which tends to zero as $k \to \infty$. This finishes the proof. □

Proposition 4.12 *Let η and η' be point processes on $\mathbb{X}$ with the same factorial moment measures α_n, $n \ge 1$. Assume that there is a sequence $B_k \in \mathcal{X}$, $k \in \mathbb{N}$, with $B_k \uparrow \mathbb{X}$ and numbers $c_k > 0$, $k \in \mathbb{N}$, such that*

$$\alpha_n(B_k^n) \le n!c_k^n, \quad k, n \in \mathbb{N}. \tag{4.30}$$

Then $\eta \overset{d}{=} \eta'$.

Proof By Proposition 2.10 and monotone convergence it is enough to prove that $L_\eta(v) = L_{\eta'}(v)$ for each bounded $v \in \mathbb{R}_+(\mathbb{X})$ such that there exists a set $B \in \{B_k : k \in \mathbb{N}\}$ with $v(x) = 0$ for all $x \notin B$. This puts us into the setting of Lemma 4.11. Let $v \in \mathbb{R}_+(\mathbb{X})$ have the upper bound $a > 0$. For each $t \in [0, -(\log(1-c^{-1}))/a)$ we can apply Lemma 4.11 with $u := 1 - e^{-tv}$. This gives us $L_\eta(tv) = L_{\eta'}(tv)$. Since $t \mapsto L_\eta(tv)$ is analytic on $(0,\infty)$, we obtain $L_\eta(tv) = L_{\eta'}(tv)$ for all $t \ge 0$ and, in particular, $L_\eta(v) = L_{\eta'}(v)$. □

4.5 Exercises

Exercise 4.1 Let η be a Poisson process on $\mathbb{X}$ with intensity measure λ and let $A \in \mathcal{N}$ have $\mathbb{P}(\eta \in A) = 0$. Use the Mecke equation to show that $\mathbb{P}(\eta + \delta_x \in A) = 0$ for λ-a.e. x.

Exercise 4.2 Let $\mu \in \mathbf{N}$ be given by (4.3) and let $m \in \mathbb{N}$. Show that

$$\mu^{(m)}(C) = \int\cdots\int \mathbf{1}_C(x_1,\dots,x_m)\Big(\mu - \sum_{j=1}^{m-1}\delta_{x_j}\Big)(dx_m)\Big(\mu - \sum_{j=1}^{m-2}\delta_{x_j}\Big)(dx_{m-1})$$
$$\cdots(\mu - \delta_{x_1})(dx_2)\,\mu(dx_1), \quad C \in \mathcal{X}^m. \tag{4.31}$$

This formula involves integrals with respect to *signed measures* of the form $\mu - \nu$, where $\mu, \nu \in \mathbf{N}$ and ν is finite. These integrals are defined as a difference of integrals in the natural way.

Exercise 4.3 Let $\mu \in \mathbf{N}$ and $x \in \mathbb{X}$. Show for all $m \in \mathbb{N}$ that

$$\int [\mathbf{1}\{(x, x_1, \dots, x_m) \in \cdot\} + \cdots + \mathbf{1}\{(x_1, \dots, x_m, x) \in \cdot\}]\, \mu^{(m)}(d(x_1, \dots, x_m)) \\ + \mu^{(m+1)} = (\mu + \delta_x)^{(m+1)}.$$

(Hint: Use Proposition A.18 to reduce to the case $\mu(\mathbb{X}) < \infty$ and then Lemma A.15 to reduce further to the case (4.3) with $k \in \mathbb{N}$.)

Exercise 4.4 Let $\mu \in \mathbf{N}$. Use the recursion (4.9) to show that (4.6), (4.7) and (4.8) hold.

Exercise 4.5 Let $\mu \in \mathbf{N}$ be given by $\mu := \sum_{j=1}^{k} \delta_{x_j}$ for some $k \in \mathbb{N}_0$ and some $x_1, \dots, x_k \in \mathbb{X}$. Let $u\colon \mathbb{X} \to \mathbb{R}$ be measurable. Show that

$$\prod_{j=1}^{k}(1 - u(x_j)) = 1 + \sum_{n=1}^{k} \frac{(-1)^n}{n!} \int u(x_1) \cdots u(x_n)\, \mu^{(n)}(d(x_1, \dots, x_n)).$$

Exercise 4.6 Let $\mu \in \mathbf{N}$ such that $\mu(\mathbb{X}) < \infty$ and let $u \in \mathbb{R}_+(\mathbb{X})$ satisfy $u < 1$. Show that

$$\exp\Big(\int \log(1 - u(x))\, \mu(dx)\Big) \\ = 1 + \sum_{n=1}^{\infty} \frac{(-1)^n}{n!} \int \prod_{j=1}^{n} u(x_j)\, \mu^{(n)}(d(x_1, \dots, x_n)).$$

(Hint: If u takes only a finite number of values, then the result follows from Lemma A.15 and Exercise 4.5.)

Exercise 4.7 (Converse to Theorem 4.4) Let $m \in \mathbb{N}$ with $m > 1$. Prove or disprove that for any σ-finite measure space $(\mathbb{X}, \mathcal{X}, \lambda)$, if η is a point process on $\mathbb{X}$ satisfying (4.11) for all $f \in \mathbb{R}_+(\mathbb{X}^m \times \mathbf{N})$, then η is a Poisson process with intensity measure λ. (For $m = 1$, this is true by Theorem 4.1.)

Exercise 4.8 Give another (inductive) proof of the multivariate Mecke identity (4.11) using the univariate version (4.2) and the recursion (4.9).

Exercise 4.9 Prove the inequalities (4.29). (Hint: Use induction.)

Exercise 4.10 Let η be a Poisson process on $\mathbb{X}$ with intensity measure λ and let $B \in \mathcal{X}$ with $0 < \lambda(B) < \infty$. Let $U_1, \dots, U_n$ be independent random elements of $\mathbb{X}$ with distribution $\lambda(B)^{-1}\lambda(B \cap \cdot)$ and assume that $(U_1, \dots, U_n)$ and η are independent. Show that the distribution of $\eta + \delta_{U_1} + \cdots + \delta_{U_n}$ is absolutely continuous with respect to $\mathbb{P}(\eta \in \cdot)$ and that $\mu \mapsto \lambda(B)^{-n}\mu^{(n)}(B^n)$ is a version of the density.

5

Mappings, Markings and Thinnings

It was shown in Chapter 3 that an independent superposition of Poisson processes is again Poisson. The properties of a Poisson process are also preserved under other operations. A mapping from the state space to another space induces a Poisson process on the new state space. A more intriguing persistence property is the Poisson nature of position-dependent markings and thinnings of a Poisson process.

5.1 Mappings and Restrictions

Consider two measurable spaces $(\mathbb{X}, \mathcal{X})$ and $(\mathbb{Y}, \mathcal{Y})$ along with a measurable mapping $T : \mathbb{X} \to \mathbb{Y}$. For any measure μ on $(\mathbb{X}, \mathcal{X})$ we define the *image* of μ under T (also known as the *push-forward* of μ), to be the measure $T(\mu)$ defined by $T(\mu) = \mu \circ T^{-1}$, i.e.

$$T(\mu)(C) := \mu(T^{-1}C), \quad C \in \mathcal{Y}. \tag{5.1}$$

In particular, if η is a point process on $\mathbb{X}$, then for any $\omega \in \Omega$, $T(\eta(\omega))$ is a measure on $\mathbb{Y}$ given by

$$T(\eta(\omega))(C) = \eta(\omega, T^{-1}(C)), \quad C \in \mathcal{Y}. \tag{5.2}$$

If η is a proper point process, i.e. one given by $\eta = \sum_{n=1}^{\kappa} \delta_{X_n}$ as in (2.4), the definition of $T(\eta)$ implies that

$$T(\eta) = \sum_{n=1}^{\kappa} \delta_{T(X_n)}. \tag{5.3}$$

Theorem 5.1 (Mapping theorem) *Let η be a point process on $\mathbb{X}$ with intensity measure λ and let $T : \mathbb{X} \to \mathbb{Y}$ be measurable. Then $T(\eta)$ is a point process with intensity measure $T(\lambda)$. If η is a Poisson process, then $T(\eta)$ is a Poisson process too.*

Proof We first note that $T(\mu) \in \mathbf{N}$ for any $\mu \in \mathbf{N}$. Indeed, if $\mu = \sum_{j=1}^{\infty} \mu_j$, then $T(\mu) = \sum_{j=1}^{\infty} T(\mu_j)$. Moreover, if the μ_j are $\mathbb{N}_0$-valued, so are the $T(\mu_j)$.

For any $C \in \mathcal{Y}$, $T(\eta)(C)$ is a random variable and by the definition of the intensity measure its expectation is

$$\mathbb{E}[T(\eta)(C)] = \mathbb{E}[\eta(T^{-1}C)] = \lambda(T^{-1}C) = T(\lambda)(C). \tag{5.4}$$

If η is a Poisson process, then it can be checked directly that $T(\eta)$ is completely independent (property (ii) of Definition 3.1), and that $T(\eta)(C)$ has a Poisson distribution with parameter $T(\lambda)(C)$ (property (i) of Definition 3.1). □

If η is a Poisson process on $\mathbb{X}$ then we may discard all of its points outside a set $B \in \mathcal{X}$ to obtain another Poisson process. Recall from (4.16) the definition of the restriction ν_B of a measure ν on $\mathbb{X}$ to a set $B \in \mathcal{X}$.

Theorem 5.2 (Restriction theorem) *Let η be a Poisson process on $\mathbb{X}$ with s-finite intensity measure λ and let $C_1, C_2, \ldots \in \mathcal{X}$ be pairwise disjoint. Then $\eta_{C_1}, \eta_{C_2}, \ldots$ are independent Poisson processes with intensity measures $\lambda_{C_1}, \lambda_{C_2}, \ldots$, respectively.*

Proof As in the proof of Proposition 3.5, it is no restriction of generality to assume that the union of the sets C_i is all of $\mathbb{X}$. (If not, add the complement of this union to the sequence (C_i).) First note that, for each $i \in \mathbb{N}$, η_{C_i} has intensity measure λ_{C_i} and satisfies the two defining properties of a Poisson process. By the existence theorem (Theorem 3.6) we can find a sequence η_i, $i \in \mathbb{N}$, of independent Poisson processes on a suitable (product) probability space, with η_i having intensity measure λ_{C_i} for each i.

By the superposition theorem (Theorem 3.3), the point process $\eta' := \sum_{i=1}^{\infty} \eta_i$ is a Poisson process with intensity measure λ. Then $\eta' \overset{d}{=} \eta$ by Proposition 3.2. Hence for any k and any $f_1, \ldots, f_k \in \mathbb{R}_+(\mathbf{N})$ we have

$$\mathbb{E}\Big[\prod_{i=1}^{k} f_i(\eta_{C_i})\Big] = \mathbb{E}\Big[\prod_{i=1}^{k} f_i(\eta'_{C_i})\Big] = \mathbb{E}\Big[\prod_{i=1}^{k} f_i(\eta_i)\Big] = \prod_{i=1}^{k} \mathbb{E}[f_i(\eta_i)].$$

Taking into account that $\eta_{C_i} \overset{d}{=} \eta_i$ for all $i \in \mathbb{N}$ (Proposition 3.2), we get the result. □

5.2 The Marking Theorem

Suppose that η is a proper point process, i.e. one that can be represented as in (2.4). Suppose that one wishes to give each of the points X_n a random *mark* Y_n with values in some measurable space $(\mathbb{Y}, \mathcal{Y})$, called the *mark space*. Given η, these marks are assumed to be independent, while their conditional distribution is allowed to depend on the value of X_n but not on any other information contained in η. This marking procedure yields a point process ξ on the product space $\mathbb{X} \times \mathbb{Y}$. Theorem 5.6 will show the remarkable fact that if η is a Poisson process then so is ξ.

To make the above marking idea precise, let K be a *probability kernel* from $\mathbb{X}$ to $\mathbb{Y}$, that is a mapping $K\colon \mathbb{X} \times \mathcal{Y} \to [0,1]$ such that $K(x, \cdot)$ is a probability measure for each $x \in \mathbb{X}$ and $K(\cdot, C)$ is measurable for each $C \in \mathcal{Y}$.

Definition 5.3 Let $\eta = \sum_{n=1}^{\kappa} \delta_{X_n}$ be a proper point process on $\mathbb{X}$. Let K be a probability kernel from $\mathbb{X}$ to $\mathbb{Y}$. Let $Y_1, Y_2, \dots$ be random elements in $\mathbb{Y}$ and assume that the conditional distribution of $(Y_n)_{n \le m}$ given $\kappa = m \in \overline{\mathbb{N}}$ and $(X_n)_{n \le m}$ is that of independent random variables with distributions $K(X_n, \cdot)$, $n \le m$. Then the point process

$$\xi := \sum_{n=1}^{\kappa} \delta_{(X_n, Y_n)} \tag{5.5}$$

is called a *K-marking* of η. If there is a probability measure $\mathbb{Q}$ on $\mathbb{Y}$ such that $K(x, \cdot) = \mathbb{Q}$ for all $x \in \mathbb{X}$, then ξ is called an *independent* $\mathbb{Q}$*-marking* of η.

For the rest of this section we fix a probability kernel K from $\mathbb{X}$ to $\mathbb{Y}$. If the random variables Y_n, $n \in \mathbb{N}$, in Definition 5.3 exist, then we say that the underlying probability space $(\Omega, \mathcal{F}, \mathbb{P})$ *supports* a K-marking of η. We now explain how $(\Omega, \mathcal{F}, \mathbb{P})$ can be modified so as to support a marking. Let $\tilde{\Omega} := \Omega \times \mathbb{Y}^\infty$ be equipped with the product σ-field. Define a probability kernel $\tilde{K}$ from Ω to $\mathbb{Y}^\infty$ by taking the infinite product

$$\tilde{K}(\omega, \cdot) := \bigotimes_{n=1}^{\infty} K(X_n(\omega), \cdot), \quad \omega \in \Omega.$$

We denote a generic element of $\mathbb{Y}^\infty$ by $\mathbf{y} = (y_n)_{n \ge 1}$. Then

$$\tilde{\mathbb{P}} := \int \mathbf{1}\{(\omega, \mathbf{y}) \in \cdot\} \tilde{K}(\omega, d\mathbf{y}) \, \mathbb{P}(d\omega) \tag{5.6}$$

is a probability measure on $\tilde{\Omega}$ that can be used to describe a K-marking of η. Indeed, for $\tilde{\omega} = (\omega, \mathbf{y}) \in \tilde{\Omega}$ we can define $\tilde{\eta}(\tilde{\omega}) := \eta(\omega)$ and, for $n \in \mathbb{N}$, $(\tilde{X}_n(\tilde{\omega}), Y_n(\tilde{\omega})) := (X_n(\omega), y_n)$. Then the distribution of $(\tilde{\eta}(\mathbb{X}), (\tilde{X}_n))$ under $\tilde{\mathbb{P}}$ coincides with that of $(\eta(\mathbb{X}), (X_n))$ under $\mathbb{P}$. Moreover, it is easy to check that under $\tilde{\mathbb{P}}$ the conditional distribution of $(Y_n)_{n \le m}$ given $\tilde{\eta}(\mathbb{X}) = m \in \overline{\mathbb{N}}$ and

$(\tilde{X}_n)_{n\le m}$ is that of independent random variables with distributions $K(\tilde{X}_n, \cdot)$, $n \le m$. This construction is known as an *extension* of a given probability space so as to support further random elements with a given conditional distribution. In particular, it is no restriction of generality to assume that our fixed probability space supports a K-marking of η.

The next proposition shows among other things that the distribution of a K-marking of η is uniquely determined by K and the distribution of η.

Proposition 5.4 *Let ξ be a K-marking of a proper point process η on $\mathbb{X}$ as in Definition 5.3. Then the Laplace functional of ξ is given by*

$$L_\xi(u) = L_\eta(u^*), \quad u \in \mathbb{R}_+(\mathbb{X} \times \mathbb{Y}), \tag{5.7}$$

where

$$u^*(x) := -\log\left[\int e^{-u(x,y)}\, K(x, dy)\right], \quad x \in \mathbb{X}. \tag{5.8}$$

Proof Recall that $\overline{\mathbb{N}}_0 := \mathbb{N}_0 \cup \{\infty\}$. For $u \in \mathbb{R}_+(\mathbb{X} \times \mathbb{Y})$ we have that

$$\begin{aligned} L_\xi(u) &= \sum_{m\in\overline{\mathbb{N}}_0} \mathbb{E}\left[\mathbf{1}\{\kappa = m\} \exp\left[-\sum_{k=1}^m u(X_k, Y_k)\right]\right] \\ &= \sum_{m\in\overline{\mathbb{N}}_0} \mathbb{E}\left[\mathbf{1}\{\kappa = m\} \int\cdots\int \exp\left[-\sum_{k=1}^m u(X_k, y_k)\right] \prod_{k=1}^m K(X_k, dy_k)\right], \end{aligned}$$

where in the case $m = 0$ empty sums are set to 0 while empty products are set to 1. Therefore

$$L_\xi(u) = \sum_{m\in\overline{\mathbb{N}}_0} \mathbb{E}\left[\mathbf{1}\{\kappa = m\}\left(\prod_{k=1}^m \int \exp[-u(X_k, y_k)] K(X_k, dy_k)\right)\right].$$

Using the function u^* defined by (5.8) this means that

$$\begin{aligned} L_\xi(u) &= \sum_{m\in\overline{\mathbb{N}}_0} \mathbb{E}\left[\mathbf{1}\{\kappa = m\}\left(\prod_{k=1}^m \exp[-u^*(X_k)]\right)\right] \\ &= \sum_{m\in\overline{\mathbb{N}}_0} \mathbb{E}\left[\mathbf{1}\{\kappa = m\} \exp\left(-\sum_{k=1}^m u^*(X_k)\right)\right], \end{aligned}$$

which is the right-hand side of the asserted identity (5.7). □

The next result says that the intensity measure of a K-marking of a point process with intensity measure λ is given by $\lambda \otimes K$, where

$$(\lambda \otimes K)(C) := \iint \mathbf{1}_C(x, y)\, K(x, dy)\, \lambda(dx), \quad C \in \mathcal{X} \otimes \mathcal{Y}. \tag{5.9}$$

In the case of an independent $\mathbb{Q}$-marking this is the product measure $\lambda \otimes \mathbb{Q}$. If λ and K are s-finite, then so is $\lambda \otimes K$.

Proposition 5.5 *Let η be a proper point process on $\mathbb{X}$ with intensity measure λ and let ξ be a K-marking of η. Then ξ is a point process on $\mathbb{X} \times \mathbb{Y}$ with intensity measure $\lambda \otimes K$.*

Proof Let $C \in \mathcal{X} \otimes \mathcal{Y}$. Similarly to the proof of Proposition 5.4 we have that

$$\mathbb{E}[\xi(C)] = \sum_{m \in \overline{\mathbb{N}}_0} \mathbb{E}\Big[\mathbf{1}\{\kappa = m\} \sum_{k=1}^{m} \mathbf{1}\{(X_k, Y_k) \in C\}\Big]$$
$$= \sum_{m \in \overline{\mathbb{N}}_0} \mathbb{E}\Big[\mathbf{1}\{\kappa = m\} \sum_{k=1}^{m} \int \mathbf{1}\{(X_k, y_k) \in C\}\, K(X_k, dy_k)\Big].$$

Using Campbell's formula (Proposition 2.7) with $u \in \mathbb{R}_+(\mathbb{X})$ defined by $u(x) := \int \mathbf{1}\{(x, y) \in C\}\, K(x, dy)$, $x \in \mathbb{X}$, we obtain the result. □

Now we formulate the previously announced behaviour of Poisson processes under marking.

Theorem 5.6 (Marking theorem) *Let ξ be a K-marking of a proper Poisson process η with s-finite intensity measure λ. Then ξ is a Poisson process with intensity measure $\lambda \otimes K$.*

Proof Let $u \in \mathbb{R}_+(\mathbb{X} \times \mathbb{Y})$. By Proposition 5.4 and Theorem 3.9,

$$L_\xi(u) = \exp\Big[-\int (1 - e^{-u^*(x)})\, \lambda(dx)\Big]$$
$$= \exp\Big[-\iint (1 - e^{-u(x,y)})\, K(x, dy)\, \lambda(dx)\Big].$$

Another application of Theorem 3.9 shows that ξ is a Poisson process. □

Under some technical assumptions we shall see in Proposition 6.16 that any Poisson process on a product space is a K-marking for some kernel K, determined by the intensity measure.

5.3 Thinnings

A thinning keeps the points of a point process η with a probability that may depend on the location and removes them otherwise. Given η, the thinning decisions are independent for different points. The formal definition can be based on a special K-marking:

Definition 5.7 Let $p\colon \mathbb{X} \to [0,1]$ be measurable and consider the probability kernel K from $\mathbb{X}$ to $\{0,1\}$ defined by

$$K_p(x,\cdot) := (1 - p(x))\delta_0 + p(x)\delta_1, \quad x \in \mathbb{X}.$$

If ξ is a K_p-marking of a proper point process η, then $\xi(\cdot \times \{1\})$ is called a *p-thinning* of η.

We shall use this terminology also in the case where $p(x) \equiv p$ does not depend on $x \in \mathbb{X}$.

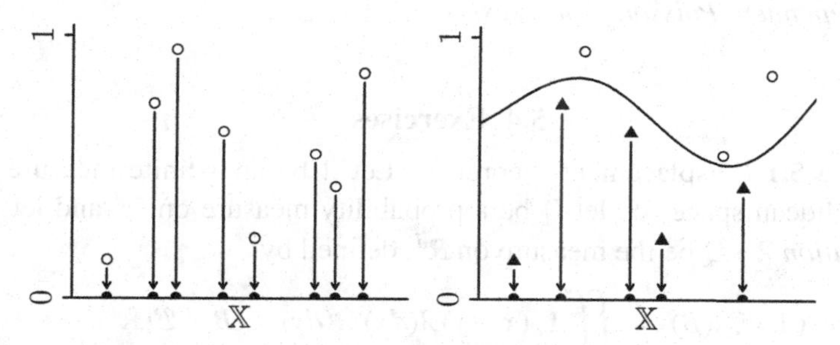

Figure 5.1 Illustration of a marking and a thinning, both based on the same set of marked points. The points on the horizontal axis represent the original point process in the first diagram, and the thinned point process in the second diagram.

More generally, let p_i, $i \in \mathbb{N}$, be a sequence of measurable functions from $\mathbb{X}$ to $[0,1]$ such that

$$\sum_{i=1}^{\infty} p_i(x) = 1, \quad x \in \mathbb{X}. \tag{5.10}$$

Define a probability kernel K from $\mathbb{X}$ to $\mathbb{N}$ by

$$K(x,\{i\}) := p_i(x), \quad x \in \mathbb{X},\, i \in \mathbb{N}. \tag{5.11}$$

If ξ is a K-marking of a point process η, then $\eta_i := \xi(\cdot \times \{i\})$ is a p_i-thinning of η for every $i \in \mathbb{N}$. By Proposition 5.5, η_i has intensity measure $p_i(x)\,\lambda(dx)$, where λ is the intensity measure of η. The following generalisation of Proposition 1.3 is consistent with the superposition theorem (Theorem 3.3).

Theorem 5.8 *Let ξ be a K-marking of a proper Poisson process η, where K is given as in* (5.11)*. Then $\eta_i := \xi(\cdot \times \{i\})$, $i \in \mathbb{N}$, are independent Poisson processes.*

Proof By Theorem 5.6, ξ is a Poisson process. Hence we can apply Theorem 5.2 with $C_i := \mathbb{X} \times \{i\}$ to obtain the result. □

If η_p is a p-thinning of a proper point process η then (according to Definitions 2.4 and 5.7) there is an $A \in \mathcal{F}$ such that $\mathbb{P}(A) = 1$ and $\eta_p(\omega) \leq \eta(\omega)$ for each $\omega \in A$. We can then define a proper point process $\eta - \eta_p$ by setting $(\eta - \eta_p)(\omega) := \eta(\omega) - \eta_p(\omega)$ for $\omega \in A$ and $(\eta - \eta_p)(\omega) := 0$, otherwise.

Corollary 5.9 (Thinning theorem) *Let $p \colon \mathbb{X} \to [0, 1]$ be measurable and let η_p be a p-thinning of a proper Poisson process η. Then η_p and $\eta - \eta_p$ are independent Poisson processes.*

5.4 Exercises

Exercise 5.1 (Displacement theorem) Let λ be an s-finite measure on the Euclidean space $\mathbb{R}^d$, let $\mathbb{Q}$ be a probability measure on $\mathbb{R}^d$ and let the *convolution* $\lambda * \mathbb{Q}$ be the measure on $\mathbb{R}^d$, defined by

$$(\lambda * \mathbb{Q})(B) := \iint \mathbf{1}_B(x + y)\, \lambda(dx)\, \mathbb{Q}(dy), \quad B \in \mathcal{B}(\mathbb{R}^d).$$

Show that $\lambda * \mathbb{Q}$ is s-finite. Let $\eta = \sum_{n=1}^{\kappa} \delta_{X_n}$ be a Poisson process with intensity measure λ and let (Y_n) be a sequence of independent random vectors with distribution $\mathbb{Q}$ that is independent of η. Show that $\eta' := \sum_{n=1}^{\kappa} \delta_{X_n + Y_n}$ is a Poisson process with intensity measure $\lambda * \mathbb{Q}$.

Exercise 5.2 Let η_1 and η_2 be independent Poisson processes with intensity measures λ_1 and λ_2, respectively. Let p be a Radon–Nikodým derivative of λ_1 with respect to $\lambda := \lambda_1 + \lambda_2$. Show that η_1 has the same distribution as a p-thinning of $\eta_1 + \eta_2$.

Exercise 5.3 Let $\xi_1, \dots, \xi_n$ be identically distributed point processes and let $\xi^{(n)}$ be an n^{-1}-thinning of $\xi := \xi_1 + \cdots + \xi_n$. Show that $\xi^{(n)}$ has the same intensity measure as ξ_1. Give examples where $\xi_1, \dots, \xi_n$ are independent and where $\xi^{(n)}$ and ξ_1 have (resp. do not have) the same distribution.

Exercise 5.4 Let $p \colon \mathbb{X} \to [0, 1]$ be measurable and let η_p be a p-thinning of a proper point process η. Using Proposition 5.4 or otherwise, show that

$$L_{\eta_p}(u) = \mathbb{E}\Big[\exp\Big(\int \log\big(1 - p(x) + p(x)e^{-u(x)}\big)\, \eta(dx) \Big) \Big], \quad u \in \mathbb{R}_+(\mathbb{X}).$$

Exercise 5.5 Let η be a proper Poisson process on $\mathbb{X}$ with σ-finite intensity measure λ. Let λ' be a σ-finite measure on $\mathbb{X}$ and let $\rho := \lambda + \lambda'$. Let $h := d\lambda/d\rho$ (resp. $h' := d\lambda'/d\rho$) be the Radon–Nikodým derivative of

λ (resp. λ') with respect to ρ; see Theorem A.10. Let $B := \{h > h'\}$ and define $p\colon \mathbb{X} \to [0,1]$ by $p(x) := h'(x)/h(x)$ for $x \in B$ and by $p(x) := 1$, otherwise. Let η' be a p-thinning of η and let η'' be a Poisson process with intensity measure $\mathbf{1}_{\mathbb{X}\setminus B}(x)(h'(x) - h(x))\,\rho(dx)$, independent of η'. Show that $\eta' + \eta''$ is a Poisson process with intensity measure λ'.

Exercise 5.6 (Poisson cluster process) Let K be a probability kernel from $\mathbb{X}$ to $\mathbf{N}(\mathbb{X})$. Let η be a proper Poisson process on $\mathbb{X}$ with intensity measure λ and let $A \in \mathcal{F}$ such that $\mathbb{P}(A) = 1$ and such that (2.4) holds on A. Let ξ be a K-marking of η and define a point process χ on $\mathbb{X}$ by setting

$$\chi(\omega, B) := \int \mu(B)\,\xi(\omega, d(x,\mu)), \quad B \in \mathcal{X}, \tag{5.12}$$

for $\omega \in A$ and $\chi(\omega, \cdot) := 0$, otherwise. Show that χ has intensity measure

$$\lambda'(B) = \iint \mu(B)\,K(x, d\mu)\,\lambda(dx), \quad B \in \mathcal{X}.$$

Show also that the Laplace functional of χ is given by

$$L_\chi(v) = \exp\left[-\int (1 - e^{-\mu(v)})\,\tilde{\lambda}(d\mu)\right], \quad v \in \mathbb{R}_+(\mathbb{X}), \tag{5.13}$$

where $\tilde{\lambda} := \int K(x, \cdot)\,\lambda(dx)$.

Exercise 5.7 Let χ be a Poisson cluster process as in Exercise 5.6 and let $B \in \mathcal{X}$. Combine Exercise 2.7 and (5.13) to show that

$$\mathbb{P}(\chi(B) = 0) = \exp\left[-\int \mathbf{1}\{\mu(B) > 0\}\,\tilde{\lambda}(d\mu)\right].$$

Exercise 5.8 Let χ be as in Exercise 5.6 and let $B \in \mathcal{X}$. Show that $\mathbb{P}(\chi(B) < \infty) = 1$ if and only if $\tilde{\lambda}(\{\mu \in \mathbf{N} : \mu(B) = \infty\}) = 0$ and $\tilde{\lambda}(\{\mu \in \mathbf{N} : \mu(B) > 0\}) < \infty$. (Hint: Use $\mathbb{P}(\chi(B) < \infty) = \lim_{t \downarrow 0} \mathbb{E}[e^{-t\chi(B)}]$.)

Exercise 5.9 Let $p \in [0,1)$ and suppose that η_p is a p-thinning of a proper point process η. Let $f \in \mathbb{R}_+(\mathbb{X} \times \mathbf{N})$ and show that

$$\mathbb{E}\left[\int f(x, \eta_p)\,\eta_p(dx)\right] = \frac{p}{1-p}\mathbb{E}\left[\int f(x, \eta_p + \delta_x)\,(\eta - \eta_p)(dx)\right].$$

6

Characterisations of the Poisson Process

A point process without multiplicities is said to be simple. For locally finite simple point processes on a metric space without fixed atoms the two defining properties of a Poisson process are equivalent. In fact, Rényi's theorem says that in this case even the empty space probabilities suffice to imply that the point process is Poisson. On the other hand, a weak (pairwise) version of the complete independence property leads to the same conclusion. A related criterion, based on the factorial moment measures, is also given.

6.1 Borel Spaces

In this chapter we assume $(\mathbb{X}, \mathcal{X})$ to be a Borel space in the sense of the following definition. In the first section we shall show that a large class of point processes is proper.

Definition 6.1 A *Borel space* is a measurable space $(\mathbb{Y}, \mathcal{Y})$ such that there is a Borel-measurable bijection φ from $\mathbb{Y}$ to a Borel subset of the unit interval $[0, 1]$ with measurable inverse.

A special case arises when $\mathbb{X}$ is a Borel subset of a complete separable metric space (CSMS) and $\mathcal{X}$ is the σ-field on $\mathbb{X}$ generated by the open sets in the inherited metric. In this case, $(\mathbb{X}, \mathcal{X})$ is called a *Borel subspace* of the CSMS; see Section A.2. By Theorem A.19, any Borel subspace $\mathbb{X}$ of a CSMS is a Borel space. In particular, $\mathbb{X}$ is then a metric space in its own right.

Recall that $\mathbf{N}_{<\infty}(\mathbb{X})$ denotes the set of all integer-valued measures on $\mathbb{X}$.

Proposition 6.2 *There exist measurable mappings* $\pi_n \colon \mathbf{N}_{<\infty}(\mathbb{X}) \to \mathbb{X}$, $n \in \mathbb{N}$, *such that for all* $\mu \in \mathbf{N}_{<\infty}(\mathbb{X})$ *we have*

$$\mu = \sum_{n=1}^{\mu(\mathbb{X})} \delta_{\pi_n(\mu)}. \tag{6.1}$$

Proof Take a measurable bijection φ from $\mathbb{X}$ onto a Borel subset U of $[0, 1]$ such that the inverse of φ is measurable. For $\mu \in \mathbf{N}_{<\infty} := \mathbf{N}_{<\infty}(\mathbb{X})$ we define a finite measure $\varphi(\mu)$ on $\mathbb{R}$ by $\varphi(\mu) := \mu \circ \varphi^{-1}$, that is,

$$\varphi(\mu)(B) = \mu(\{x : \varphi(x) \in B\}), \quad B \in \mathcal{B}(\mathbb{R}).$$

Here we interpret φ as a mapping from $\mathbb{X}$ to $\mathbb{R}$, so that $\varphi^{-1}(B) = \emptyset$ whenever $B \cap U = \emptyset$. Hence $\varphi(\mu)$ is concentrated on U, that is $\varphi(\mu)(\mathbb{R} \setminus U) = 0$. For $n \in \mathbb{N}$, set

$$Y_n(\mu) := \inf\{x \in \mathbb{R} : \varphi(\mu)((-\infty, x]) \geq n\}, \quad \mu \in \mathbf{N}_{<\infty},$$

where $\inf \emptyset := \infty$. For $n > \mu(\mathbb{X})$ we have $Y_n(\mu) = \infty$. For $n \leq \mu(\mathbb{X})$ we have $Y_n(\mu) \in U$. Indeed, in this case $\varphi(\mu)\{Y_n(\mu)\} > 0$.

For $x \in \mathbb{R}$ we have

$$\begin{aligned}\{\mu \in \mathbf{N}_{<\infty} : Y_n(\mu) \leq x\} &= \{\mu \in \mathbf{N}_{<\infty} : \varphi(\mu)((-\infty, x]) \geq n\} \\ &= \{\mu \in \mathbf{N}_{<\infty} : \mu(\varphi^{-1}((-\infty, x])) \geq n\},\end{aligned}$$

so Y_n is a measurable mapping on $\mathbf{N}_{<\infty}$. Also

$$\varphi(\mu)(B) = \sum_{n=1}^{\mu(\mathbb{X})} \delta_{Y_n(\mu)}(B), \quad \mu \in \mathbf{N}_{<\infty}, \tag{6.2}$$

for all B of the form $B = (-\infty, x]$ with $x \in \mathbb{R}$ (a π-system of sets), and hence for all Borel sets $B \subset \mathbb{R}$ (by Theorem A.5). Fix $x_0 \in \mathbb{X}$ and define

$$X_n(\mu) := \begin{cases} \varphi^{-1}(Y_n(\mu)), & \text{if } n \leq \mu(\mathbb{X}), \\ x_0, & \text{otherwise.} \end{cases}$$

By (6.2) we have for all $B \in \mathcal{X}$ that

$$\mu(B) = \mu(\varphi^{-1}(\varphi(B))) = \sum_{n=1}^{\mu(\mathbb{X})} \mathbf{1}\{Y_n(\mu) \in \varphi(B)\} = \sum_{n=1}^{\mu(\mathbb{X})} \mathbf{1}\{X_n(\mu) \in B\}$$

and hence $\mu = \sum_{n=1}^{\mu(\mathbb{X})} \delta_{X_n(\mu)}$. Then (6.1) holds with $\pi_n(\mu) = X_n(\mu)$. □

In the case where $\mathbb{X}$ is a Borel subspace of a CSMS, recall from Definition 2.11 that $\mathbf{N}_l(\mathbb{X})$ denotes the class of all measures from $\mathbf{N}(\mathbb{X})$ that are locally finite, that is finite on bounded Borel sets. The preceding proposition implies a measurable decomposition of these measures.

Proposition 6.3 *Suppose that $\mathbb{X}$ is a Borel subspace of a CSMS. Then there are measurable mappings $\pi_n : \mathbf{N}(\mathbb{X}) \to \mathbb{X}$, $n \in \mathbb{N}$, such that for all $\mu \in \mathbf{N}_l(\mathbb{X})$ we have $\mu = \sum_{n=1}^{\mu(\mathbb{X})} \delta_{\pi_n(\mu)}$.*

Proof Let $B_1, B_2, \dots$ be a sequence of disjoint bounded sets in $\mathcal{X}$, forming a partition of $\mathbb{X}$. Recall from (4.16) the definition of the restriction μ_{B_i} of μ to B_i. By Theorem A.19, each B_i is a Borel space. Hence we can apply Proposition 6.2 to obtain for each $i \in \mathbb{N}$ measurable mappings $\pi_{i,j}\colon \mathbf{N}_l(\mathbb{X}) \to B_i$, $j \in \mathbb{N}$, such that

$$\mu_{B_i} = \sum_{j=1}^{\mu(B_i)} \delta_{\pi_{i,j}(\mu)}, \quad \mu \in \mathbf{N}_l(\mathbb{X}).$$

Fix $x_0 \in \mathbb{X}$ and let $\mu \in \mathbf{N}_l(\mathbb{X})$. If $\mu = 0$ is the zero measure, for all $n \in \mathbb{N}$ we set $\pi_n(\mu) := x_0$. Otherwise let $k_1 = k_1(\mu)$ be the smallest $i \in \mathbb{N}$ such that $\mu(B_i) > 0$ and define $\pi_n(\mu) := \pi_{k_1,n}(\mu)$ for $1 \le n \le \mu(B_{k_1})$. If $\mu(\mathbb{X}) = \mu(B_{k_1})$ let $\pi_n(\mu) := x_0$ for $n > \mu(B_1)$. Otherwise we define $\pi_{k_1+m}(\mu) := \pi_{k_2,m}(\mu)$ for $1 \le m \le \mu(B_{k_2})$, where $k_2 \equiv k_2(\mu)$ is the smallest $i > k$ such that $\mu(B_i) > 0$.

It is now clear how to construct a sequence $\pi_n\colon \mathbf{N}_l(\mathbb{X}) \to \mathbb{X}$, $n \in \mathbb{N}$, inductively, such that (6.1) holds. Measurability can be proved by induction, using the fact that the $\pi_{i,j}$ are measurable. Since $\mathbf{N}_l(\mathbb{X})$ is a measurable subset of $\mathbf{N}(\mathbb{X})$ (see the discussion after Definition 2.11) the mappings π_n can be extended to measurable mappings on $\mathbf{N}(\mathbb{X})$. □

The following definition generalises the concept of a locally finite point process (see Definition 2.13) to point processes on an arbitrary (not necessarily Borel) phase space.

Definition 6.4 A point process η on a measurable space $(\mathbb{Y}, \mathcal{Y})$ is said to be *uniformly σ-finite* if there exist $B_n \in \mathcal{Y}$, $n \in \mathbb{N}$, such that $\cup_{n=1}^{\infty} B_n = \mathbb{Y}$ and

$$\mathbb{P}(\eta(B_n) < \infty) = 1, \quad n \in \mathbb{N}. \tag{6.3}$$

We note that Poisson processes with σ-finite intensity measure and locally finite point processes on a metric space are uniformly σ-finite.

It follows from Proposition 6.2 that every uniformly σ-finite point process on the Borel space $\mathbb{X}$ is proper. As mentioned just after Definition 2.4 this shows in particular that all locally finite point processes are proper.

Corollary 6.5 *Let η be a uniformly σ-finite point process on $\mathbb{X}$. Then η is a proper point process. That is, there exist random elements $X_1, X_2, \dots$ in $\mathbb{X}$ and an $\overline{\mathbb{N}}_0$-valued random variable κ such that almost surely*

$$\eta = \sum_{n=1}^{\kappa} \delta_{X_n}. \tag{6.4}$$

Proof Choose the sets B_k, $k \in \mathbb{N}$, as in Definition 6.4 and assume without loss of generality that these sets are pairwise disjoint. Let $\eta_k := \eta_{B_k}$ be the restriction of η to B_k. By definition (see also the discussion after Definition 2.13) there are random elements $\tilde{\eta}_k$ of $\mathbf{N}_{<\infty}(\mathbb{X})$ such that η_k and $\tilde{\eta}_k$ are almost surely equal. For each k we can now use Proposition 6.2 to define $\kappa := \tilde{\eta}(\mathbb{X})$ and, for $n \in \mathbb{N}$, $X_n := \pi_n(\tilde{\eta}_k)$, to see that η_k is proper. Since $\eta = \sum_k \eta_k$, Exercise 2.4 shows that η is proper. □

6.2 Simple Point Processes

In this section we discuss point processes without multiplicities.

Definition 6.6 A measure $\mu \in \mathbf{N}(\mathbb{X})$ is said to be *simple* if $\mu\{x\} \leq 1$ for all $x \in \mathbb{X}$. Let $\mathbf{N}_s(\mathbb{X})$ denote the set of all simple measures in $\mathbf{N}(\mathbb{X})$. If $(\mathbb{X}, \mathcal{X})$ is a metric space then let $\mathbf{N}_{ls}(\mathbb{X}) := \mathbf{N}_l(\mathbb{X}) \cap \mathbf{N}_s(\mathbb{X})$; see Definition 2.11.

There is a convenient description of $\mathbf{N}_s(\mathbb{X})$ based on the *diagonal* in $\mathbb{X}^2$, defined by

$$D_{\mathbb{X}} := \{(x, y) \in \mathbb{X}^2 : x = y\}. \tag{6.5}$$

Proposition 6.7 *Let $\mu \in \mathbf{N}(\mathbb{X})$. Then $\mu \in \mathbf{N}_s(\mathbb{X})$ if and only if $\mu^{(2)}(D_{\mathbb{X}}) = 0$. Moreover, $\mathbf{N}_s(\mathbb{X})$ is measurable, i.e. $\mathbf{N}_s(\mathbb{X}) \in \mathcal{N}(\mathbb{X})$.*

Proof We first note that $D_{\mathbb{X}}$ is measurable, that is $D_{\mathbb{X}} \in \mathcal{X} \otimes \mathcal{X}$. Indeed, this holds if $\mathbb{X}$ is a Borel subset of $[0, 1]$. Using the definition of a Borel space, the measurability can be extended to the general case.

By definition, there is a sequence μ_n, $n \in \mathbb{N}$, of finite measures in $\mathbf{N}(\mathbb{X})$ such that $\mu = \sum_n \mu_n$. By Proposition 6.2, each of the μ_n and hence also μ is of the form (4.3). Therefore (4.5) implies for each $x \in \mathbb{X}$ that $\mu^{(2)}\{(x, x)\} = 0$ if and only if $\mu\{x\} \leq 1$. This proves the first assertion. The measurability of $\mathbf{N}_s(\mathbb{X})$ is then a consequence of Proposition 4.3. □

Point processes without multiplicities deserve a special name:

Definition 6.8 A point process η is said to be *simple* if $\mathbb{P}(\eta \in \mathbf{N}_s(\mathbb{X})) = 1$.

If η is a simple point process on $\mathbb{X}$ and $\eta' \overset{d}{=} \eta$, then η' is also simple.

Similarly to Section 2.4 we say that a measure ν on $\mathbb{X}$ is *diffuse* if $\nu\{x\} := \nu(\{x\}) = 0$ for each $x \in \mathbb{X}$.

Proposition 6.9 *Let η be a Poisson process on $\mathbb{X}$ with s-finite intensity measure λ. Then η is simple if and only if λ is diffuse.*

Proof Suppose λ is not diffuse. Let $x \in \mathbb{X}$ with $c := \lambda\{x\} > 0$. Then

$$\mathbb{P}(\eta\{x\} \geq 2) = 1 - e^{-c} - ce^{-c} > 0,$$

so that η is not simple.

Conversely, suppose that λ is diffuse. We need to show that η is simple. By Proposition 6.7 this amounts to proving that $\mathbb{P}(\eta^{(2)}(D_{\mathbb{X}}) = 0) = 1$ or, equivalently, $\mathbb{E}[\eta^{(2)}(D_{\mathbb{X}})] = 0$. By Corollary 4.10 we have that

$$\mathbb{E}[\eta^{(2)}(D_{\mathbb{X}})] = \iint \mathbf{1}\{x = y\}\, \lambda(dx)\, \lambda(dy) = \int \lambda(\{y\})\, \lambda(dy) = 0,$$

and the proof is complete. □

6.3 Rényi's Theorem

The following (at first glance surprising) result shows that the two defining properties of a Poisson process are not independent of each other. In fact, this result, together with Theorem 6.12, shows that under certain extra conditions, either of the defining properties of the Poisson process implies the other. We base the proof on a more general result for simple point processes.

Theorem 6.10 (Rényi's theorem) *Suppose that λ is a diffuse s-finite measure on $\mathbb{X}$, and that η is a simple point process on $\mathbb{X}$ satisfying*

$$\mathbb{P}(\eta(B) = 0) = \exp[-\lambda(B)], \quad B \in \mathcal{X}. \tag{6.6}$$

Then η is a Poisson process with intensity measure λ.

Proof Let η' be a Poisson process with intensity measure λ. Then assumption (6.6) implies (6.7) below. Proposition 6.9 shows that η' is simple. Theorem 6.11 shows that η and η' have the same distribution. □

Theorem 6.11 *Let η and η' be simple point processes on $\mathbb{X}$ such that*

$$\mathbb{P}(\eta(B) = 0) = \mathbb{P}(\eta'(B) = 0), \quad B \in \mathcal{X}. \tag{6.7}$$

Then $\eta \overset{d}{=} \eta'$.

Proof Take a measurable bijection φ from $\mathbb{X}$ onto a Borel subset of $I := [1/4, 3/4]$ such that the inverse of φ is measurable. We interpret φ as a mapping from $\mathbb{X}$ to I. Define a point process ξ on I by

$$\xi(B) := \eta \circ \varphi^{-1}(B) = \eta(\{x \in \mathbb{X} : \varphi(x) \in B\}), \quad B \in \mathcal{B}(I). \tag{6.8}$$

Since φ is one-to-one it follows that $\xi\{x\} = \eta(\varphi^{-1}(\{x\})) \leq 1$ for all $x \in I$, provided $\eta \in \mathbf{N}_s$. Hence ξ is simple. The same holds for $\xi' := \eta' \circ \varphi^{-1}$. Since φ is one-to-one we have $\eta = \xi \circ \varphi$ and $\eta' = \xi' \circ \varphi$. Furthermore, since $\mu \mapsto \mu(\varphi(B))$ is measurable for all $B \in \mathcal{X}$, $\mu \mapsto \mu \circ \varphi$ is a measurable mapping from $\mathbf{N}(I)$ to $\mathbf{N}(\mathbb{X})$. Since equality in distribution is preserved under measurable mappings, it now suffices to prove that $\xi \overset{d}{=} \xi'$.

Let $\mathcal{N}^*$ denote the sub-σ-field of $\mathcal{N}(I)$ generated by the system

$$\mathcal{H} := \{\{\mu \in \mathbf{N}(I) : \mu(B) = 0\} : B \in \mathcal{B}(I)\}.$$

Since, for any measure μ on I and any two sets $B, B' \in \mathcal{B}(I)$, the equation $\mu(B \cup B') = 0$ is equivalent to $\mu(B) = \mu(B') = 0$, $\mathcal{H}$ is a π-system. By assumption (6.7), $\mathbb{P}_\xi$ agrees with $\mathbb{P}_{\xi'}$ on $\mathcal{H}$, and therefore, by Theorem A.5, $\mathbb{P}_\xi$ agrees with $\mathbb{P}_{\xi'}$ on $\mathcal{N}^*$.

For $n \in \mathbb{N}$ and $j \in \{1, \dots, 2^n\}$ let $I_{n,j} := ((j-1)2^{-n}, j2^{-n}]$. Given $B \in \mathcal{B}(I)$, define

$$g_{n,B}(\mu) := \sum_{j=1}^{2^n} \mu(I_{n,j} \cap B) \wedge 1, \quad \mu \in \mathbf{N}(I),\ n \in \mathbb{N}, \tag{6.9}$$

where $a \wedge b := \min\{a, b\}$ denotes the minimum of $a, b \in \overline{\mathbb{R}}$. Define the function $g_B \colon \mathbf{N}(I) \to \overline{\mathbb{R}}_+$ by

$$g_B(\mu) := \lim_{n\to\infty} g_{n,B}(\mu), \quad \mu \in \mathbf{N}(I). \tag{6.10}$$

Then g_B is an $\mathcal{N}^*$-measurable function on $\mathbf{N}(I)$. Moreover, if $\mu \in \mathbf{N}_s(\mathbb{X})$, then $g_B(\mu) = \mu(B)$. To see this, one can represent μ in the form (4.3) (justified by Proposition 6.2 and the definition of $\mathbf{N}(\mathbb{X})$) and distinguish the cases $\mu(B) < \infty$ and $\mu(B) = \infty$. Since ξ is simple we obtain $\xi(B) = g_B(\xi)$, almost surely, and therefore, for any $m \in \mathbb{N}$, $B_1, \dots, B_m \in \mathcal{B}(I)$ and $k_1, \dots, k_m \in \mathbb{N}_0$, we have

$$\mathbb{P}\big(\cap_{i=1}^m \{\xi(B_i) = k_i\}\big) = \mathbb{P}\big(\xi \in \cap_{i=1}^m g_{B_i}^{-1}(\{k_i\})\big),$$

and, since $\cap_{i=1}^m g_{B_i}^{-1}(\{k_i\}) \in \mathcal{N}^*$, the corresponding probability for ξ' is the same. Therefore, by Proposition 2.10, $\xi' \overset{d}{=} \xi$. □

A point process η on $\mathbb{X}$ satisfies

$$\eta\{x\} = 0, \quad \mathbb{P}\text{-a.s., } x \in \mathbb{X}, \tag{6.11}$$

if and only if its intensity measure is diffuse. If, in addition, η is uniformly σ-finite and simple, then the following result shows that we need only a weak version of the complete independence property to ensure that η is a Poisson process. This complements Theorem 6.10.

Theorem 6.12 *Suppose that η is a uniformly σ-finite simple point process on $\mathbb{X}$ satisfying* (6.11). *Assume also that $\{\eta(B) = 0\}$ and $\{\eta(B') = 0\}$ are independent whenever $B, B' \in \mathcal{X}$ are disjoint. Then η is a Poisson process.*

Proof Let the sets B_n, $n \in \mathbb{N}$, be as in Definition 6.4 and assume without loss of generality that $B_n \subset B_{n+1}$. Suppose for each $n \in \mathbb{N}$ that η_{B_n} is a Poisson process. Then it follows from Theorem 3.9 and monotone convergence that η is a Poisson process. Hence we can assume that $\mathbb{P}(\eta(\mathbb{X}) < \infty) = 1$. Furthermore, we can (and do) assume $\mathbb{X} = \mathbb{R}$ and $\eta(\mathbb{R} \setminus [0, 1]) = 0$, cf. the proof of Theorem 6.11.

For $t \in \mathbb{R}$ set $f(t) := \mathbb{P}(\eta((-\infty, t]) = 0)$, which is clearly non-increasing. Clearly $f(-1) = 1$. Suppose $f(1) = 0$. Let $t_0 := \inf\{t \in \mathbb{R} : f(t) = 0\}$. By continuity of $\mathbb{P}$, (6.11) and the assumption $\mathbb{P}(\eta(\mathbb{R}) < \infty) = 1$, we have $\mathbb{P}(\eta((t_0 - 1/n, t_0 + 1/n)) = 0) \to 1$ as $n \to \infty$. Hence we can choose n with

$$c := \mathbb{P}(\eta((t_0 - 1/n, t_0 + 1/n]) = 0) > 0.$$

Then by our assumption we have

$$f(t_0 + 1/n) = cf(t_0 - 1/n) > 0$$

which is a contradiction, so $f(1) > 0$.

Define

$$\lambda(B) := -\log \mathbb{P}(\eta(B) = 0), \quad B \in \mathcal{B}(\mathbb{R}). \tag{6.12}$$

Then $\lambda(\emptyset) = 0$ and $\lambda(\mathbb{R}) < \infty$. We show that λ is a measure. By our assumption λ is additive and hence also finitely additive. Let C_n, $n \in \mathbb{N}$, be an increasing sequence of Borel sets with union C. Then the events $\{\eta(C_n) = 0\}$ are decreasing and have intersection $\{\eta(C) = 0\}$. Therefore $\lambda(C_n) \to \lambda(C)$ as $n \to \infty$, showing that λ is indeed a measure. Furthermore, (6.11) implies for any $x \in \mathbb{R}$ that $\lambda\{x\} = -\log \mathbb{P}(\eta\{x\} = 0) = 0$, so that λ is diffuse. Now we can apply Rényi's theorem (Theorem 6.10) to conclude that η is a Poisson process. □

6.4 Completely Orthogonal Point Processes

For simple point processes satisfying (4.30) the assumptions of Proposition 4.12 can be relaxed as follows.

Theorem 6.13 *Suppose that η and η' are simple point processes on $\mathbb{X}$*

such that, for each $m \in \mathbb{N}$ and each collection $B_1, \dots, B_m$ of pairwise disjoint measurable sets,

$$\mathbb{E}[\eta(B_1) \cdots \eta(B_m)] = \mathbb{E}[\eta'(B_1) \cdots \eta'(B_m)]. \tag{6.13}$$

Suppose also that the factorial moment measures of η satisfy (4.30). Then $\eta \overset{d}{=} \eta'$.

Proof As in the proof of Theorem 6.12 we can assume that $\mathbb{X} = \mathbb{R}$.

We wish to apply Proposition 4.12. Let $m \in \mathbb{N}$ with $m \geq 2$ and let

$$D_m := \{(x_1, \dots, x_m) \in \mathbb{X}^m : \text{there exist } i < j \text{ with } x_i = x_j\} \tag{6.14}$$

denote the *generalised diagonal* in $\mathbb{X}^m$. Let $\mathcal{H}$ be the class of all Borel sets in $\mathbb{R}^m$ which are either of the form $B_1 \times \cdots \times B_m$ with the $B_1, \dots, B_m$ Borel and pairwise disjoint, or are contained in the generalised diagonal D_m. Then $\mathcal{H}$ is a π-system and the m-th factorial moment measure of η agrees with that of η' on all sets in $\mathcal{H}$. Indeed, this is true by assumption for the first kind of set in $\mathcal{H}$, and by Exercise 6.9 and our assumption both factorial moment measures are zero on the diagonal D_m. Then by Theorem A.5 and Proposition 4.12 we are done if we can show that $\mathcal{H}$ generates the product σ-field $\mathcal{B}(\mathbb{R})^m = \mathcal{B}(\mathbb{R}^m)$; see Lemma A.24. The latter is generated by all sets of the form $B_1 \times \cdots \times B_m$, where $B_1, \dots, B_m$ are open intervals. Let us fix such intervals. For all $n \in \mathbb{N}$ and $j \in \mathbb{Z}$ let $I_{n,j} := ((j-1)/n, j/n]$. Define

$$J_{n,i} := \{j \in \mathbb{Z} : I_{n,j} \subset B_i\}, \quad i \in \{1, \dots, m\},$$

and $J_n := J_{n,1} \times \cdots \times J_{n,m}$. Let Δ_m denote the generalised diagonal in $\mathbb{Z}^m$. We leave it to the reader to check that

$$B_1 \times \cdots \times B_m \setminus D_m = \bigcup_{n=1}^{\infty} \bigcup_{(i_1,\dots,i_m) \in J_n \setminus \Delta_m} I_{n,i_1} \times \cdots \times I_{n,i_m}.$$

It therefore follows that $B_1 \times \cdots \times B_m \in \sigma(\mathcal{H})$, finishing the proof. □

We say that a point process η on $\mathbb{X}$ is *completely orthogonal* if

$$\mathbb{E}[\eta(B_1) \cdots \eta(B_m)] = \prod_{j=1}^{m} \mathbb{E}[\eta(B_j)] \tag{6.15}$$

for all $m \in \mathbb{N}$ and all pairwise disjoint $B_1, \dots, B_m \in \mathcal{X}$.

Theorem 6.13 implies the following characterisation of simple Poisson processes.

Theorem 6.14 *Let η be a simple, completely orthogonal point process on $\mathbb{X}$ with a σ-finite diffuse intensity measure λ. Then η is a Poisson process.*

Proof Let η' be a Poisson process with intensity measure λ. Proposition 6.9 shows that η' is simple. Corollary 4.10, (4.6) and assumption (6.15) show that the hypothesis (6.13) of Theorem 6.13 is satisfied. It remains to note that η' satisfies (4.30). □

6.5 Turning Distributional into Almost Sure Identities

In this section we prove a converse of Theorem 5.6. Consider a Poisson process ξ on $\mathbb{X} \times \mathbb{Y}$ with intensity measure λ_ξ, where $(\mathbb{X}, \mathcal{X})$ and $(\mathbb{Y}, \mathcal{Y})$ are Borel subspaces of a CSMS. Assuming that $\lambda := \lambda_\xi(\cdot \times \mathbb{Y})$ is σ-finite, we can apply Theorem A.14 to obtain $\lambda_\xi = \lambda \otimes K$, where K is a probability kernel from $\mathbb{X}$ to $\mathbb{Y}$. Since $\xi(\cdot \times \mathbb{Y})$ is a Poisson process with intensity measure λ (Theorem 5.1), Theorem 5.6 shows that ξ has the same distribution as a K-marking of $\xi(\cdot \times \mathbb{Y})$. Moreover, if λ is locally finite, then it turns out that the second coordinates of the points of ξ have the conditional independence properties of Definition 5.3.

First we refine Proposition 6.3 in a special case. Let $\mathbf{N}^*$ be the measurable set of all $\mu \in \mathbf{N}(\mathbb{X} \times \mathbb{Y})$ with $\mu(\cdot \times \mathbb{Y}) \in \mathbf{N}_{ls}(\mathbb{X})$; see Definition 6.6.

Lemma 6.15 *There is a measurable mapping $T \colon \mathbb{X} \times \mathbf{N}^* \to \mathbb{Y}$ such that*

$$\mu = \sum_{n=1}^{\bar{\mu}(\mathbb{X})} \delta_{(\pi_n(\bar{\mu}), T(\pi_n(\bar{\mu}), \mu))}, \qquad \mu \in \mathbf{N}^*, \tag{6.16}$$

where $\bar{\mu} := \mu(\cdot \times \mathbb{Y})$.

Proof Let $\mu \in \mathbf{N}^*$. If $\bar{\mu}\{x\} = 0$ we set $T(x, \mu) := y_0$ for some fixed value $y_0 \in \mathbb{Y}$. If $\bar{\mu}\{x\} > 0$ then $\nu := \mu(\{x\} \times \cdot)$ is an integer-valued measure on $\mathbb{Y}$ with $\nu(\mathbb{X}) = 1$. By Proposition 6.2 there exists a unique $y \in \mathbb{Y}$ such that $\nu\{y\} = 1$, so that we can define $T(x, \mu) := y$. Then (6.16) holds.

It remains to show that the mapping T is measurable. Let $C \in \mathcal{Y}$. Then we have for all $(x, \mu) \in \mathbb{X} \times \mathbf{N}^*$ that

$$\mathbf{1}\{T(x, \mu) \in C\} = \mathbf{1}\{\bar{\mu}\{x\} = 0, y_0 \in C\} + \mathbf{1}\{\mu(\{x\} \times C) > 0\}.$$

Since $\mathbb{X} \times \mathbb{Y}$ is a Borel subspace of a CSMS, it follows from Proposition 6.3 that $(x, \mu) \mapsto (\mathbf{1}\{\bar{\mu}\{x\} = 0\}, \mathbf{1}\{\mu(\{x\} \times C) > 0\})$ is measurable. This shows that T is measurable. □

Proposition 6.16 *Let ξ be a Poisson process on $\mathbb{X} \times \mathbb{Y}$, where $(\mathbb{X}, \mathcal{X})$ and*

$(\mathbb{Y}, \mathcal{Y})$ are Borel subspaces of a CSMS. Suppose that the intensity measure of ξ is given by $\lambda\otimes K$, where λ is a locally finite diffuse measure on $\mathbb{X}$ and K is a probability kernel from $\mathbb{X}$ to $\mathbb{Y}$. Then ξ is a K-marking of $\eta := \xi(\cdot\times\mathbb{Y})$.

Proof Since λ is locally finite and diffuse, we can apply Proposition 6.9 to the Poisson process η to obtain $\mathbb{P}(\xi \in \mathbf{N}^*) = 1$. It is then no restriction of generality to assume that $\xi \in \mathbf{N}^*$ everywhere on Ω. By Lemma 6.15 we have the representation

$$\xi = \sum_{n=1}^{\kappa} \delta_{(X_n,Y_n)},$$

where $\kappa := \xi(\mathbb{X}\times\mathbb{Y})$, and for each $n \in \mathbb{N}$, $X_n := \pi_n(\bar{\xi})$, $Y_n := T(X_n, \xi)$. (Recall that $\bar{\xi} = \xi(\cdot\times\mathbb{Y})$.) We wish to show that the sequence (Y_n) has the properties required in Definition 5.3. Since κ has a Poisson distribution, we have $\mathbb{P}(\kappa = \infty) \in \{0, 1\}$. Let us first assume that $\mathbb{P}(\kappa = \infty) = 1$. Let $n \in \mathbb{N}$, $k \in \{1, \dots, n\}$, $A \in \mathcal{X}^n$ and $B_1, \dots, B_k \in \mathcal{Y}$. Set $B := B_1 \times\cdots\times B_k$ and

$$C := \{((x_1, y_1), \dots, (x_n, y_n)) \in (\mathbb{X}\times\mathbb{Y})^n : (x_1, \dots, x_n) \in A, (y_1, \dots, y_k) \in B\}.$$

Then

$$\begin{aligned}\mathbb{P}((X_1, \dots, X_n) \in A, (Y_1, \dots, Y_k) \in B)& \\ = \mathbb{E}\Big[\int_C g_n(x_1, \dots, x_n, \bar{\xi})\,\xi^{(n)}(d((x_1, y_1), \dots, (x_n, y_n)))\Big],&\end{aligned}$$

where, for $x_1, \dots, x_n \in \mathbb{X}$ and $\mu \in \mathbf{N}_{ls}$,

$$g_n(x_1, \dots, x_n, \mu) := \mathbf{1}\{\pi_1(\mu) = x_1, \dots, \pi_n(\mu) = x_n\}.$$

By the multivariate Mecke equation (Theorem 4.4) and the assumed form of the intensity measure of ξ, this equals

$$\mathbb{E}\Big[\int_A g_n(x_1, \dots, x_n, \eta + \delta_{x_1} + \cdots + \delta_{x_n}) \prod_{i=1}^{k} K(x_i, B_i)\,\lambda^n(d(x_1, \dots, x_n))\Big].$$

By the multivariate Mecke identity for the Poisson process η, this comes to

$$\mathbb{E}\Big[\mathbf{1}\{(X_1, \dots, X_n) \in A\} \prod_{i=1}^{k} K(X_i, B_i)\Big].$$

Since n is arbitrary it follows from Theorem A.5 that

$$\mathbb{P}((X_n)_{n\geq 1} \in \cdot, (Y_1, \dots, Y_k) \in B) = \mathbb{E}\Big[\mathbf{1}\{(X_n)_{n\geq 1} \in \cdot\} \prod_{i=1}^{k} K(X_i, B_i)\Big].$$

Since k is arbitrary this implies the assertion.

Assume now that $\mathbb{P}(\kappa < \infty) = 1$ and let $n \in \mathbb{N}$. Then, using similar notation to above (for the case $k = n$),

$$\mathbb{P}((X_1, \dots, X_n) \in A, (Y_1, \dots, Y_n) \in B, \eta(\mathbb{X}) = n)$$
$$= \mathbb{E}\Big[\int_C h_n(x_1, \dots, x_n, \bar{\xi})\, \xi^{(n)}(d((x_1, y_1), \dots, (x_n, y_n)))\Big],$$

where $h_n(x_1, \dots, x_n, \mu) := \mathbf{1}\{\pi_1(\mu) = x_1, \dots, \pi_n(\mu) = x_n, \mu(\mathbb{X}) = n\}$, and we can argue as above to conclude the proof. □

Exercise 6.12 shows that Proposition 6.16 remains true for possibly non-diffuse λ, provided that the underlying probability space supports a K-marking of ξ.

6.6 Exercises

Exercise 6.1 Suppose that $\mathbb{X}$ is a Borel subspace of a CSMS. Show that the mapping $(x, \mu) \mapsto \mu\{x\}$ from $\mathbb{X} \times \mathbf{N}_l(\mathbb{X})$ to $\mathbb{N}_0$ is $\mathcal{B}(\mathbb{X}) \otimes \mathcal{N}_l(\mathbb{X})$-measurable. (Hint: Use Proposition 6.3.)

Exercise 6.2 Give an example to show that if the word "simple" is omitted from the hypothesis of Rényi's theorem, then the conclusion need not be true. (This can be done by taking a simple point process and modifying it to make it "complicated", i.e. not simple.)

Exercise 6.3 Give an example to show that if the word "diffuse" is omitted from the hypothesis of Rényi's theorem, then the conclusion need not be true. (This can be done by taking a "complicated" point process and modifying it to make it simple.)

Exercise 6.4 Let $(\mathbb{X}, \mathcal{X})$ be a Borel space. A measure λ on $\mathbb{X}$ is said to be *purely discrete*, if $\lambda = \sum_{i \in I} c_i \delta_{x_i}$, for some $I \subset \mathbb{N}$, $x_i \in \mathbb{X}$ and $c_i > 0$. Let λ be a σ-finite measure on $\mathbb{X}$ and let $A := \{x \in \mathbb{X} : \lambda\{x\} = 0\}$. Show that λ_A is diffuse and that $\lambda_{\mathbb{X} \setminus A}$ is purely discrete.

Exercise 6.5 Give an example to show that if we drop the assumption (6.11) from the conditions of Theorem 6.12, then we cannot always conclude that η is a Poisson process.

Exercise 6.6 Suppose that $(\mathbb{X}, \mathcal{X})$ is a Borel subspace of a CSMS. Define for each $\mu \in \mathbf{N}_l(\mathbb{X})$ the measure $\mu^* \in \mathbf{N}_s(\mathbb{X})$ by

$$\mu^* := \int \mu\{x\}^{\oplus} \mathbf{1}\{x \in \cdot\}\, \mu(dx), \tag{6.17}$$

where $a^{\oplus} := \mathbf{1}\{a \neq 0\}a^{-1}$ is the generalised inverse of $a \in \mathbb{R}$. Prove that the mapping $\mu \mapsto \mu^*$ from $\mathbf{N}_l(\mathbb{X})$ to $\mathbf{N}_l(\mathbb{X})$ is measurable. Prove also that the system of all sets $\{\mu \in \mathbf{N}_l(\mathbb{X}) : \mu(B) = 0\}$, where B is a bounded Borel set, is a π-system generating the σ-field

$$\mathcal{N}^* = \{\{\mu \in \mathbf{N}_l(\mathbb{X}) : \mu^* \in A\} : A \in \mathcal{N}_l(\mathbb{X})\}.$$

(Hint: Check the proof of Theorem 6.11.)

Exercise 6.7 Suppose that $(\mathbb{X}, \mathcal{X})$ is a Borel subspace of a CSMS. Recall from Definition 6.6 the notation $\mathbf{N}_{ls}(\mathbb{X}) = \mathbf{N}_l(\mathbb{X}) \cap \mathbf{N}_s(\mathbb{X})$ and let

$$\mathcal{N}_{ls}(\mathbb{X}) := \{A \cap \mathbf{N}_{ls}(\mathbb{X}) : A \in \mathcal{N}(\mathbb{X})\}.$$

Show that the system of all sets $\{\mu \in \mathbf{N}_{ls}(\mathbb{X}) : \mu(B) = 0\}$, where B is a bounded Borel set, is a π-system generating $\mathcal{N}_{ls}(\mathbb{X})$.

Exercise 6.8 Let η and η' be point processes on an arbitrary measurable space $(\mathbb{X}, \mathcal{X})$. Assume that there are $B_n \in \mathcal{X}$, $n \in \mathbb{N}$, such that $\cup_{n=1}^{\infty} B_n = \mathbb{X}$ and such that (6.3) holds for both η and η'. Prove that $\eta \overset{d}{=} \eta'$ if and only if $\eta_{B_n} \overset{d}{=} \eta'_{B_n}$ for each $n \in \mathbb{N}$.

Exercise 6.9 Let $(\mathbb{X}, \mathcal{X})$ be a Borel space, $\mu \in \mathbf{N}_s(\mathbb{X})$ and $m \in \mathbb{N}$ with $m \geq 2$. Show that $\mu^{(m)}(D_m) = 0$, where the generalised diagonal D_m is given by (6.14). Why is D_m a measurable set? (Hint: Use Proposition 6.2 and (4.5).)

Exercise 6.10 Let ν be the measure on $[0, 1]$ defined by $\nu(B) = 0$ if $\lambda_1(B) = 0$ and $\nu(B) = \infty$ otherwise. (Here λ_1 denotes Lebesgue measure.) Show that ν is s-finite but does not belong to $\mathbf{N}([0, 1])$. (Hint: To prove the second assertion you can use the fact that each $\mu \in \mathbf{N}([0, 1])$ is an at most countably infinite sum of Dirac measures.)

Exercise 6.11 (Uniform randomisation) Let η be a proper Poisson process on a Borel space $(\mathbb{X}, \mathcal{X})$ and let ξ be an independent λ_1-marking of η, where λ_1 is Lebesgue measure on $[0, 1]$. Show that the Mecke identity for η can be derived from that for ξ.

Exercise 6.12 Suppose the assumptions of Proposition 6.16 are all satisfied except for the assumption that λ is diffuse. Assume that the probability space supports uniform randomisation of ξ (see Exercise 6.11) and show that then the assertion of Proposition 6.16 remains valid.

7

Poisson Processes on the Real Line

A Poisson process on the real half-line is said to be homogeneous if its intensity measure is a multiple of Lebesgue measure. Such a process is characterised by the fact that the distances between consecutive points are independent and identically exponentially distributed. Using conditional distributions this result can be generalised to position-dependent markings of non-homogeneous Poisson processes. An interesting example of a non-homogeneous Poisson process is given by the consecutive record values in a sequence of independent and identically distributed non-negative random variables.

7.1 The Interval Theorem

In this chapter we study point processes on the real half-line $\mathbb{R}_+ := [0, \infty)$. We shall consider point processes that are simple with at most one accumulation point of their atoms.

Given a measure μ on $\mathbb{R}_+$ (or on $\mathbb{R}$) and an interval $I \subset \mathbb{R}_+$ (resp. $I \subset \mathbb{R}$), we shall write $\mu I := \mu(I)$. For $\mu \in \mathbf{N}(\mathbb{R}_+)$ set

$$T_n(\mu) := \inf\{t \geq 0 : \mu[0, t] \geq n\}, \quad n \in \overline{\mathbb{N}},$$

where $\inf \emptyset := \infty$, $\overline{N} := \mathbb{N} \cup \{\infty\}$ and where we interpret $\mu[0, t] \geq \infty$ as $\mu[0, t] = \infty$. Let $\mathbf{N}^+$ be the space of all measures $\mu \in \mathbf{N}(\mathbb{R}_+)$ such that $\mu[T_\infty(\mu), \infty) = 0$ and $T_n(\mu) < T_{n+1}(\mu)$ for all $n \in \mathbb{N}$ such that $T_n(\mu) < \infty$. In Exercise 7.1 the reader is asked to show that $\mathbf{N}^+ \in \mathcal{N}(\mathbb{R}_+)$.

Definition 7.1 We say that a point process η on $\mathbb{R}_+$ is *ordinary* if it satisfies $\mathbb{P}(\eta \in \mathbf{N}^+) = 1$.

If η is an ordinary point process we can almost surely write

$$\eta = \sum_{n=1}^{\infty} \mathbf{1}\{T_n < \infty\}\delta_{T_n}, \tag{7.1}$$

where $T_n := T_n(\eta)$ for $n \in \overline{\mathbb{N}}$. Sometimes T_n is called n-th *arrival time* of η. In the case $T_n = \infty$ the measure $\mathbf{1}\{T_n < \infty\}\delta_{T_n}$ is interpreted as the zero measure on $\mathbb{R}_+$. If $T_\infty < \infty$ we say that *explosion* occurs.

An important example of an ordinary point process is a *homogeneous Poisson process* of *rate* (or *intensity*) $\gamma > 0$. This is a Poisson process on $\mathbb{R}_+$ with intensity measure $\gamma\lambda_+$, where λ_+ is Lebesgue measure on $\mathbb{R}_+$. (More generally, for $d \in \mathbb{N}$ and $B \in \mathcal{B}(\mathbb{R}^d)$, a *homogeneous Poisson process* on B is a Poisson process η on B whose intensity measure is a multiple of Lebesgue measure on B. This multiple is called the *intensity* of η.) Given $B \subset \mathbb{R}$ and $t \in \mathbb{R}$ we set $B + t := \{s + t : s \in B\}$. A point process η on $\mathbb{R}_+$ is said to be *stationary* if

$$\theta_t^+ \eta \overset{d}{=} \eta, \quad t \in \mathbb{R}_+,$$

where, for any measure μ on $\mathbb{R}_+$ and $t \in \mathbb{R}_+$, the measure $\theta_t^+\mu$ on $\mathbb{R}_+$ is defined by

$$\theta_t^+ \mu(B) := \mu(B + t), \quad B \in \mathcal{B}(\mathbb{R}_+). \tag{7.2}$$

Any homogeneous Poisson process on $\mathbb{R}_+$ is stationary.

Our first aim in this chapter is to characterise homogeneous Poisson processes in terms of the inter-point distances $T_n - T_{n-1}$, where $T_0 := 0$.

Theorem 7.2 (Interval theorem) *Let η be a point process on $\mathbb{R}_+$. Then η is a homogeneous Poisson process with rate $\gamma > 0$ if and only if the $T_n - T_{n-1}$, $n \geq 1$, are independent and exponentially distributed with parameter γ.*

Proof Suppose first that η is a Poisson process as stated. Let $n \in \mathbb{N}$. Since η is locally finite we have

$$\{T_n \leq t\} = \{\eta[0, t] \geq n\}, \quad \mathbb{P}\text{-a.s.}, \ t \in \mathbb{R}_+. \tag{7.3}$$

Since $\mathbb{P}(\eta(\mathbb{R}_+) = \infty) = 1$ we have $\mathbb{P}(T_n < \infty) = 1$. Let $f \in \mathbb{R}_+(\mathbb{R}_+^n)$. Then

$$\mathbb{E}[f(T_1, T_2 - T_1, \dots, T_n - T_{n-1})] = \mathbb{E}\Big[\int \mathbf{1}\{t_1 < \cdots < t_n\}$$
$$\times f(t_1, t_2 - t_1, \dots, t_n - t_{n-1})\mathbf{1}\{\eta[0, t_n) = n - 1\}\, \eta^{(n)}(d(t_1, \dots, t_n))\Big]. \tag{7.4}$$

Now we use the multivariate Mecke theorem (Theorem 4.4). Since, for $0 \leq t_1 < \cdots < t_n$,

$$\{(\eta + \delta_{t_1} + \cdots + \delta_{t_n})[0, t_n) = n - 1\} = \{\eta[0, t_n) = 0\},$$

the right-hand side of (7.4) equals

$$\gamma^n \int \mathbf{1}\{0 < t_1 < \cdots < t_n\} f(t_1, t_2 - t_1, \dots, t_n - t_{n-1}) \exp[-\gamma t_n]\, d(t_1, \dots, t_n),$$

where the integration is with respect to Lebesgue measure on $\mathbb{R}^n$. After the change of variables $s_1 := t_1, s_2 := t_2 - t_1, \dots, s_n := t_n - t_{n-1}$ this yields

$$\mathbb{E}[f(T_1, T_2 - T_1, \dots, T_n - T_{n-1})] = \int_0^\infty \cdots \int_0^\infty f(s_1, \dots, s_n)\gamma^n \exp[-\gamma(s_1 + \cdots + s_n)]\, ds_1 \cdots ds_n.$$

Therefore $T_1, T_2 - T_1, \dots, T_n - T_{n-1}$ are independent and exponentially distributed with parameter γ. Since $n \in \mathbb{N}$ is arbitrary, the asserted properties of the sequence (T_n) follow.

Suppose, conversely, that (T_n) has the stated properties. Let η' be a homogeneous Poisson process of intensity $\gamma > 0$. Then η' has a representation as in (7.1) with random variables T'_n instead of T_n. We have just proved that $(T_n) \overset{d}{=} (T'_n)$. Since, for any $B \in \mathcal{B}(\mathbb{R}_+)$,

$$\eta(B) = \sum_{n=1}^\infty \mathbf{1}\{T_n \in B\}$$

is a measurable function of the sequence (T_n), we can use Proposition 2.10 ((ii) implies (i)) to conclude that $\eta \overset{d}{=} \eta'$ and hence η is a homogeneous Poisson process. □

A Poisson process on $\mathbb{R}_+$ whose intensity measure is not a multiple of λ_+ is said to be *non-homogeneous*. Such a process can be constructed from a homogeneous Poisson process by a suitable *time transform*. This procedure is a special case of the mapping theorem (Theorem 5.1). Let ν be a locally finite measure on $\mathbb{R}_+$ and define a function $\nu^{\leftarrow} : \mathbb{R}_+ \to [0, \infty]$ by

$$\nu^{\leftarrow}(t) := \inf\{s \geq 0 : \nu[0, s] \geq t\}, \quad t \geq 0, \tag{7.5}$$

where $\inf \emptyset := \infty$. This function is increasing, left-continuous and, in particular, measurable.

Proposition 7.3 *Let ν be a locally finite measure on $\mathbb{R}_+$, let η be a homogeneous Poisson process on $\mathbb{R}_+$ with rate 1 and let (T_n) be given by (7.1). Then*

$$\eta' := \sum_{n=1}^\infty \mathbf{1}\{\nu^{\leftarrow}(T_n) < \infty\}\delta_{\nu^{\leftarrow}(T_n)} \tag{7.6}$$

is a Poisson process on $\mathbb{R}_+$ with intensity measure ν.

Proof By the mapping theorem (Theorem 5.1) $\sum_{n=1}^\infty \delta_{\nu^{\leftarrow}(T_n)}$ is a Poisson process on $\overline{\mathbb{R}}_+$ with intensity measure

$$\lambda = \int \mathbf{1}\{\nu^{\leftarrow}(t) \in \cdot\}\, dt.$$

Proposition A.31 shows that $\lambda = \nu$ (on $\mathbb{R}_+$), and the assertion follows. □

7.2 Marked Poisson Processes

In this section we consider Poisson processes on $\mathbb{R}_+ \times \mathbb{Y}$, where $(\mathbb{Y}, \mathcal{Y})$ (the mark space) is a Borel space. Let $\mathbf{N}^+(\mathbb{Y})$ be the space of all $\mu \in \mathbf{N}(\mathbb{R}_+ \times \mathbb{Y})$ such that $\mu(\cdot \times \mathbb{Y}) \in \mathbf{N}^+$. Exercise 7.4 shows that there are measurable mappings $T'_n\colon \mathbf{N}^+(\mathbb{Y}) \to [0, \infty]$, $n \in \mathbb{N}$, and $Y'_n\colon \mathbf{N}^+(\mathbb{Y}) \to \mathbb{Y}$, such that $T'_n \le T'_{n+1}$ for each $n \in \mathbb{N}$ and

$$\mu = \sum_{n=1}^{\infty} \mathbf{1}\{T'_n(\mu) < \infty\}\delta_{(T'_n(\mu), Y'_n(\mu))}, \quad \mu \in \mathbf{N}^+(\mathbb{Y}). \tag{7.7}$$

Let ξ be a point process on $\mathbb{R}_+ \times \mathbb{Y}$ such that $\eta := \xi(\cdot \times \mathbb{Y})$ is ordinary. By (7.7) we have almost surely that

$$\xi = \sum_{n=1}^{\infty} \mathbf{1}\{T_n < \infty\}\delta_{(T_n, Y_n)}, \tag{7.8}$$

where $T_n := T_n(\eta)$, $n \in \mathbb{N}$, and where the Y_n are random elements of $\mathbb{Y}$ such that almost surely $\xi\{(T_n, Y_n)\} = 1$ for $T_n < \infty$.

If ξ is a Poisson process, then our next result (Theorem 7.4) provides a formula for the distribution of $(T_1, Y_1, \dots, T_n, Y_n)$ in terms of the intensity measure of ξ. Corollary 7.5 will then extend Theorem 7.2 by allowing both for marks and for non-homogeneity of the Poisson process.

Given a measure μ on $\mathbb{R}_+ \times \mathbb{Y}$ and given $t \ge 0$, we define another measure $\vartheta^+_t \mu$ on $\mathbb{R}_+ \times \mathbb{Y}$ by

$$\vartheta^+_t \mu(B) := \int \mathbf{1}\{(s - t, y) \in B\}\, \mu(d(s, y)), \quad B \in \mathcal{B}(\mathbb{R}_+) \otimes \mathcal{Y}.$$

This definition generalises (7.2). If $\mu \in \mathbf{N}^+(\mathbb{Y})$ then (7.7) implies that

$$\vartheta^+_t \mu = \sum_{n=1}^{\infty} \mathbf{1}\{t \le T'_n(\mu) < \infty\}\delta_{(T'_n(\mu) - t, Y'_n(\mu))}, \quad \mu \in \mathbf{N}^+(\mathbb{Y}).$$

This shows that $(t, \mu) \mapsto \vartheta^+_t \mu$ is measurable on $\mathbb{R}_+ \times \mathbf{N}^+(\mathbb{Y})$. Indeed, for each $n \in \mathbb{N}$ and each $B \in \mathcal{B}(\mathbb{R}_+) \otimes \mathcal{Y}$ the expression $\delta_{(T'_n(\mu) - t, Y'_n(\mu))}(B)$ is a measurable function of (t, μ).

Theorem 7.4 (Memoryless property) *Suppose that ξ is a Poisson process on $\mathbb{R}_+ \times \mathbb{Y}$ with a σ-finite intensity measure λ such that $\mathbb{P}(\xi(\cdot \times \mathbb{Y}) \in \mathbf{N}^+) = 1$. For $n \in \mathbb{N}$ let $T_n := T_n(\eta)$, where $\eta := \xi(\cdot \times \mathbb{Y})$. Let the sequence (Y_n) be as in (7.8). Then the following hold for all $n \in \mathbb{N}$.*

(i) *For any $f \in \mathbb{R}_+((\mathbb{R}_+ \times \mathbb{Y})^n)$,*

$$\mathbb{E}[\mathbf{1}\{T_n < \infty\} f(T_1, Y_1, \dots, T_n, Y_n)] = \int \mathbf{1}\{0 < t_1 < \cdots < t_n\}$$
$$\times f(t_1, y_1, \dots, t_n, y_n) \exp[-\lambda((0, t_n] \times \mathbb{Y})]\, \lambda^n(d(t_1, y_1, \dots, t_n, y_n)).$$

(ii) *The conditional distribution of $\vartheta^+_{T_n} \xi$ given $(T_1, Y_1, \dots, T_n, Y_n)$ and $T_n < \infty$ is almost surely that of a Poisson process with intensity measure $\vartheta^+_{T_n} \lambda$.*

Proof We interpret ξ as a random element of the space $\mathbf{N}^+(\mathbb{Y})$ introduced at the beginning of this section. The assumptions and Proposition 6.9 imply that the measure $\lambda(\cdot \times \mathbb{Y})$ is diffuse. We now use the same idea as in the proof of Theorem 7.2. Let f be as in (i) and let $g \in \mathbb{R}_+(\mathbf{N}^+(\mathbb{Y}))$. Then

$$\mathbb{E}[\mathbf{1}\{T_n < \infty\} f(T_1, Y_1, \dots, T_n, Y_n) g(\vartheta^+_{T_n} \xi)] = \mathbb{E}\Big[\int \mathbf{1}\{t_1 < \cdots < t_n\}$$
$$\times f(t_1, y_1, \dots, t_n, y_n) g(\vartheta^+_{t_n} \xi) \mathbf{1}\{\eta[0, t_n) = n - 1\}\, \xi^{(n)}(d(t_1, y_1, \dots, t_n, y_n))\Big].$$

By the Mecke equation this equals

$$\int \mathbf{1}\{t_1 < \cdots < t_n\} f(t_1, y_1, \dots, t_n, y_n)$$
$$\times \mathbb{E}[g(\vartheta^+_{t_n} \xi) \mathbf{1}\{\eta[0, t_n) = 0\}]\, \lambda^n(d(t_1, y_1, \dots, t_n, y_n)).$$

By Theorem 5.2, $\mathbf{1}\{\eta[0, t_n) = 0\}$ and $g(\vartheta^+_{t_n} \xi)$ are independent for any fixed t_n. Moreover, $\vartheta^+_{t_n} \xi$ is a Poisson process with intensity measure $\vartheta^+_{t_n} \lambda$. Therefore we obtain both (i) and (ii). □

If, in the situation of Theorem 7.4, $\lambda(\mathbb{R}_+ \times \mathbb{Y}) < \infty$, then ξ has only finitely many points and $T_n = \infty$ for $n > \xi(\mathbb{R}_+ \times \mathbb{Y})$. In fact, the theorem shows that, $\mathbb{P}$-a.s. on the event $\{T_n < \infty\}$,

$$\mathbb{P}(T_{n+1} = \infty \mid T_0, Y_0, \dots, T_n, Y_n) = \exp[-\lambda([T_n, \infty) \times \mathbb{Y})].$$

If ν, ν' are measures on a measurable space $(\mathbb{X}, \mathcal{X})$ and $f \in \mathbb{R}_+(\mathbb{X})$, we write $\nu'(dx) = f(x)\nu(dx)$ if f is a density of ν' with respect to ν, that is $\nu'(B) = \nu(\mathbf{1}_B f)$ for all $B \in \mathcal{X}$.

Corollary 7.5 *Under the hypotheses of Theorem 7.4, we have for every* $n \in \mathbb{N}$ *that*

$$\begin{aligned}&\mathbf{1}\{t < \infty\}\mathbb{P}((T_n, Y_n) \in d(t,y) \mid T_0, Y_0, \dots, T_{n-1}, Y_{n-1})\\&\quad = \mathbf{1}\{T_{n-1} < t\} \exp[-\lambda([T_{n-1}, t] \times \mathbb{Y})]\lambda(d(t,y)), \quad \mathbb{P}\text{-a.s. on } \{T_{n-1} < \infty\},\end{aligned}$$

where $T_0 = 0$ *and* Y_0 *is chosen as a constant function.*

Proof The result is an immediate consequence of Theorem 7.4 and the definition of conditional distributions given in Section B.4. □

Independent markings of homogeneous Poisson processes can be characterised as follows.

Theorem 7.6 *Let the point process* ξ *on* $\mathbb{R}_+ \times \mathbb{Y}$ *be given by* (7.8) *and define* η *by* (7.1)*. Let* $\gamma > 0$ *and let* $\mathbb{Q}$ *be a probability measure on* $\mathbb{Y}$*. Then* ξ *is an independent* $\mathbb{Q}$*-marking of a homogeneous Poisson process with rate* $\gamma > 0$ *if and only if* $T_1, Y_1, T_2 - T_1, Y_2, \dots$ *are independent, the* $T_n - T_{n-1}$ *have an exponential distribution with parameter* γ *and the* Y_n *have distribution* $\mathbb{Q}$*.*

Proof If η is a homogeneous Poisson process and ξ is an independent $\mathbb{Q}$-marking of η, then by Theorem 5.6, ξ is a Poisson process with intensity measure $\gamma\lambda_+ \otimes \mathbb{Q}$. Hence the properties of the sequence $((T_n, Y_n))_{n\geq 1}$ follow from Corollary 7.5 (or from Theorem 7.4). The converse is an immediate consequence of the interval theorem (Theorem 7.2). □

7.3 Record Processes

Here we discuss how non-homogeneous Poisson processes describe the occurrence of *record levels* in a sequence $X_1, X_2, \dots$ of independent random variables with values in $\mathbb{R}_+$ and common distribution $\mathbb{Q}$. Let $N_1 := 1$ be the first record time and $R_1 := X_1$ the first record. The further record times $N_2, N_3, \dots$ are defined inductively by

$$N_{k+1} := \inf\{n > N_k : X_n > X_{N_k}\}, \quad k \in \mathbb{N},$$

where $\inf \emptyset := \infty$. The k-th record level is $R_k := X_{N_k}$. We consider the following point process on $\mathbb{R}_+ \times \mathbb{N}$:

$$\chi := \sum_{n=1}^{\infty} \mathbf{1}\{N_{n+1} < \infty\}\delta_{(R_n, N_{n+1}-N_n)}. \tag{7.9}$$

Proposition 7.7 *Let $\mathbb{Q}$ be a diffuse probability measure on $\mathbb{R}_+$ and let $(X_n)_{n\geq 1}$ be a sequence of independent $\mathbb{R}_+$-valued random variables with common distribution $\mathbb{Q}$. Then the point process χ on $\mathbb{R}_+ \times \mathbb{N}$ defined by (7.9) is a Poisson process whose intensity measure λ is given by*

$$\lambda(dt \times \{k\}) = \mathbb{Q}(0, t]^{k-1}\mathbb{Q}(dt), \quad k \in \mathbb{N}.$$

Proof Let $n \in \mathbb{N}$, $k_1, \ldots, k_n \in \mathbb{N}$ and $f \in \mathbb{R}_+(\mathbb{R}_+^n)$. We assert that

$$\begin{aligned}
&\mathbb{E}[\mathbf{1}\{N_2 - N_1 = k_1, \ldots, N_{n+1} - N_n = k_n\} f(R_1, \ldots, R_n)] \\
&\quad = \int \mathbf{1}\{t_1 < \cdots < t_n\} f(t_1, \ldots, t_n) \\
&\qquad \times \mathbb{Q}[0, t_1]^{k_1-1} \cdots \mathbb{Q}[0, t_n]^{k_n-1} \mathbb{Q}(t_n, \infty)\, \mathbb{Q}^n(d(t_1, \ldots, t_n)). \qquad (7.10)
\end{aligned}$$

To prove this let A denote the event inside the indicator in the left-hand side of (7.10). Set $Y_1 := X_1$ and $Y_i := X_{1+k_1+\cdots+k_{i-1}}$ for $2 \leq i \leq n$, and let $B := \{Y_1 < \cdots < Y_n\}$. Then the left-hand side of (7.10) equals

$$\mathbb{E}[\mathbf{1}_A \mathbf{1}_B f(Y_1, \ldots, Y_n)] = \mathbb{E}[f(Y_1, \ldots, Y_n)\mathbf{1}_B\, \mathbb{P}(A \mid Y_1, \ldots, Y_n)],$$

where the identity follows from independence and Fubini's theorem (or, equivalently, by conditioning on $Y_1, \ldots, Y_n$). This equals the right-hand side of (7.10).

Effectively, the range of integration in (7.10) can be restricted to $t_n < t_\infty$, where

$$t_\infty := \sup\{t \in \mathbb{R}_+ : \mathbb{Q}[0, t] < 1\}.$$

Indeed, since $\mathbb{Q}$ is diffuse, we have $\mathbb{Q}[t_\infty, \infty) = 0$. Summing in (7.10) over $k_1, \ldots, k_n \in \mathbb{N}$, we obtain

$$\begin{aligned}
&\mathbb{E}[\mathbf{1}\{N_{n+1} < \infty\} f(R_1, \ldots, R_n)] \\
&\quad = \int \mathbf{1}\{t_1 < \cdots < t_n\} f(t_1, \ldots, t_n) \\
&\qquad \times \mathbb{Q}(t_1, \infty)^{-1} \cdots \mathbb{Q}(t_{n-1}, \infty)^{-1}\, \mathbb{Q}^n(d(t_1, \ldots, t_n)).
\end{aligned}$$

Taking $f \equiv 1$ and performing the integration yields $\mathbb{P}(N_{n+1} < \infty) = 1$.

Next we note that

$$\lambda(dt \times \mathbb{N}) = (\mathbb{Q}[t, \infty))^{\oplus}\mathbb{Q}(dt) \qquad (7.11)$$

is the *hazard measure* of $\mathbb{Q}$, where $a^{\oplus} := \mathbf{1}\{a \neq 0\}a^{-1}$ is the generalised inverse of $a \in \mathbb{R}$. Therefore by Proposition A.32

$$\mathbb{Q}[t, \infty) = \exp[-\lambda([0, t] \times \mathbb{N})]. \qquad (7.12)$$

Hence we obtain, for all $n \in \mathbb{N}$ and $g \in \mathbb{R}_+((\mathbb{R}_+ \times \mathbb{N})^n)$, from (7.10) that

$$\begin{aligned}\mathbb{E}[g(R_1, N_2 - N_1, \dots, R_n, N_{n+1} - N_n)] \\ = \int \mathbf{1}\{t_1 < \cdots < t_n\} g(t_1, k_1, \dots, t_n, k_n) \\ \times \exp[-\lambda([0, t_n] \times \mathbb{N})]\, \lambda^n(d(t_1, k_1, \dots, t_n, k_n)). \quad (7.13)\end{aligned}$$

Now let ξ be a Poisson process as in Theorem 7.4 (with $\mathbb{Y} = \mathbb{N}$) with intensity measure λ. The identity (7.12) implies that $\lambda([0,\infty) \times \mathbb{N}) = \infty$, so that $\mathbb{P}(T_n < \infty) = 1$ holds for all $n \in \mathbb{N}$. Comparing (7.13) and Theorem 7.4(i) yields

$$(R_1, N_2 - N_1, \dots, R_n, N_{n+1} - N_n) \stackrel{d}{=} (T_1, Y_1, \dots, T_n, Y_n), \quad n \in \mathbb{N}.$$

As in the final part of the proof of Theorem 7.2 we obtain $\xi \stackrel{d}{=} \chi$ and hence the assertion of the theorem. □

Proposition 7.7, (7.11) and the mapping theorem (Theorem 5.1) together show that the point process $\chi(\cdot \times \mathbb{N})$ of successive record levels is a Poisson process with the hazard measure of $\mathbb{Q}$ as intensity measure. Further consequences of the proposition are discussed in Exercise 7.14.

7.4 Polar Representation of Homogeneous Poisson Processes

In this section we discuss how Poisson processes on $\mathbb{R}_+$ naturally show up in a spatial setting. For $d \in \mathbb{N}$ let ν_{d-1} denote the uniform distribution on the *unit sphere* $\mathbb{S}^{d-1} := \{x \in \mathbb{R}^d : \|x\| = 1\}$, where $\|\cdot\|$ denotes the Euclidean norm on $\mathbb{R}^d$. This normalised *spherical Lebesgue measure* on $\mathbb{S}^{d-1}$ is the probability measure defined by

$$\nu_{d-1}(C) := \kappa_d^{-1} \int_{B^d} \mathbf{1}\{x/\|x\| \in C\}\, dx, \quad C \in \mathcal{B}(\mathbb{S}^{d-1}), \tag{7.14}$$

where $B^d := \{x \in \mathbb{R}^d : \|x\| \le 1\}$ is the *unit ball* and $\kappa_d := \lambda_d(B^d)$ is its volume. For $x = 0$ we let $x/\|x\|$ equal some fixed point in $\mathbb{S}^{d-1}$.

Proposition 7.8 *Let ζ be a homogeneous Poisson process on $\mathbb{R}^d$ with intensity $\gamma > 0$. Then the point process ξ on $\mathbb{R}^d \times \mathbb{S}^{d-1}$ defined by*

$$\xi(A) := \int \mathbf{1}\{(\kappa_d \|x\|^d, x/\|x\|) \in A\}\, \zeta(dx), \quad A \in \mathcal{B}(\mathbb{R}_+ \times \mathbb{S}^{d-1}), \tag{7.15}$$

is an independent ν_{d-1}-marking of a homogeneous Poisson process with intensity γ.

Proof By Theorem 5.1 (mapping theorem) and Proposition 6.16 it is sufficient to show for each $A \in \mathcal{B}(\mathbb{R}_+ \times \mathbb{S}^{d-1})$ that

$$\mathbb{E}[\xi(A)] = \gamma \int_{\mathbb{S}^{d-1}} \int_0^\infty \mathbf{1}\{(r, u) \in A\} \, dr \, \nu_{d-1}(du). \tag{7.16}$$

To this end, we need the *polar representation* of Lebesgue measure, which says that

$$\int g(x) \, dx = d\kappa_d \int_{\mathbb{S}^{d-1}} \int_0^\infty r^{d-1} g(ru) \, dr \, \nu_{d-1}(du), \tag{7.17}$$

for all $g \in \mathbb{R}_+(\mathbb{R}^d)$. Indeed, if $g(x) = \mathbf{1}\{\|x\| \le s, x/\|x\| \in C\}$, for $s \ge 0$ and $C \in \mathcal{B}(\mathbb{S}^{d-1})$, then (7.17) follows from definition (7.14) and the scaling properties of Lebesgue measure. Using first Campbell's formula for ζ and then (7.17) yields

$$\begin{aligned}\mathbb{E}[\xi(A)] &= \gamma \int \mathbf{1}\{(\kappa_d \|x\|^d, x/\|x\|) \in A\} \, dx \\ &= \gamma d\kappa_d \int_{\mathbb{S}^{d-1}} \int_0^\infty \mathbf{1}\{(\kappa_d r^d, u) \in A\} r^{d-1} \, dr \, \nu_{d-1}(du).\end{aligned}$$

Hence (7.16) follows upon a change of variables. □

Proposition 7.8 can be used along with the interval theorem (Theorem 7.2) to simulate the points of a homogeneous Poisson process in order of increasing distance from the origin.

7.5 Exercises

Exercise 7.1 Let $\mu \in \mathbf{N}(\mathbb{R}_+)$ and define the function $f_\mu \in \overline{\mathbb{R}}_+(\mathbb{R}_+)$ as the right-continuous version of $t \mapsto \mu[0, t]$, that is

$$f_\mu(t) := \lim_{s \downarrow t} \mu[0, s], \quad t \ge 0.$$

(If μ is locally finite, then $f_\mu(t) = \mu[0, t]$.) Let $n \in \overline{\mathbb{N}}$ and $t \in \mathbb{R}_+$. Show that $T_n(\mu) \le t$ if and only if $f_\mu(t) \ge n$. Show that this implies that the T_n are measurable mappings on $\mathbf{N}(\mathbb{R}_+)$ and that $\mathbf{N}^+$ (see Definition 7.1) is a measurable subset of $\mathbf{N}(\mathbb{R}_+)$.

Exercise 7.2 Let η be a Poisson process on $\mathbb{R}_+$ with intensity measure λ. Show that η is ordinary if and only if λ is diffuse and $\lambda([\lambda^\leftarrow(\infty), \infty)) = \infty$, where $\lambda^\leftarrow(\infty) := \inf\{s \ge 0 : \nu[0, s] = \infty\}$ and $[\infty, \infty) := \emptyset$.

Exercise 7.3 Let T_n be the n-th point of a homogeneous Poisson process η on $\mathbb{R}_+$ with intensity γ. Use (7.3) to show that

$$\mathbb{P}(T_n \in dt) = \frac{\gamma^n}{(n-1)!} t^{n-1} e^{-\gamma t}\, dt.$$

This is a Gamma distribution with scale parameter γ and shape parameter n; see Section 1.4.

Exercise 7.4 Let $(\mathbb{Y}, \mathcal{Y})$ be a Borel space and let $C \in \mathcal{Y}$. Show that $(t, \mu) \mapsto \mu(\{t\} \times C)$ is a measurable mapping on $\mathbb{R}_+ \times \mathbf{N}^+(\mathbb{Y})$. Show also that there are measurable mappings $Y'_n \colon \mathbf{N}^+(\mathbb{Y}) \to \mathbb{Y}$, $n \in \mathbb{N}$, such that (7.7) holds with $T'_n(\mu) := T_n(\mu(\cdot \times \mathbb{Y}))$. (Hint: To prove the first assertion it suffices to show that $(t, \mu) \mapsto \mathbf{1}\{t < T_\infty(\mu)\}\mu(\{t\} \times C)$ is measurable, which can be done by a limit procedure. Check the proof of Lemma 6.15 to see the second assertion.)

Exercise 7.5 Suppose that the hypotheses of Theorem 7.4 apply, and that there is a probability kernel J from $\mathbb{R}_+$ to $(\mathbb{Y}, \mathcal{Y})$ such that

$$\lambda(d(t, y)) = J(t, dy)\, \lambda(dt \times \mathbb{Y}).$$

(By Theorem A.14 this is no restriction of generality.) Show for all $n \in \mathbb{N}$ that

$$\mathbb{P}(Y_n \in dy \mid T_1, Y_1, \dots, Y_{n-1}, T_n) = J(T_n, dy), \quad \mathbb{P}\text{-a.s. on } \{T_n < \infty\}.$$

Exercise 7.6 Suppose that X is a Poisson distributed random variable and let $k, i \in \mathbb{N}$ with $i \le k$. Show that $\mathbb{P}(X \ge k \mid X \ge i) \le \mathbb{P}(X \ge k - i)$. (Hint: Use the interval theorem.)

Exercise 7.7 Let η be a homogeneous Poisson process of intensity $\gamma > 0$. Prove that $\eta[0, t]/t \to \gamma$ as $t \to \infty$. (Hint: Use the fact that $\eta[0, n]/n$ satisfies a law of large numbers; see Theorem B.6.)

Exercise 7.8 Let η be a Poisson process on $\mathbb{R}_+$, whose intensity measure ν satisfies $0 < \nu[0, t] < \infty$ for all sufficiently large t and $\nu[0, \infty) = \infty$. Use Exercise 7.7 to prove that

$$\lim_{t \to \infty} \frac{\eta[0, t]}{\nu[0, t]} = 1, \quad \mathbb{P}\text{-a.s.}$$

Exercise 7.9 Let η_+ and η_- be two independent homogeneous Poisson processes on $\mathbb{R}_+$ with intensity γ. Define a point process η on $\mathbb{R}$ by

$$\eta(B) := \eta_+(B \cap \mathbb{R}_+) + \eta_-(B^* \cap \mathbb{R}_+), \quad B \in \mathcal{B}^1,$$

where $B^* := \{-t : t \in B\}$. Show that η is a homogeneous Poisson process on $\mathbb{R}$.

Exercise 7.10 Let η be a point process on $\mathbb{R}_+$ with intensity measure λ_+ and let ν be a locally finite measure on $\mathbb{R}_+$. Show that the point process η' defined by (7.6) has intensity measure ν. (Hint: Use Theorem 5.1 and the properties of generalised inverses; see the proof of Proposition 7.3.)

Exercise 7.11 Show that Proposition 7.7 remains valid for a sequence $(X_n)_{n\geq 1}$ of independent and identically distributed random elements of $\overline{\mathbb{R}}_+$, provided that the distribution of X_1 is diffuse on $\mathbb{R}_+$.

Exercise 7.12 Let T be a random element of $\overline{\mathbb{R}}_+$. Show that there is a Poisson process η on $\mathbb{R}_+$ such that $T \overset{d}{=} T_1(\eta)$. (Hint: Use Exercise 7.11.)

Exercise 7.13 Suppose the assumptions of Proposition 7.7 hold. For $n \in \mathbb{N}$ let $M_n := \max\{X_1, \dots, X_n\}$ (running maximum) and for $t \in \mathbb{R}_+$ let

$$L_t = \int \mathbf{1}\{s \leq t\} k \chi(d(s,k)).$$

Show that $\mathbb{P}$-a.s. $\inf\{n \in \mathbb{N} : M_n > t\} = 1 + L_t$, provided $\mathbb{Q}[0,t] > 0$. (Hence $L_t + 1$ is the first time the running maximum exceeds the level t.)

Exercise 7.14 Suppose the assumptions of Proposition 7.7 hold and for $t \in \mathbb{R}_+$ define L_t as in Exercise 7.13. Show that L_a and $L_b - L_a$ are independent whenever $0 \leq a < b$. Show also that

$$\mathbb{E}[w^{L_b - L_a}] = \frac{\mathbb{Q}(b,\infty)(1 - w\,\mathbb{Q}(b,\infty))}{\mathbb{Q}(a,\infty)(1 - w\,\mathbb{Q}(a,\infty))}, \quad w \in [0,1], \tag{7.18}$$

whenever $\mathbb{Q}[0,b] < 1$. Use this formula to prove that

$$\mathbb{P}(L_b - L_a = n) = \frac{\mathbb{Q}(b,\infty)}{\mathbb{Q}(a,\infty)} \mathbb{Q}(a,b]\mathbb{Q}(b)^{n-1}, \quad n \in \mathbb{N}. \tag{7.19}$$

(Hint: Use Theorem 3.9, Proposition A.31 and the logarithmic series to prove (7.18). Then compare the result with the probability generating function of the right-hand side of (7.19).)

Exercise 7.15 Let the assumptions of Proposition 7.7 hold. For $j \in \mathbb{N}$ let I_j be the indicator of the event that X_j is a record. Use a direct combinatorial argument to show that $I_1, I_2, \dots$ are independent with $\mathbb{P}(I_j = 1) = 1/j$.

Exercise 7.16 By setting $g(x) := e^{-\|x\|^2/2}$ in (7.17), show that the volume of the unit ball $B^d \subset \mathbb{R}^d$ is given by $\kappa_d = 2\pi^{d/2}/\Gamma(1 + d/2)$, where the Gamma function $\Gamma(\cdot)$ is defined by (1.23).

8

Stationary Point Processes

A point process η on $\mathbb{R}^d$ is said to be stationary if it looks statistically the same from all sites of $\mathbb{R}^d$. In this case the intensity measure is a multiple of Lebesgue measure. The reduced second factorial moment measure of a stationary point process can be used to express variances and covariances of point process integrals. Its density (when existing) is called the pair correlation function of η. A stationary point process is ergodic if its distribution is degenerate on translation invariant sets. In this case it satisfies a spatial ergodic theorem.

8.1 Stationarity

In this chapter we fix $d \in \mathbb{N}$ and consider point processes on the Euclidean space $\mathbb{X} = \mathbb{R}^d$. To distinguish between points of the point process and elements of $\mathbb{R}^d$ we call the latter *sites*. Stationarity is an important invariance concept in probability theory. Our aim here is to discuss a few basic properties of stationary point processes, using the Poisson process as illustration. Throughout the chapter we abbreviate $(\mathbf{N}, \mathcal{N}) := (\mathbf{N}(\mathbb{R}^d), \mathcal{N}(\mathbb{R}^d))$.

The formal definition of stationarity is based on the family of *shifts* $\theta_y : \mathbf{N} \to \mathbf{N}$, $y \in \mathbb{R}^d$, defined by

$$\theta_y \mu(B) := \mu(B + y), \quad \mu \in \mathbf{N},\ B \in \mathcal{B}^d, \tag{8.1}$$

where $B + y := \{x + y : x \in B\}$ and $\mathcal{B}^d := \mathcal{B}(\mathbb{R}^d)$ is the Borel σ-field on $\mathbb{R}^d$. We write $B - y := B + (-y)$. A good way to memorise (8.1) is the formula $\theta_y \delta_y = \delta_0$, where 0 is the origin in $\mathbb{R}^d$. Definition (8.1) is equivalent to

$$\int g(x)\, (\theta_y \mu)(dx) = \int g(x - y)\, \mu(dx), \quad \mu \in \mathbf{N},\ g \in \mathbb{R}_+(\mathbb{R}^d). \tag{8.2}$$

We note that θ_0 is the identity on $\mathbf{N}$ and the *flow property* $\theta_y \circ \theta_x = \theta_{x+y}$ for all $x, y \in \mathbb{R}^d$. For any fixed $y \in \mathbb{R}^d$, the mapping θ_y is measurable.

Definition 8.1 A point process η on $\mathbb{R}^d$ is said to be *stationary* if $\theta_x \eta \overset{d}{=} \eta$ for all $x \in \mathbb{R}^d$.

Let λ_d denote Lebesgue measure on $\mathbb{R}^d$. Under a natural integrability assumption the intensity measure of a stationary point process is a multiple of λ_d.

Proposition 8.2 *Let η be a stationary point process on $\mathbb{R}^d$ such that the quantity*

$$\gamma := \mathbb{E}[\eta([0,1]^d)] \tag{8.3}$$

is finite. Then the intensity measure of η equals $\gamma\lambda_d$.

Proof The stationarity of η implies that its intensity measure λ is *translation invariant*, that is $\lambda(B + x) = \lambda(B)$ for all $B \in \mathcal{B}^d$ and all $x \in \mathbb{R}^d$. Moreover, $\lambda([0,1]^d) = \gamma < \infty$. It is a fundamental result from measure theory that $\gamma\lambda_d$ is the only measure with these two properties. □

The number γ given by (8.3) is called the *intensity* of η. Proposition 8.2 shows that a stationary point process with a finite intensity is locally finite. For a stationary Poisson process the intensity determines the distribution:

Proposition 8.3 *Let η be a Poisson process on $\mathbb{R}^d$ such that the quantity γ defined by* (8.3) *is finite. Then η is stationary if and only if the intensity measure λ of η equals $\gamma\lambda_d$.*

Proof In view of Proposition 8.2 we need only to show that $\lambda = \gamma\lambda_d$ implies that $\theta_x\eta$ has the same distribution as η for all $x \in \mathbb{R}^d$. Since θ_x preserves Lebesgue measure, this follows from Theorem 5.1 (the mapping theorem) or by a direct check of Definition 3.1. □

For examples of non-Poisson stationary point processes we refer to Exercises 8.2 and 8.3, Section 14.2 and Section 16.5.

The next result says that a stationary point process cannot have a positive but finite number of points. Given $\mu \in \mathbf{N}$ we define the *support* of μ by

$$\operatorname{supp}\mu := \{x : \mu\{x\} > 0\}. \tag{8.4}$$

Given $x = (x_1, \dots, x_d) \in \mathbb{R}^d$ and $y = (y_1, \dots, y_d) \in \mathbb{R}^d$ one says that x is *lexicographically smaller* than y if there exists $i \in \{1, \dots, d-1\}$ such that $x_j = y_j$ for $j \in \{1, \dots, i-1\}$ and $x_i < y_i$. In this case we write $x < y$. Every non-empty finite set $B \subset \mathbb{R}^d$ has a unique *lexicographic minimum* $l(B)$; note that $l(B + x) = l(B) + x$ for all $x \in \mathbb{R}^d$. If $B \subset \mathbb{R}^d$ is empty or infinite we set $l(B) := 0$.

Proposition 8.4 *Suppose that η is a stationary point process on $\mathbb{R}^d$. Then* $\mathbb{P}(0 < \eta(\mathbb{R}^d) < \infty) = 0$.

Proof Assume on the contrary that $\mathbb{P}(0 < \eta(\mathbb{R}^d) < \infty) > 0$ and consider the conditional probability measure $\mathbb{P}' := \mathbb{P}(\cdot \mid 0 < \eta(\mathbb{R}^d) < \infty)$. Since η is stationary under $\mathbb{P}$ and $\eta(\mathbb{R}^d) = \theta_x\eta(\mathbb{R}^d)$, η is stationary under $\mathbb{P}'$. By Proposition 6.2, $\mu \mapsto l(\operatorname{supp}\mu)$ is a measurable mapping on $\mathbf{N}_{<\infty}(\mathbb{R}^d)$. For $x \in \mathbb{R}^d$ the random variable $l(\operatorname{supp}\eta)$ has under $\mathbb{P}'$ the same distribution as

$$l(\operatorname{supp}\theta_x\eta) = l((\operatorname{supp}\eta) - x) = l(\operatorname{supp}\eta) - x,$$

where the second identity holds $\mathbb{P}'$-a.s. This contradicts the fact that there is no translation invariant probability measure on $\mathbb{R}^d$. □

8.2 The Pair Correlation Function

In this section we deal with certain second order properties of stationary point processes. We say that a point process η on $\mathbb{R}^d$ is *locally square integrable* if

$$\mathbb{E}[\eta(B)^2] < \infty, \quad B \in \mathcal{B}^d_b, \tag{8.5}$$

where $\mathcal{B}^d_b$ denotes the system of all bounded Borel subsets of $\mathbb{R}^d$. For a stationary point process it suffices to check (8.5) for only one bounded set:

Lemma 8.5 *Let η be a stationary point process on $\mathbb{R}^d$ and assume that $\mathbb{E}[\eta([0,1]^d)^2] < \infty$. Then η is locally square integrable.*

Proof Let $B \in \mathcal{B}^d_b$. Then there exist $n \in \mathbb{N}$ and $x_1, \dots, x_n \in \mathbb{R}^d$ such that $B \subset \cup_{i=1}^n B_i$, where $B_i := [0,1]^d + x_i$. Then $\eta(B) \le \sum_{i=1}^n \eta(B_i)$ and Minkowski's inequality together with stationarity shows that $\mathbb{E}[\eta(B)^2]$ is finite. □

If η is locally square integrable, then

$$\mathbb{E}[\eta^{(2)}(C)] < \infty, \quad C \subset \mathbb{R}^d \times \mathbb{R}^d \text{ bounded and measurable.} \tag{8.6}$$

Indeed, for such C there exists $B \in \mathcal{B}^d_b$ such that $C \subset B \times B$, so that (8.6) follows from (4.7) (see also Exercise 4.4) and Lemma 8.5. Recall the definition (4.22) of the factorial moment measures.

Definition 8.6 Let η be a stationary point process on $\mathbb{R}^d$ with second factorial moment measure α_2. The measure $\alpha_2^!$ on $\mathbb{R}^d$ defined by

$$\alpha_2^!(B) := \int \mathbf{1}\{x \in [0,1]^d, y - x \in B\}\,\alpha_2(d(x,y)), \quad B \in \mathcal{B}^d, \tag{8.7}$$

is called the *reduced second factorial moment measure* of η.

Loosely speaking, $\alpha_2^!(dx)$ measures the intensity of pairs of points at a relative displacement of x. The reduced second factorial moment measure $\alpha_2^!$ determines α_2:

Proposition 8.7 *Let η be a stationary point process on $\mathbb{R}^d$ with second factorial moment measure α_2. Then η is locally square integrable if and only if its reduced second factorial moment measure $\alpha_2^!$ is locally finite. In this case,*

$$\int f(x,y)\,\alpha_2(d(x,y)) = \iint f(x,x+y)\,\alpha_2^!(dy)\,dx, \quad f \in \mathbb{R}_+(\mathbb{R}^d \times \mathbb{R}^d). \tag{8.8}$$

Proof Assume that η is locally square integrable and let $B \subset \mathbb{R}^d$ be compact. Then $B' := \{x + y : x \in [0,1]^d, y \in B\}$ is a compact set and, if $y \in [0,1]^d$ and $y - x \in B$, then $y \in B'$, so that $\alpha_2^!(B) \le \alpha_2([0,1]^d \times B') < \infty$ by (8.6). Hence $\alpha_2^!$ is locally finite. Conversely, the set inclusion

$$[0,1]^d \times [0,1]^d \subset \{(x,y) \in \mathbb{R}^d \times \mathbb{R}^d : x \in [0,1]^d, y - x \in [-1,1]^d\}$$

shows that $\mathbb{E}[\eta([0,1]^d)(\eta([0,1]^d) - 1)] \le \alpha_2^!([-1,1]^d)$, and by Exercise 8.12 and Lemma 8.5 this proves the other direction of the asserted equivalence.

The proof of (8.8) is based on the fact that Lebesgue measure is the only translation invariant locally finite measure on $\mathbb{R}^d$, up to a multiplicative factor. Let $B \in \mathcal{B}_b^d$ and define a measure ν_B on $\mathbb{R}^d$ by

$$\nu_B(C) := \int \mathbf{1}\{x \in C, y - x \in B\}\,\alpha_2(d(x,y)), \quad C \in \mathcal{B}^d.$$

It follows, as in the first step of the proof, that this measure is locally finite. To show that it is also translation invariant we let $z \in \mathbb{R}^d$ and $C \in \mathcal{B}^d$. Then

$$\begin{aligned}\nu_B(C+z) &= \mathbb{E}\Big[\int \mathbf{1}\{x - z \in C, y - z - (x - z) \in B\}\,\eta^{(2)}(d(x,y))\Big]\\ &= \mathbb{E}\int \mathbf{1}\{x \in C, y - x \in B\}\,(\theta_z\eta)^{(2)}(d(x,y)) = \nu_B(C),\end{aligned}$$

where we have used the stationarity of η. Hence ν_B is translation invariant, so that $\nu_B = c_B\lambda_d$ for some $c_B \in \mathbb{R}_+$. Then $c_B = \nu_B([0,1]^d)$ and

$$\nu_B(C) = \nu_B([0,1]^d)\lambda_d(C) = \alpha_2^!(B)\lambda_d(C).$$

Therefore

$$\int g(x, y - x)\,\alpha_2(d(x,y)) = \int g(x,z)\,\alpha_2^!(dz)\,dx,$$

first for $g = \mathbf{1}_{C\times B}$ and then for all $g \in \mathbb{R}_+(\mathbb{R}^d \times \mathbb{R}^d)$. Applying this with $g(x, y) := f(x, y + x)$ yields the result. □

In principle, (8.8) can be used to compute second moments of integrals with respect to locally square integrable point processes. If η is stationary and locally square integrable and $v, w \in \mathbb{R}_+(\mathbb{R}^d)$, then by (4.25) and (8.8)

$$\mathbb{E}[\eta(v)\eta(w)] = \iint v(x)w(x + y)\, \alpha_2^!(dy)\, dx + \gamma \int v(x)w(x)\, dx,$$

where γ is the intensity of η. This formula remains true for $v, w \in \mathbb{R}(\mathbb{R}^d)$, whenever the integrals on the right are finite.

Proposition 8.8 *Let η be a stationary locally square integrable point process on $\mathbb{R}^d$ with second (resp. reduced second) factorial moment measure α_2 (resp. $\alpha_2^!$). Then $\alpha_2^! \ll \lambda_d$ with Radon–Nikodým derivative $\rho \in \mathbb{R}_+(\mathbb{R}^d)$ if and only if*

$$\int f(x, y)\, \alpha_2(d(x, y)) = \int f(x, y)\rho(y - x)\, d(x, y), \quad f \in \mathbb{R}_+(\mathbb{R}^d \times \mathbb{R}^d). \tag{8.9}$$

Proof If $\alpha_2^! \ll \lambda_d$ with density ρ, then Proposition 8.7 and a change of variables show that (8.9) holds. Conversely, if (8.9) holds, then Proposition 8.7 shows for each $f \in \mathbb{R}_+(\mathbb{R}^d \times \mathbb{R}^d)$ that

$$\iint f(x, x + y)\, \alpha_2^!(dy)\, dx = \int f(x, x + y)\rho(y)\, d(x, y),$$

or, for each $g \in \mathbb{R}_+(\mathbb{R}^d \times \mathbb{R}^d)$,

$$\iint g(x, y)\, \alpha_2^!(dy)\, dx = \int g(x, y)\rho(y)\, d(x, y).$$

Choosing $g(x, y) = \mathbf{1}\{x \in [0, 1]^d\}h(y)$ for $h \in \mathbb{R}_+(\mathbb{R}^d)$ gives $\alpha_2^!(dy) = \rho(y)\, dy$ and hence the asserted result. □

Definition 8.9 Let η be a locally square integrable stationary point process on $\mathbb{R}^d$ with positive intensity γ and assume that (8.9) holds for some $\rho \in \mathbb{R}_+(\mathbb{R}^d)$. Then $\rho_2 := \gamma^{-2}\rho$ is called the *pair correlation function* of η.

If η is not simple, that is $\mathbb{P}(\eta \in \mathbf{N}_s(\mathbb{R}^d)) < 1$, Exercise 8.10 shows that the pair correlation function cannot exist. In such a case one may apply Definition 8.9 to the simple point process η^* defined in Exercise 8.7.

Example 8.10 Let η be a stationary Poisson process on $\mathbb{R}^d$ with positive intensity. Then Corollary 4.10 and Proposition 8.8 show that the pair correlation function of η is given by $\rho_2 \equiv 1$.

On a heuristic level, (4.24) and (8.9) (or (8.7)) imply that

$$\mathbb{E}[\eta(dx)\eta(dy)] = \rho(y - x)\,dx\,dy.$$

Hence the pair correlation function ρ_2 of a stationary point process with intensity γ satisfies the heuristic equation

$$\mathbb{C}\mathrm{ov}(\eta(dx), \eta(dy)) = \gamma^2(\rho_2(y - x) - 1)\,dx\,dy. \tag{8.10}$$

Therefore the inequality $\rho_2(y-x) < 1$ indicates attraction between potential points at x and y, while $\rho_2(y-x) > 1$ indicates repulsion. Later in this book we shall encounter several examples of pair correlation functions.

8.3 Local Properties

In this section we assume that η is a stationary simple point process on $\mathbb{R}^d$ with intensity $\gamma \in (0, \infty)$. We might expect that both $\mathbb{P}(\eta(B) = 1)$ and $\mathbb{P}(\eta(B) \geq 1)$ behave like $\gamma\lambda_d(B)$ for small $B \in \mathcal{B}^d$. This is made precise by the following result.

Proposition 8.11 *Let $B \subset \mathbb{R}^d$ be a bounded Borel set with $\lambda_d(B) > 0$ and let $r_n > 0$, $n \in \mathbb{N}$, with $\lim_{n\to\infty} r_n = 0$. For each $n \in \mathbb{N}$ let $B_n := r_n B$. Then*

$$\lim_{n\to\infty} \frac{\mathbb{P}(\eta(B_n) \geq 1)}{\lambda_d(B_n)} = \lim_{n\to\infty} \frac{\mathbb{P}(\eta(B_n) = 1)}{\lambda_d(B_n)} = \gamma. \tag{8.11}$$

Proof Let $c > 0$. For each $n \in \mathbb{N}$ let $C_n := [0, c/n)^d$ and let $C_{n,1}, \dots, C_{n,n^d}$ be a collection of disjoint translates of C_n with union C_1. Since η is simple (and locally finite), we have almost surely that

$$\eta(C_1) = \lim_{n\to\infty} \sum_{i=1}^{n^d} \mathbf{1}\{\eta(C_{n,i}) \geq 1\} = \lim_{n\to\infty} \sum_{i=1}^{n^d} \mathbf{1}\{\eta(C_{n,i}) = 1\}.$$

The sums in the above limits are bounded by $\eta(C_1)$, so taking expectations we obtain from dominated convergence and stationarity that

$$\gamma c^d = \lim_{n\to\infty} n^d\, \mathbb{P}(\eta(C_n) \geq 1) = \lim_{n\to\infty} n^d\, \mathbb{P}(\eta(C_n) = 1),$$

which is (8.11) in the case where $B = [0, c)^d$ and $r_n = 1/n$.

Next we note that for each bounded $B \in \mathcal{B}^d$ we have

$$\mathbb{E}[\eta(B)] - \mathbb{E}[\mathbf{1}\{\eta(B) \geq 2\}\eta(B)] = \mathbb{P}(\eta(B) = 1) \leq \mathbb{P}(\eta(B) \geq 1) \leq \mathbb{E}[\eta(B)].$$

Therefore (8.11) is equivalent to

$$\lim_{n\to\infty} \frac{\mathbb{E}[\mathbf{1}\{\eta(B_n) \geq 2\}\eta(B_n)]}{\lambda_d(B_n)} = 0. \tag{8.12}$$

Consider now a general sequence (r_n) but still assume that B is a half-open cube containing the origin. Let $n \in \mathbb{N}$ and let $m_n \in \mathbb{N}$ satisfy the inequalities $1/(m_n + 1) < r_n \leq 1/m_n$ (assuming without loss of generality that $r_n \leq 1$). Set $B'_n := m_n^{-1}B$, so that $B_n \subset B'_n$. Then

$$\frac{\lambda_d(B'_n)}{\lambda_d(B_n)} = \frac{1}{m_n^d r_n^d} \leq \frac{(m_n + 1)^d}{m_n^d} \leq 2^d.$$

Hence

$$\frac{\mathbb{E}[\mathbf{1}\{\eta(B_n) \geq 2\}\eta(B_n)]}{\lambda_d(B_n)} \leq \frac{\mathbb{E}[\mathbf{1}\{\eta(B'_n) \geq 2\}\eta(B'_n)]}{\lambda_d(B_n)} \leq 2^d \frac{\mathbb{E}[\mathbf{1}\{\eta(B'_n) \geq 2\}\eta(B'_n)]}{\lambda_d(B'_n)}$$

which converges to zero by the case considered previously.

For a general B, choose a half-open cube C with $B \subset C$ and set $B'_n := r_n C$, $n \in \mathbb{N}$. Then $B_n \subset B'_n$ and $\lambda_d(B'_n)/\lambda_d(B_n) = \lambda_d(C)/\lambda_d(B)$, so that the assertion follows as before. □

8.4 Ergodicity

Sometimes stationary point processes satisfy a useful zero-one law. The *invariant σ-field* is defined by

$$\mathcal{I} := \{A \in \mathcal{N} : \theta_x A = A \text{ for all } x \in \mathbb{R}^d\}, \tag{8.13}$$

where $\theta_x A := \{\theta_x \mu : \mu \in A\}$. A stationary point process η is said to be *ergodic* if $\mathbb{P}(\eta \in A) \in \{0, 1\}$ for all $A \in \mathcal{I}$. Recall that a function $h\colon \mathbb{R}^d \to \mathbb{R}$ satisfies $\lim_{\|x\|\to\infty} h(x) = a$ for some $a \in \mathbb{R}$ if, for each $\varepsilon > 0$, there exists $c > 0$ such that each $x \in \mathbb{R}^d$ with $\|x\| > c$ satisfies $|h(x) - a| < \varepsilon$.

Proposition 8.12 *Suppose that η is a stationary point process on $\mathbb{R}^d$. Assume that there exists a π-system $\mathcal{H}$ generating $\mathcal{N}$ such that*

$$\lim_{\|x\|\to\infty} \mathbb{P}(\eta \in A, \theta_x \eta \in A') = \mathbb{P}(\eta \in A)\mathbb{P}(\eta \in A') \tag{8.14}$$

for all $A, A' \in \mathcal{H}$. Then (8.14) *holds for all $A, A' \in \mathcal{N}$.*

Proof We shall use the monotone class theorem (Theorem A.1). First fix $A' \in \mathcal{H}$. Let $\mathcal{D}$ be the class of sets $A \in \mathcal{N}$ satisfying (8.14). Then $\mathbf{N} \in \mathcal{D}$ and $\mathcal{D}$ is closed with respect to proper differences. Let $A_n \in \mathcal{H}$, $n \in \mathbb{N}$, be such that $A_n \uparrow A$ for some $A \in \mathcal{N}$. Then we have for all $x \in \mathbb{R}^d$ that

$$\begin{aligned}
&|\mathbb{P}(\eta \in A)\mathbb{P}(\eta \in A') - \mathbb{P}(\eta \in A, \theta_x\eta \in A')| \\
&\quad \le |\mathbb{P}(\eta \in A)\mathbb{P}(\eta \in A') - \mathbb{P}(\eta \in A_n)\mathbb{P}(\eta \in A')| \\
&\quad\quad + |\mathbb{P}(\eta \in A_n)\mathbb{P}(\eta \in A') - \mathbb{P}(\eta \in A_n, \theta_x\eta \in A')| \\
&\quad\quad + |\mathbb{P}(\eta \in A_n, \theta_x\eta \in A') - \mathbb{P}(\eta \in A, \theta_x\eta \in A')|.
\end{aligned}$$

For large n the first and third terms are small uniformly in $x \in \mathbb{R}^d$. For any fixed $n \in \mathbb{N}$ the second term tends to 0 as $\|x\| \to \infty$. It follows that $A \in \mathcal{D}$. Hence $\mathcal{D}$ is a Dynkin system and Theorem A.1 shows that $\sigma(\mathcal{H}) \subset \mathcal{D}$, that is $\mathcal{N} = \mathcal{D}$. Therefore (8.14) holds for all $A \in \mathcal{N}$ and $A' \in \mathcal{H}$.

Now fix $A \in \mathcal{N}$ and let $\mathcal{D}'$ be the class of sets $A' \in \mathcal{N}$ satisfying (8.14). It follows as before that $\mathcal{D}'$ is a Dynkin system. When checking that $\mathcal{D}'$ contains any monotone union A' of sets $A'_n \in \mathcal{D}'$ one has to use the fact that

$$\begin{aligned}
|\mathbb{P}(\eta \in A, \theta_x\eta \in A'_n) - \mathbb{P}(\eta \in A, \theta_x\eta \in A')| &\le \mathbb{P}(\theta_x\eta \in A') - \mathbb{P}(\theta_x\eta \in A'_n) \\
&= \mathbb{P}(\eta \in A') - \mathbb{P}(\eta \in A'_n),
\end{aligned}$$

where the equality comes from the stationarity of η. Theorem A.1 shows that $\mathcal{D}'$ contains all $A' \in \mathcal{N}$, implying the assertion. □

A stationary point process η satisfying (8.14) for all $A, A' \in \mathcal{N}$ is said to be *mixing*. Any point process with this property is ergodic. Indeed, if $A \in \mathcal{N}$ satisfies $\theta_x A = A$ for all $x \in \mathbb{R}^d$, then we can take $A = A'$ in (8.14) and conclude that $\mathbb{P}(\eta \in A) = (\mathbb{P}(\eta \in A))^2$.

Proposition 8.13 *Let η be a stationary Poisson process with finite intensity. Then η is mixing and in particular is ergodic.*

Proof Let $\mathbf{N}_{ls} := \mathbf{N}_{ls}(\mathbb{R}^d)$; see Definition 6.6. Proposition 6.9 shows that η is simple, that is $\mathbb{P}(\eta \in \mathbf{N}_{ls}) = 1$. Define $\mathcal{N}_{ls} := \{A \cap \mathbf{N}_{ls} : A \in \mathcal{N}\}$. Let $\mathcal{H}$ denote the system of all sets of the form $\{\mu \in \mathbf{N}_{ls} : \mu(B) = 0\}$ for some bounded $B \in \mathcal{B}^d$. By Exercise 6.7, $\mathcal{H}$ is a π-system and generates $\mathcal{N}_{ls}$. Let $B, B' \in \mathcal{B}^d$ be bounded. Let $A = \{\mu \in \mathbf{N}_{ls} : \mu(B) = 0\}$ and define A' similarly in terms of B'. For $x \in \mathbb{R}^d$ we have by (8.1) that

$$\{\theta_x\eta \in A'\} = \{\eta(B' + x) = 0\}.$$

If $\|x\|$ is sufficiently large, then $B \cap (B' + x) = \emptyset$, so that the events $\{\eta \in A\}$ and $\{\theta_x\eta \in A'\}$ are independent. Therefore,

$$\mathbb{P}(\eta \in A, \theta_x\eta \in A') = \mathbb{P}(\eta \in A)\,\mathbb{P}(\theta_x\eta \in A') = \mathbb{P}(\eta \in A)\,\mathbb{P}(\eta \in A'),$$

implying (8.14). Since $\mathbb{P}(\eta \in \mathbf{N}_{ls}) = \mathbb{P}(\theta_x\eta \in \mathbf{N}_{ls}) = 1$ for all $x \in \mathbb{R}^d$, the proof of Proposition 8.12 shows that (8.14) holds for all $A, A' \in \mathcal{N}$. □

For an extension of Proposition 8.13 to marked Poisson processes see Exercise 10.1.

8.5 A Spatial Ergodic Theorem

In this section we let η be a stationary point process on $\mathbb{R}^d$ with finite intensity γ. Define

$$\mathcal{I}_\eta := \{\eta^{-1}(A) : A \in \mathcal{I}\}, \tag{8.15}$$

where the invariant σ-field $\mathcal{I}$ is given by (8.13). The following result is derived from the univariate version of the mean ergodic theorem, proved in Section B.2; see Theorem B.11.

Theorem 8.14 (Mean ergodic theorem) *Let $W \subset \mathbb{R}^d$ be compact and convex with non-empty interior, and set $W_n := a_n W$ for some sequence $(a_n)_{n\geq 1}$ with $a_n \to \infty$ as $n \to \infty$. Then, as $n \to \infty$,*

$$\frac{\eta(W_n)}{\lambda_d(W_n)} \to \mathbb{E}[\eta([0,1]^d) \mid \mathcal{I}_\eta] \quad \text{in } L^1(\mathbb{P}). \tag{8.16}$$

Proof We consider only the case with $d = 2$, leaving the generalisation to other values of d to the reader.

For $i, j \in \mathbb{N}$, set $X_{i,j} := \eta([i, i+1) \times [j, j+1))$. Then for each $j \in \mathbb{N}$ the sequence $(X_{i,j})_{i\in\mathbb{N}}$ is stationary and satisfies $0 < \mathbb{E}[X_{1,j}] < \infty$. By Theorem B.11, for each $j \in \mathbb{N}$ we have L^1-convergence

$$n^{-1} \sum_{i=1}^{n} X_{i,j} \to Y_j$$

for some integrable Y_j. Moreover, $(Y_j)_{j\geq 1}$ is a stationary sequence and $\mathbb{E}[|Y_1|] < \infty$, so by applying Theorem B.11 again we have

$$m^{-1} \sum_{j=1}^{m} Y_j \to Z_1 \quad \text{in } L^1(\mathbb{P})$$

for some integrable Z_1. Writing $\|X\|_1$ for $\mathbb{E}[|X|]$ for any random variable X, we have for $n, m \in \mathbb{N}$ that

$$\left\| (nm)^{-1} \sum_{i=1}^{n} \sum_{j=1}^{m} X_{i,j} - Z_1 \right\|_1$$
$$\leq \left\| m^{-1} \sum_{j=1}^{m} \left(n^{-1} \sum_{i=1}^{n} X_{i,j} - Y_j \right) \right\|_1 + \left\| m^{-1} \sum_{j=1}^{m} Y_j - Z_1 \right\|_1.$$

Therefore,

$$\left\|(nm)^{-1}\sum_{i=1}^{n}\sum_{j=1}^{m}X_{i,j}-Z_1\right\|_1$$
$$\leq m^{-1}\sum_{j=1}^{m}\left\|n^{-1}\sum_{i=1}^{n}X_{i,j}-Y_j\right\|_1+\left\|m^{-1}\sum_{j=1}^{m}Y_j-Z_1\right\|_1$$
$$=\left\|n^{-1}\sum_{i=1}^{n}X_{i,1}-Y_1\right\|_1+\left\|m^{-1}\sum_{j=1}^{m}Y_j-Z_1\right\|_1,$$

which tends to zero as $n, m \to \infty$.

Let $\mathcal{R}_0$ be the class of all rectangles R of the form $R = [0, s) \times [0, t)$ with $s, t > 0$. Then we assert that

$$a_n^{-2}\eta(a_n R) \to \lambda_2(R)Z_1 \quad \text{in } L^1(\mathbb{P}), \quad R \in \mathcal{R}_0. \tag{8.17}$$

To show this we use (here and later in the proof) for $a \in \mathbb{R}$ the notation

$$\lfloor a\rfloor := \max\{k \in \mathbb{Z} : k \leq a\}, \quad \lceil a\rceil := \min\{k \in \mathbb{Z} : k \geq a\}.$$

Given $s, t > 0$ and setting $R = [0, s) \times [0, t)$ we have

$$\|a_n^{-2}\eta(a_nR) - stZ_1\|_1 \leq a_n^{-2}\left\|\sum_{i=1}^{\lfloor a_n s\rfloor}\sum_{j=1}^{\lfloor a_n t\rfloor}X_{i,j} - \lfloor a_n s\rfloor\lfloor a_n t\rfloor Z_1\right\|_1$$
$$+ a_n^{-2}\|\eta(a_nR \setminus [0, \lfloor a_n s\rfloor) \times [0, \lfloor a_n t\rfloor))\|_1 + a_n^{-2}\|(st - \lfloor a_n s\rfloor\lfloor a_n t\rfloor)Z_1\|_1$$

and the first term in the expression on the right-hand side tends to zero by the preceding argument, while the second and the third term are bounded by $a_n^{-2}\gamma\lambda_2(a_nR \setminus [0, \lfloor a_n s\rfloor) \times [0, \lfloor a_n t\rfloor))$, which tends to zero.

Now let $\mathcal{R}$ be the class of all rectangles R of the form $R = [a, b) \times [c, d)$ with $0 \leq a < b$ and $0 \leq c < d$. Let $R = [a, b) \times [c, d) \in \mathcal{R}$. Defining

$$R_1 := [0, b) \times [0, d), \quad R_2 := [0, a) \times [0, d),$$
$$R_3 := [0, b) \times [0, c), \quad R_4 := [0, a) \times [0, c),$$

by additivity of η we have for any $t \in \mathbb{R}_+$ that

$$\eta(tR) = \eta(tR_1) - \eta(tR_2) - \eta(tR_3) + \eta(tR_4).$$

Therefore, applying (8.17) to $R_1, \ldots, R_4$, we obtain

$$a_n^{-2}\eta(a_nR) \to \lambda_2(R)Z_1 \quad \text{in } L^1(\mathbb{P}), \quad R \in \mathcal{R}. \tag{8.18}$$

Now we assume that $W \subset \mathbb{R}_+^2$. In this case, given $\varepsilon > 0$, since W is assumed convex we can choose disjoint rectangles $R_1, \dots, R_k \in \mathcal{R}$ such that $R := \cup_{i=1}^k R_i \subset W$ and $\lambda_2(W \setminus R) < \varepsilon$. Then

$$\begin{aligned} a_n^{-2} \| \eta(W_n) - Z_1 \lambda_2(W_n) \|_1 &\le a_n^{-2} \| \eta(a_n(W \setminus R)) \|_1 + a_n^{-2} \| \eta(a_n R) - a_n^2 \lambda_2(R) Z_1 \|_1 \\ &\quad + a_n^{-2} \| \lambda_2(a_n R) Z_1 - \lambda_2(a_n W) Z_1 \|_1. \end{aligned}$$

On the right-hand side the first and third terms are each bounded by $\gamma\varepsilon$, while the middle term tends to zero by (8.18) and the additivity of η. Therefore, since ε is arbitrarily small, the left-hand side tends to zero as $n \to \infty$, and this gives us

$$a_n^{-2} \eta(W_n) \to \lambda_2(W) Z_1 \quad \text{in } L^1(\mathbb{P}).$$

Here we have not used that $\lambda_2(W) > 0$.

Let Q_1 be the upper right (closed) quadrant of $\mathbb{R}^2$ and let Q_2, Q_3, Q_4 denote the other quadrants of $\mathbb{R}^2$. Just as in the case $i = 1$ we see that there are integrable random variables Z_2, Z_3, Z_4 such that

$$a_n^{-2} \eta(W_n) \to Z_i \quad \text{in } L^1(\mathbb{P}), \tag{8.19}$$

whenever $i \in \{1, 2, 3, 4\}$ is such that $W \subset Q_i$.

For general W we may write $W = \cup_{i=1}^4 W_i$ with each of the (convex) sets $W_1, \dots, W_4$ contained in a single quadrant of $\mathbb{R}^2$. Using (8.19) and the fact that $\mathbb{E}[\eta(W_i \cap W_j)] = 0$ for $i \ne j$, we have as $n \to \infty$ that

$$\lambda_2(a_n W)^{-1} \eta(a_n W) \to Z, \quad \text{in } L^1(\mathbb{P}), \tag{8.20}$$

where $Z := \lambda(W)^{-1}(\lambda_2(W_1)Z_1 + \cdots + \lambda_2(W_4)Z_4)$.

Next we show that the random variable Z can be chosen $\mathcal{I}_\eta$-measurable. Set $Q := [-1/2, 1/2]^2$ and let

$$Z'(\mu) := \liminf_{n \to \infty} n^{-2} \mu(nQ), \quad \mu \in \mathbf{N}.$$

Let $x = (u, v) \in \mathbb{R}^2$. Then we have for each $n \ge 2 \max\{\lceil |u| \rceil, \lceil |v| \rceil\}$ that

$$(n - 2 \max\{\lceil |u| \rceil, \lceil |v| \rceil\}) Q \subset nQ + x \subset (n + 2 \max\{\lceil |u| \rceil, \lceil |v| \rceil\}) Q,$$

so that $Z'(\theta_x \mu) = Z'(\mu)$. Hence the mapping $Z'(\cdot)$ is $\mathcal{I}$-measurable. By (8.20) with $W = Q$ and $a_n = n$, and since L^1-convergence implies convergence in probability (see Proposition B.8), we have that $\mathbb{P}(Z = Z'(\eta)) = 1$. Hence we can assume that Z is $\mathcal{I}_\eta$-measurable.

It remains to prove that $Z = \mathbb{E}[\eta[0, 1]^d \mid \mathcal{I}_\eta]$. Let $A \in \mathcal{I}$. For $n \in \mathbb{N}$

let $S_n := n^{-2}\eta(nQ)$, where again $Q := [-1/2, 1/2]^d$. Then $\mathbf{1}\{\eta \in A\}S_n$ converges to $\mathbf{1}\{\eta \in A\}Z$ in $L^1(\mathbb{P})$. Therefore

$$\begin{aligned}\mathbb{E}[\mathbf{1}\{\eta \in A\}Z] &= \lim_{n\to\infty} \mathbb{E}[\mathbf{1}\{\eta \in A\}S_n] = \lim_{n\to\infty} n^{-2}\mathbb{E}[\mathbf{1}\{\eta \in A\}\eta(nQ)]\\ &= \mathbb{E}[\mathbf{1}\{\eta \in A\}\mathbb{E}[\eta([0,1]^d) \mid \mathcal{I}_\eta]],\end{aligned}$$

where we have used Exercise 8.11. Thus Z has the defining properties of $\mathbb{E}[\eta([0,1]^d) \mid \mathcal{I}_\eta]$, so $Z = \mathbb{E}[\eta([0,1]^d) \mid \mathcal{I}_\eta]$ $\mathbb{P}$-a.s. □

A bounded set $B \subset \mathbb{R}^d$ is said to be *Riemann measurable* if $\mathbf{1}_B$ is Riemann integrable, that is, if B can be approximated in measure from below and above by finite unions of hypercubes. For example, a convex set has this property. The proof of Theorem 8.14 shows that it is enough to assume that W is Riemann measurable.

Theorem 8.14 justifies calling

$$\hat{\eta} := \mathbb{E}[\eta([0,1]^d) \mid \mathcal{I}_\eta] \tag{8.21}$$

the *sample intensity* of η. If η is ergodic, then $\mathbb{P}(\hat{\eta} = \gamma) = 1$. The theorem remains valid for stationary random measures to be introduced in Exercise 13.12; see Exercise 13.13.

8.6 Exercises

Exercise 8.1 Let η be a stationary point process on $\mathbb{R}$. Prove that

$$\mathbb{P}(\eta \neq 0, \eta((-\infty, 0]) < \infty) = \mathbb{P}(\eta \neq 0, \eta([0, \infty)) < \infty) = 0.$$

Exercise 8.2 Let $\mathbb{Q}$ be a probability measure on $\mathbf{N}(\mathbb{R}^d)$ and let K be the probability kernel from $\mathbb{R}^d$ to $\mathbf{N}(\mathbb{R}^d)$ defined by $K(x, A) := \mathbb{Q}(\theta_x A)$. Let η be a stationary Poisson process on $\mathbb{R}^d$ with a (finite) intensity $\gamma \geq 0$ and let χ be a Poisson cluster process as in Exercise 5.6. Show that χ is stationary with intensity $\gamma \int \mu(\mathbb{R}^d)\, \mathbb{Q}(d\mu)$. (Hint: Use the Laplace functional in Exercise 5.6 to prove the stationarity.)

Exercise 8.3 Let $C := [0, 1)^d$ be a half-open unit cube. Let $\mu \in \mathbf{N}_{<\infty}(\mathbb{R}^d)$ such that $\mu(\mathbb{R}^d \setminus C) = 0$ and let X be uniformly distributed on C. Show that $\eta := \sum_{y\in\mathbb{Z}^d} \theta_{y+X}\mu$ is a stationary point process with intensity $\mu(C)$. Show also that η is ergodic. (Hint: To check ergodicity use that $\eta = \theta_X \mu_0$ for some $\mu_0 \in \mathbf{N}_l(\mathbb{R}^d)$.)

Exercise 8.4 Let $T\colon \mathbf{N} \to \mathbf{N}$ be measurable such that $T(\theta_x\mu) = \theta_x T(\mu)$ for all $(x, \mu) \in \mathbb{R}^d \times \mathbf{N}$. Show that if η is a stationary (resp. stationary and ergodic) point process on $\mathbb{R}^d$, then so is $T(\eta)$.

Exercise 8.5 Give an example of a stationary locally finite point process η with $\mathbb{E}[\eta([0,1]^d)] = \infty$.

Exercise 8.6 Give an example of a stationary point process that is not locally finite. (Hint: Such a point process has to have infinitely many accumulation points.)

Exercise 8.7 Let η be a locally finite stationary point process on $\mathbb{R}^d$. Use Exercise 8.4 to show that

$$\eta^* := \int \eta\{x\}^{\oplus} \mathbf{1}\{x \in \cdot\}\, \eta(dx) \tag{8.22}$$

is a stationary point process; see also Exercise 6.6.

Exercise 8.8 Let η be a stationary locally finite point process on $\mathbb{R}^d$ with finite intensity but do not assume that η is simple. Show that the first limit in (8.11) exists and equals $\mathbb{E}[\eta^*([0,1]^d)]$, where η^* is given by (8.22).

Exercise 8.9 Let η be a locally square integrable stationary point process on $\mathbb{R}^d$ with intensity γ. Assume that η has a pair correlation function ρ_2. Let $W \subset \mathbb{R}^d$ be a bounded Borel set and show that

$$\mathbb{V}\mathrm{ar}[\eta(W)] = \gamma^2 \int \lambda_d(W \cap (W+x))(\rho_2(x) - 1)\, dx + \gamma \lambda_d(W). \tag{8.23}$$

Exercise 8.10 Let η be a stationary point process on $\mathbb{R}^d$ with reduced second factorial moment measure $\alpha_2^!$. Show that η is simple if and only if $\alpha_2^!(\{0\}) = 0$. (Hint: Assume without loss of generality that η is a random element of $\mathbf{N}_l(\mathbb{R}^d)$ and note that η is simple if and only if the stationary point process $\eta' := \int \mathbf{1}\{x \in \cdot\}(\eta\{x\} - 1)\, \eta(dx)$ has intensity zero.)

Exercise 8.11 Let η be a stationary point process with finite intensity. Show that

$$\mathbb{E}[\eta(B) \mid \mathcal{I}_\eta] = \lambda_d(B)\mathbb{E}[\eta([0,1]^d) \mid \mathcal{I}_\eta] \quad \mathbb{P}\text{-a.s}$$

holds for each $B \in \mathcal{B}^d$. (Hint: Take $A \in \mathcal{I}$ and use the proof of Proposition 8.2 to show that $\mathbb{E}[\mathbf{1}\{\eta \in A\}\eta(B)] = \lambda_d(B)\mathbb{E}[\mathbf{1}\{\eta \in A\}\eta([0,1]^d)]$.)

Exercise 8.12 Let $X \geq 0$ be a random variable with $\mathbb{E}[X(X-1)] < \infty$. Show that $\mathbb{E}[X^2] < \infty$.

Exercise 8.13 Let η be a locally finite stationary point process on $\mathbb{R}^d$, interpreted as a random element of $\mathbf{N}_l := \mathbf{N}_l(\mathbb{R}^d)$. Let $f \in \mathbb{R}_+(\mathbf{N}_l)$ and let ν be a σ-finite measure on $\mathbb{R}^d$ such that $\nu(\mathbb{R}^d) > 0$ and $\nu(\{x : f(\theta_x \eta) \neq 0\}) = 0$ $\mathbb{P}$-a.s. Show that $\mathbb{E}[f(\eta)] = 0$. (Hint: Use Fubini's theorem.)

9

The Palm Distribution

The Palm distribution of a stationary point process can be introduced via a refined Campbell theorem. It can be interpreted as a conditional distribution given that there is a point at the origin. A stationary point process is Poisson if and only if its Palm distribution is the distribution of the original process with an extra point added at the origin. For a given simple point process, Voronoi tessellations partition the space into regions based on the nearest neighbour principle. They are an important model of stochastic geometry but also provide a convenient setting for formulating the close relationship between the stationary distribution and the Palm distribution of a stationary simple point process. The latter is a volume-debiased version of the first, while, conversely, the former is a volume-biased version of the latter.

9.1 Definition and Basic Properties

Throughout this chapter η denotes a locally finite point process on $\mathbb{R}^d$. It is convenient (and no loss of generality) to assume that $\eta(\omega) \in \mathbf{N}_l$ for all $\omega \in \Omega$, where $\mathbf{N}_l := \mathbf{N}_l(\mathbb{R}^d) \in \mathcal{N}(\mathbb{R}^d)$ is the space of locally finite measures from $\mathbf{N}(\mathbb{R}^d)$ as in Definition 2.11. The advantage is that Lemma 9.2 below shows that the mapping $(x, \mu) \mapsto \theta_x \mu$ is measurable on $\mathbb{R}^d \times \mathbf{N}_l$, where we choose $\mathcal{N}_l := \mathcal{N}_l(\mathbb{R}^d) = \{A \in \mathcal{N}(\mathbb{R}^d) : A \subset \mathbf{N}_l\}$ as the σ-field on $\mathbf{N}_l$.

If η is stationary, then the distribution $\mathbb{P}_\eta = \mathbb{P}(\eta \in \cdot)$ of η does not change under a shift of the origin. We then also refer to $\mathbb{P}_\eta$ as the *stationary distribution* of η. We now introduce another distribution that describes η as seen from a *typical point* of η. In Chapter 10 we shall make this precise under an additional ergodicity hypothesis.

Theorem 9.1 (Refined Campbell theorem) *Suppose that η is a stationary point process on $\mathbb{R}^d$ with finite strictly positive intensity γ. Then there exists* *a unique probability measure* $\mathbb{P}^0_\eta$ *on* $\mathbf{N}_l$ *such that*

$$\mathbb{E}\Big[\int f(x,\theta_x\eta)\,\eta(dx)\Big] = \gamma \iint f(x,\mu)\,\mathbb{P}^0_\eta(d\mu)\,dx, \quad f \in \mathbb{R}_+(\mathbb{R}^d \times \mathbf{N}_l). \tag{9.1}$$

Proof The proof generalises that of Proposition 8.7. For each $A \in \mathcal{N}_l$ we define a measure ν_A on $\mathbb{R}^d$ by

$$\nu_A(B) := \mathbb{E}\Big[\int \mathbf{1}_B(x)\mathbf{1}_A(\theta_x\eta)\,\eta(dx)\Big], \quad B \in \mathcal{B}^d,$$

where the integrations are justified by Lemma 9.2. By definition of $\nu_A(\cdot)$ and (8.2) we have for all $y \in \mathbb{R}^d$ that

$$\begin{aligned}\nu_A(B+y) &= \mathbb{E}\Big[\int \mathbf{1}_B(x-y)\mathbf{1}_A(\theta_x\eta)\,\eta(dx)\Big]\\ &= \mathbb{E}\Big[\int \mathbf{1}_B(x)\mathbf{1}_A(\theta_{x+y}\eta)\,(\theta_y\eta)(dx)\Big].\end{aligned}$$

Because of the flow property $\theta_{x+y}\eta = \theta_x(\theta_y\eta)$, we can use stationarity to conclude that

$$\nu_A(B+y) = \mathbb{E}\Big[\int \mathbf{1}_B(x)\mathbf{1}_A(\theta_x\eta)\,\eta(dx)\Big] = \nu_A(B),$$

so that ν_A is translation invariant. Furthermore, $\nu_A(B) \le \mathbb{E}[\eta(B)] = \gamma\lambda_d(B)$, so that ν_A is locally finite. Hence there is a number $\gamma_A \ge 0$ such that

$$\nu_A(B) = \gamma_A\lambda_d(B), \quad B \in \mathcal{B}^d. \tag{9.2}$$

Choosing $B = [0,1]^d$ shows that $\gamma_A = \nu_A([0,1]^d)$ is a measure in A. Since $\gamma_{\mathbf{N}_l} = \gamma$, the definition $\mathbb{P}^0_\eta(A) := \gamma_A/\gamma$ yields a probability measure $\mathbb{P}^0_\eta$, and it follows from (9.2) that

$$\mathbb{E}\Big[\int \mathbf{1}_B(x)\mathbf{1}_A(\theta_x\eta)\,\eta(dx)\Big] = \gamma\,\mathbb{P}^0_\eta(A)\lambda_d(B), \quad A \in \mathcal{N}_l,\ B \in \mathcal{B}^d. \tag{9.3}$$

Hence (9.1) holds for functions of the form $f(x,\mu) = \mathbf{1}_B(x)\mathbf{1}_A(\mu)$ and then also for general measurable indicator functions by the monotone class theorem (Theorem A.1). Linearity of the integral and monotone convergence yield (9.1) for general $f \in \mathbb{R}_+(\mathbb{R}^d \times \mathbf{N}_l)$.

Conversely, (9.1) yields (9.3) and therefore

$$\mathbb{P}^0_\eta(A) = \frac{1}{\gamma\lambda_d(B)}\mathbb{E}\Big[\int \mathbf{1}_B(x)\mathbf{1}_A(\theta_x\eta)\,\eta(dx)\Big], \tag{9.4}$$

provided that $0 < \lambda_d(B) < \infty$. □

Clearly (9.4) extends to

$$\int f(\mu)\,\mathbb{P}^0_\eta(d\mu) = \frac{1}{\gamma\lambda_d(B)}\mathbb{E}\Big[\int \mathbf{1}_B(x) f(\theta_x\eta)\,\eta(dx)\Big] \tag{9.5}$$

whenever $0 < \lambda_d(B) < \infty$ and $f \in \mathbb{R}_+(\mathbf{N}_l)$. Multiplying this identity by $\gamma\lambda_d(B)$ yields a special case of (9.1).

We still need to prove the following measurability assertion.

Lemma 9.2 *The mapping* $(x,\mu) \mapsto \theta_x\mu$ *from* $\mathbb{R}^d \times \mathbf{N}_l$ *to* $\mathbf{N}_l$ *is measurable.*

Proof By Proposition 6.3 each $\mu \in \mathbf{N}_l$ can be written as $\mu = \sum_{n=1}^{\mu(\mathbb{X})} \delta_{\pi_n(\mu)}$, where the mappings $\pi_n : \mathbf{N}_l \to \mathbb{R}^d$, $n \in \mathbb{N}$, are measurable. For $x \in \mathbb{R}^d$ we then have $\theta_x\mu = \sum_{n=1}^{\mu(\mathbb{X})} \delta_{\pi_n(\mu)-x}$. Hence it remains to note that for each $B \in \mathcal{B}^d$ the mapping $(x,\mu) \mapsto \delta_{\pi_n(\mu)-x}(B) = \mathbf{1}\{\pi_n(\mu) - x \in B\}$ is measurable. □

Definition 9.3 Under the assumptions of Theorem 9.1 the measure $\mathbb{P}^0_\eta$ is called the *Palm distribution* of η.

Sometimes we shall use the refined Campbell theorem in the equivalent form

$$\mathbb{E}\Big[\int f(x,\eta)\,\eta(dx)\Big] = \gamma \iint f(x,\theta_{-x}\mu)\,\mathbb{P}^0_\eta(d\mu)\,dx, \quad f \in \mathbb{R}_+(\mathbb{R}^d \times \mathbf{N}_l). \tag{9.6}$$

Indeed, for any $f \in \mathbb{R}_+(\mathbb{R}^d \times \mathbf{N}_l)$ we can apply (9.1) with $\tilde{f} \in \mathbb{R}_+(\mathbb{R}^d \times \mathbf{N}_l)$ defined by $\tilde{f}(x,\mu) := f(x,\theta_{-x}\mu)$.

It follows from Proposition 6.3 that the mapping $(x,\mu) \mapsto \mathbf{1}\{\mu\{x\} > 0\}$ is measurable on $\mathbb{R}^d \times \mathbf{N}_l$. Indeed, we have

$$\mathbf{1}\{\mu\{x\} = 0\} = \prod_{n=1}^{\mu(\mathbb{X})} \mathbf{1}\{\pi_n(\mu) \neq x\}.$$

By (9.4) we have for a stationary point process η that

$$\mathbb{P}^0_\eta(\{\mu \in \mathbf{N}_l : \mu\{0\} > 0\}) = \gamma^{-1}\mathbb{E}\Big[\int \mathbf{1}_{[0,1]^d}(x)\mathbf{1}\{\eta\{x\} > 0\}\,\eta(dx)\Big] = 1, \tag{9.7}$$

so that $\mathbb{P}^0_\eta$ is concentrated on those $\mu \in \mathbf{N}_l$ having an atom at the origin. If η is simple, we shall see in Proposition 9.5 that $\mathbb{P}^0_\eta$ can be interpreted as the conditional distribution of η given that $\eta\{0\} > 0$.

9.2 The Mecke–Slivnyak Theorem

Our next result is a stationary version of Mecke's characterisation of the Poisson process (Theorem 4.1).

Theorem 9.4 *Let η be a stationary point process on $\mathbb{R}^d$ with intensity $\gamma \in (0,\infty)$. Then η is a Poisson process if and only if*

$$\mathbb{P}^0_\eta = \mathbb{P}(\eta + \delta_0 \in \cdot). \tag{9.8}$$

Proof Assume first that η is a Poisson process. For any $A \in \mathcal{N}_l$ we then obtain from the Mecke equation (4.2) that

$$\begin{aligned}\mathbb{E}\Big[\int \mathbf{1}_{[0,1]^d}(x)\mathbf{1}_A(\theta_x\eta)\,\eta(dx)\Big] &= \gamma\,\mathbb{E}\Big[\int \mathbf{1}_{[0,1]^d}(x)\mathbf{1}_A(\theta_x(\eta+\delta_x))\,dx\Big]\\ &= \gamma \int \mathbf{1}_{[0,1]^d}(x)\mathbb{P}(\eta+\delta_0 \in A)\,dx = \gamma\,\mathbb{P}(\eta+\delta_0\in A),\end{aligned}$$

where we have used Fubini's theorem and stationarity for the second identity. Hence (9.8) follows from (9.4) with $B = [0,1]^d$.

Conversely, if (9.8) holds, we take $f \in \mathbb{R}_+(\mathbb{R}^d \times \mathbf{N}_l)$ to obtain from (9.6) that

$$\mathbb{E}\Big[\int f(x,\eta)\,\eta(dx)\Big] = \gamma \int \mathbb{E}[f(x,\theta_{-x}(\eta+\delta_0))]\,dx.$$

Stationarity yields that the Mecke equation (4.2) holds with $\lambda = \gamma\lambda_d$. Theorem 4.1 then shows that η is a Poisson process. □

9.3 Local Interpretation of Palm Distributions

Let η be a stationary simple point process on $\mathbb{R}^d$ with intensity $\gamma \in (0,\infty)$. Let η^0 denote a *Palm version* of η, that is a point process (defined on the basic probability space $(\Omega, \mathcal{F}, \mathbb{P})$) with distribution $\mathbb{P}^0_\eta$. By Exercise 9.1 η^0 is simple, while (9.7) implies that $\mathbb{P}(0 \in \eta^0) = 1$. If $\eta(\mathbb{R}^d) > 0$ we let X denote the point of η with smallest Euclidean norm, taking the lexicographically smallest such point if there is more than one. In the case $\eta(\mathbb{R}^d) = 0$ we set $X := 0$. We now give a local interpretation of the Palm distribution.

Proposition 9.5 *For $n \in \mathbb{N}$ let $r_n > 0$ and let B_n be the closed ball with centre 0 and radius r_n. Assume that $r_n \to 0$ as $n \to \infty$. Let $g \in \mathbb{R}(\mathbf{N})$ be bounded. Then*

$$\lim_{n\to\infty} \mathbb{E}[g(\theta_X\eta) \mid \eta(B_n) \geq 1] = \mathbb{E}[g(\eta^0)]. \tag{9.9}$$

If, moreover, $x \mapsto g(\theta_x\mu)$ is continuous for all $\mu \in \mathbf{N}_{ls}(\mathbb{R}^d)$, then

$$\lim_{n\to\infty} \mathbb{E}[g(\eta) \mid \eta(B_n) \geq 1] = \mathbb{E}[g(\eta^0)]. \tag{9.10}$$

Proof By the triangle inequality

$$\left|\mathbb{E}[g(\theta_X\eta) \mid \eta(B_n) \geq 1] - \mathbb{E}[g(\eta^0)]\right| \leq I_{1,n} + I_{2,n},$$

where we set

$$I_{1,n} := \gamma^{-1}\lambda_d(B_n)^{-1}\left|\mathbb{E}[\mathbf{1}\{\eta(B_n) \geq 1\}g(\theta_X\eta)] - \gamma\lambda_d(B_n)\mathbb{E}[g(\eta^0)]\right|$$

and

$$\begin{aligned} I_{2,n} &:= \left|(\mathbb{P}(\eta(B_n) \geq 1)^{-1} - \gamma^{-1}\lambda_d(B_n)^{-1})\mathbb{E}[\mathbf{1}\{\eta(B_n) \geq 1\}g(\theta_X\eta)]\right| \\ &\leq \frac{\mathbb{E}[\mathbf{1}\{\eta(B_n) \geq 1\}|g(\theta_X\eta)|]}{\gamma\lambda_d(B_n)}\left|\frac{\gamma\lambda_d(B_n)}{\mathbb{P}(\eta(B_n) \geq 1)} - 1\right|. \end{aligned}$$

Since g is bounded, Proposition 8.11 shows that $I_{2,n} \to 0$ as $n \to \infty$.

By the refined Campbell theorem (Theorem 9.1),

$$I_{1,n} = \gamma^{-1}\lambda_d(B_n)^{-1}\left|\mathbb{E}\left[\mathbf{1}\{\eta(B_n) \geq 1\}g(\theta_X\eta) - \int_{B_n} g(\theta_x\eta)\,\eta(dx)\right]\right|.$$

Since B_n is a ball, $X \in B_n$ whenever $\eta(B_n) \geq 1$. In the last expectation, distinguish the cases $\eta(B_n) = 1$ and $\eta(B_n) \geq 2$; then we obtain

$$\begin{aligned} I_{1,n} &\leq \gamma^{-1}\lambda_d(B_n)^{-1}\mathbb{E}\left[\mathbf{1}\{\eta(B_n) \geq 2\}\int \mathbf{1}_{B_n}(x)|g(\theta_x\eta)|\,\eta(dx)\right] \\ &\leq c\gamma^{-1}\lambda_d(B_n)^{-1}\mathbb{E}[\mathbf{1}\{\eta(B_n) \geq 2\}\eta(B_n)], \end{aligned}$$

where c is an upper bound of $|g|$. Therefore

$$I_{1,n} \leq \frac{c(\mathbb{E}[\eta(B_n)] - \mathbb{P}(\eta(B_n) = 1))}{\gamma\lambda_d(B_n)} = c\left(1 - \frac{\mathbb{P}(\eta(B_n) = 1)}{\gamma\lambda_d(B_n)}\right),$$

which tends to zero by Proposition 8.11.

To prove (9.10) it is now sufficient to show that

$$\lim_{n\to\infty} \mathbb{E}[|g(\eta) - g(\theta_X\eta)| \mid \eta(B_n) \geq 1] = 0.$$

Given $\varepsilon > 0$ and $\mu \in \mathbf{N}_{ls} := \mathbf{N}_{ls}(\mathbb{R}^d)$, define

$$g_\varepsilon(\mu) := \sup\{|g(\mu) - g(\theta_x\mu)| : \|x\| \leq \varepsilon\}.$$

For $\mu \in \mathbf{N} \setminus \mathbf{N}_{ls}$ we set $g_\varepsilon(\mu) := 0$. Assuming that g has the stated additional continuity property, the supremum can be taken over a countable dense subset of $B(0, \varepsilon)$. Since $\mu \mapsto \theta_x\mu$ is measurable for each $x \in \mathbb{R}^d$ (Lemma 9.2), we see that g_ε is a measurable function. Fixing $\varepsilon > 0$ we note that

$$\mathbb{E}[|g(\eta) - g(\theta_X\eta)| \mid \eta(B_n) \geq 1] \leq \mathbb{E}[g_\varepsilon(\theta_X\eta) \mid \eta(B_n) \geq 1]$$

for sufficiently large n, where we have again used the fact that $X \in B_n$ if $\eta(B_n) \geq 1$. Applying (9.9) to g_ε we therefore obtain

$$\limsup_{n\to\infty} \mathbb{E}[|g(\eta) - g(\theta_X\eta)| \mid \eta(B_n) \geq 1] \leq \mathbb{E}[g_\varepsilon(\eta^0)].$$

The assumption on g implies that $g_\varepsilon(\eta^0) \to 0$ as $\varepsilon \to 0$, so that $\mathbb{E}[g_\varepsilon(\eta^0)] \to 0$ by dominated convergence. This concludes the proof. □

9.4 Voronoi Tessellations and the Inversion Formula

In this section we shall assume that η is a stationary simple point process on $\mathbb{R}^d$ with finite intensity γ. We also assume that $\mathbb{P}(\eta(\mathbb{R}^d) = 0) = 0$, so that $\mathbb{P}(\eta(\mathbb{R}^d) = \infty) = 1$ by Proposition 8.4. In particular, $\gamma > 0$. We shall discuss some basic relationships between the stationary distribution and the Palm distribution of η. For simplicity we assume for all $\omega \in \Omega$ that $\eta(\omega)$ is a simple locally finite counting measure, that is $\eta(\omega) \in \mathbf{N}_{ls} := \mathbf{N}_{ls}(\mathbb{R}^d)$; see Definition 6.6. As in the preceding section, we let η^0 denote a Palm version of η.

In what follows we take advantage of the geometric idea of a *Voronoi tessellation*. For $\mu \in \mathbf{N}_{ls}$ with $\mu(\mathbb{R}^d) > 0$ and for $x \in \mathbb{R}^d$, let $\tau(x, \mu) \in \mu$ be the nearest neighbour of x in $\operatorname{supp}\mu$, i.e. the point in μ of minimal Euclidean distance from x. If there is more than one such point we take the lexicographically smallest. In the (exceptional) case $\mu(\mathbb{R}^d) = 0$ we put $\tau(x, \mu) := x$ for all $x \in \mathbb{R}^d$. The mapping $\tau\colon \mathbb{R}^d \times \mathbf{N}_{ls} \to \mathbb{R}^d$ is *covariant* under translations, that is,

$$\tau(x - y, \theta_y\mu) = \tau(x, \mu) - y, \quad x, y \in \mathbb{R}^d,\ \mu \in \mathbf{N}_{ls}. \tag{9.11}$$

For $x \in \mu \in \mathbf{N}_{ls}$ the *Voronoi cell* of x (with respect to μ) is defined by

$$C(x, \mu) := \{y \in \mathbb{R}^d : \tau(y, \mu) = x\}. \tag{9.12}$$

If $\mu(\mathbb{R}^d) \neq 0$ these cells are pairwise disjoint and cover $\mathbb{R}^d$; see Figure 9.1 for an illustration. In the following formulae we shall frequently use the abbreviation

$$C_0 := C(0, \eta^0).$$

This random set is also called the *typical cell* of the Voronoi tessellation. As before we denote by

$$X := \tau(0, \eta) \tag{9.13}$$

the point of η closest to the origin, and we set $X := 0$ if $\eta(\mathbb{R}^d) = 0$.

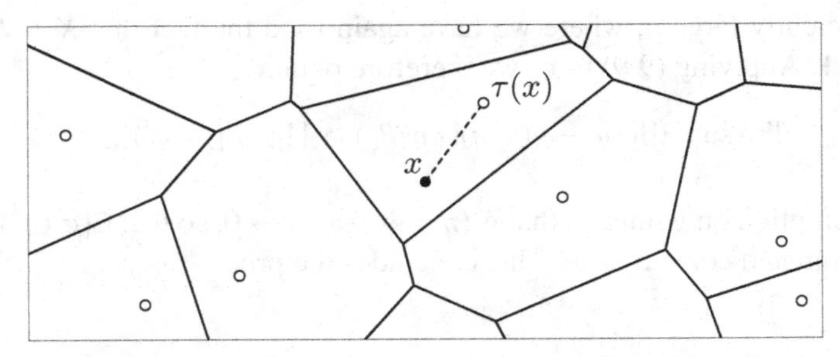

Figure 9.1 Voronoi tessellation based on a planar pattern of points. The dashed line connects the site x with its closest point.

Theorem 9.6 *For all $h \in \mathbb{R}_+(\mathbb{R}^d \times \mathbf{N}_{ls})$ it is the case that*

$$\mathbb{E}[h(X, \eta)] = \gamma\, \mathbb{E}\left[\int_{C_0} h(-x, \theta_x \eta^0)\, dx\right]. \tag{9.14}$$

Proof Equation (9.6) (the refined Campbell theorem) and a change of variables yield

$$\mathbb{E}\left[\int f(x, \eta)\, \eta(dx)\right] = \gamma\, \mathbb{E}\left[\int f(-x, \theta_x \eta^0)\, dx\right] \tag{9.15}$$

for all $f \in \mathbb{R}_+(\mathbb{R}^d \times \mathbf{N}_{ls})$. We apply this formula with

$$f(x, \mu) := h(x, \mu)\mathbf{1}\{\tau(0, \mu) = x\}. \tag{9.16}$$

Then the left-hand side of (9.15) reduces to the left-hand side of (9.14). Since, by the covariance property (9.11), $\tau(0, \theta_x \eta^0) = -x$ if and only if $\tau(x, \eta^0) = 0$ (that is, $x \in C_0$), the right-hand side of (9.15) coincides with the right-hand side of (9.14). □

Let $f \in \mathbb{R}_+(\mathbf{N}_{ls})$. Taking $h(x, \mu) := f(\mu)$ in (9.14) yields the *inversion formula*

$$\mathbb{E}[f(\eta)] = \gamma\, \mathbb{E}\left[\int_{C_0} f(\theta_x \eta^0)\, dx\right], \quad f \in \mathbb{R}_+(\mathbf{N}_{ls}), \tag{9.17}$$

expressing the stationary distribution in terms of the Palm distribution. The choice $f \equiv 1$ yields the intuitively obvious formula

$$\mathbb{E}[\lambda_d(C_0)] = \gamma^{-1}. \tag{9.18}$$

Let $g \in \mathbb{R}_+(\mathbf{N}_{ls})$. Taking $h(x, \mu) := g(\theta_x \mu)$ in (9.14) yields

$$\gamma\, \mathbb{E}[\lambda_d(C_0) g(\eta^0)] = \mathbb{E}[g(\theta_X \eta)], \tag{9.19}$$

showing that the distribution of $\theta_X\eta$ is absolutely continuous with respect to the Palm distribution. The formula says that the stationary distribution is (up to a shift) a *volume-biased* version of the Palm distribution.

We define the (stationary) *zero-cell* of η by

$$V_0 := C(X, \eta) = \{x \in \mathbb{R}^d : \tau(x, \eta) = \tau(0, \eta)\}.$$

In the exceptional case where $\eta(\mathbb{R}^d) = 0$, we have defined $\tau(x, \mu) := x$ for all $x \in \mathbb{R}^d$ so that $V_0 = \{0\}$. The next result shows that the Palm distribution can be derived from the stationary distribution by *volume debiasing* and shifting X to 0.

Proposition 9.7 *We have for all $f \in \mathbb{R}_+(\mathbf{N}_{ls})$ that*

$$\gamma\,\mathbb{E}[f(\eta^0)] = \mathbb{E}[\lambda_d(V_0)^{-1} f(\theta_X\eta)]. \tag{9.20}$$

Proof We apply (9.19) with $g(\mu) := f(\mu) \cdot \lambda_d(C(0, \mu))^{-1}$ to obtain

$$\gamma\,\mathbb{E}[f(\eta^0)] = \mathbb{E}[\lambda_d(C(0, \theta_X\eta))^{-1} f(\theta_X\eta)].$$

Since $C(0, \theta_X\eta) = C(X, \eta) - X = V_0 - X$ the result follows. □

Given $\alpha \in \mathbb{R}$, putting $f(\mu) := \lambda_d(C(0, \mu))^{\alpha+1}$ in equation (9.20) yields

$$\gamma\,\mathbb{E}[\lambda_d(C_0)^{\alpha+1}] = \mathbb{E}[\lambda_d(V_0)^{\alpha}]. \tag{9.21}$$

In particular,

$$\mathbb{E}[\lambda_d(V_0)^{-1}] = \gamma. \tag{9.22}$$

By Jensen's inequality (Proposition B.1), $\mathbb{E}[\lambda_d(V_0)^{-1}] \geq (\mathbb{E}[\lambda_d(V_0)])^{-1}$. Hence by (9.22) and (9.18) we obtain

$$\mathbb{E}[\lambda_d(C_0)] \leq \mathbb{E}[\lambda_d(V_0)]. \tag{9.23}$$

9.5 Exercises

Exercise 9.1 Let η be a stationary simple point process on $\mathbb{R}^d$ with positive finite intensity. Show that $\mathbb{P}^0_\eta(\mathbf{N}_{ls}) = 1$.

Exercise 9.2 Let η be a stationary simple point process with finite intensity and with $\mathbb{P}(\eta = 0) = 0$. Let X be given by (9.13). Show that the conditional distribution of $-X$ given $\theta_X\eta$ is the uniform distribution on $V_0 - X$. (Hint: Use Theorem 9.6, then Proposition 9.7 and then again Theorem 9.6.)

Exercise 9.3 Let $h \in \mathbb{R}(\mathbb{R}^d)$ be continuous with compact support. Show that $x \mapsto \int h\, d(\theta_x\mu)$ is continuous for all $\mu \in \mathbf{N}_{ls}$.

Exercise 9.4 (Palm–Khinchin equations) Let η be a stationary simple point process on $\mathbb{R}$ with finite intensity γ such that $\mathbb{P}(\eta = 0) = 0$. Let $x \geq 0$ and $j \in \mathbb{N}_0$. Show that $\mathbb{P}$-almost surely

$$\mathbf{1}\{\eta(0, x] \leq j\} = \int \mathbf{1}\{t \leq 0\}\mathbf{1}\{\eta(t, x] = j\}\, \eta(dt).$$

Then use the refined Campbell theorem to prove that

$$\mathbb{P}(\eta(0, x] \leq j) = \gamma \int_x^\infty \mathbb{P}(\eta^0(0, t] = j)\, dt,$$

where η^0 has distribution $\mathbb{P}^0_\eta$.

Exercise 9.5 Let η be a stationary locally finite point process with finite positive intensity γ, and let $A \in \mathbf{N}_l$. Show that $\mathbb{P}^0_\eta(A) = 1$ if and only if

$$\eta(\{x \in \mathbb{R}^d : \theta_x \eta \notin A\}) = 0, \quad \mathbb{P}\text{-a.s.}$$

(Hint: Use (9.4) to prove in the case $\mathbb{P}^0_\eta(A) = 1$ that

$$\eta(\{x \in B : \theta_x \eta \in A\}) = \eta(B), \quad \mathbb{P}\text{-a.s.},$$

for any $B \in \mathcal{B}^d$ with $0 < \lambda_d(B) < \infty$.)

Exercise 9.6 Let η be a stationary simple point process with finite intensity γ. Let the sequence (B_n) be as in Proposition 8.11 and let X_n be a random vector such that X_n is the point of η in B_n whenever $\eta(B_n) = 1$. Let $g \in \mathbb{R}(\mathbf{N}_l)$ be bounded. Show that

$$\lim_{n\to\infty} \mathbb{E}[g(\theta_{X_n}\eta) \mid \eta(B_n) = 1] = \mathbb{E}[g(\eta^0)].$$

Exercise 9.7 In the setting of Exercise 9.6, assume that $x \mapsto g(\theta_x \mu)$ is continuous for all $\mu \in \mathbf{N}_{ls}$. Show that

$$\lim_{n\to\infty} \mathbb{E}[g(\eta) \mid \eta(B_n) = 1] = \mathbb{E}[g(\eta^0)].$$

(Hint: Use Exercise 9.6 and the proof of (9.10).)

Exercise 9.8 Let χ be the stationary Poisson cluster process defined in Exercise 8.2. Assume that $\gamma_{\mathbb{Q}} := \int \mu(\mathbb{R}^d)\, \mathbb{Q}(d\mu) \in (0, \infty)$. Show that the Palm distribution of χ is given by

$$\mathbb{P}^0_\chi = \int \mathbb{P}(\chi + \mu \in \cdot)\, \mathbb{Q}^0(d\mu),$$

where the probability measure $\mathbb{Q}^0$ is given by

$$\mathbb{Q}^0 := \gamma_{\mathbb{Q}}^{-1} \iint \mathbf{1}\{\theta_x \mu \in \cdot\}\, \mu(dx)\, \mathbb{Q}(d\mu).$$

(Hint: Use the definition (5.12) of χ and the Mecke equation for ξ.)

Exercise 9.9 Let η be a stationary point process with finite positive intensity γ. Show that the reduced second factorial moment measure $\alpha_2^!$ of η (see Definition 8.6) is given by

$$\alpha_2^!(B) = \gamma \int (\mu(B) - \mathbf{1}\{0 \in B\})\, \mathbb{P}_\eta^0(d\mu). \quad B \in \mathcal{B}^d.$$

Exercise 9.10 Let χ be a stationary Poisson cluster process as in Exercise 9.8. Show that the reduced second factorial moment measure $\alpha_2^!$ of χ is given by

$$\alpha_2^!(B) = \gamma_\chi^2 \lambda_d(B) + \gamma \alpha_{\mathbb{Q}}^!(B), \quad B \in \mathcal{B}^d,$$

where $\gamma_\chi := \gamma\gamma_{\mathbb{Q}}$ is the intensity of χ and the measure $\alpha_{\mathbb{Q}}^!$ on $\mathbb{R}^d$ is given by

$$\alpha_{\mathbb{Q}}^!(B) := \iint \mu(B + x)\mu(dx)\, \mathbb{Q}(d\mu) - \gamma_{\mathbb{Q}} \mathbf{1}\{0 \in B\}. \tag{9.24}$$

(Hint: You may combine Exercises 9.8 and 9.9.)

Exercise 9.11 Let $\mathbb{Q}$ be a probability measure on $\mathbf{N}(\mathbb{R}^d)$ such that $\gamma_{\mathbb{Q}} = \int \mu(B)\, \mathbb{Q}(d\mu) < \infty$. Define a measure $\alpha_{\mathbb{Q}}^!$ by (9.24) and show that

$$\alpha_{\mathbb{Q}}^!(B) := \iint \mathbf{1}\{y - x \in B\}\, \mu^{(2)}(d(x, y))\, \mathbb{Q}(d\mu), \quad B \in \mathcal{B}^d.$$

Exercise 9.12 Let χ be a stationary Poisson cluster process as in Exercise 9.8 but assume in addition that $\mathbb{Q}(\mathbf{N}_{ls}(\mathbb{R}^d)) = 1$. Assume that $\alpha_{\mathbb{Q}}^!$ is locally finite and absolutely continuous with respect to λ_d. Let $\rho_{\mathbb{Q}}$ denote the density. Show that the pair correlation function ρ_2 of χ exists and can be chosen as $\rho_2 = 1 + \gamma^{-1}\gamma_{\mathbb{Q}}^{-2}\rho_{\mathbb{Q}}$. (As expected this function is at least one.)

Exercise 9.13 Suppose η has distribution $\mathbb{P}_\eta(A) = (1/2)\Pi_\gamma(A)$, $A \in \mathcal{N}_l$, and $\mathbb{P}_\eta(\{0\}) = 1/2$, where Π_γ is the distribution of a stationary Poisson process on $\mathbb{R}^d$ with intensity $\gamma \in (0, \infty)$. Find the Palm distribution of η.

10

Extra Heads and Balanced Allocations

Is it possible to choose a point of a stationary Poisson process such that after removing this point and centring the process around its location (i.e. shifting so that this location goes to the origin) the resulting point process is still Poisson? More generally, one can ask whether it is possible to choose a point of a stationary simple point process such that the re-centred process has the Palm distribution. This question can be answered using balanced allocations, which partition the space into regions of equal volume in a translation invariant way, such that each point is associated with exactly one region. Under an ergodicity assumption a spatial version of the Gale–Shapley algorithm of economics provides an important example of such an allocation. These results show that the Palm version of a stationary ergodic simple point process η can be constructed by a random shift of η.

10.1 The Extra Head Problem

Let $d \in \mathbb{N}$. Throughout this chapter, η denotes a stationary simple point process on $\mathbb{R}^d$ with $\mathbb{P}(\eta = 0) = 0$ and with intensity $\gamma \in (0, \infty)$. Recall from Section A.1 that the σ-field $\sigma(\eta)$ generated by η consists of all sets $\{\eta \in A\}$ with $A \in \mathcal{N}(\mathbb{R}^d)$. Let T be a $\sigma(\eta)$-measurable random element of $\mathbb{R}^d$ such that $\mathbb{P}(\eta\{T\} = 1) = 1$. Thus T picks one of the points of η using only the information contained in η. The shifted point process $\theta_T\eta$ (that is, the point process $\omega \mapsto \theta_{T(\omega)}\eta(\omega)$) has a point at the origin, and one might ask whether

$$\theta_T\eta \overset{d}{=} \eta^0, \tag{10.1}$$

where η^0 is a Palm version of η, that is a point process with the Palm distribution $\mathbb{P}^0_\eta$, as in Section 9.3. If η is a Poisson process, then the Mecke–Slivnyak theorem (Theorem 9.4) shows that (10.1) is equivalent to

$$\theta_T\eta \overset{d}{=} \eta + \delta_0. \tag{10.2}$$

This (as well as the more general version (10.1)) is known as the *extra head problem*. The terminology comes from the analogous discrete problem, given a doubly infinite sequence of independent and identically distributed coin tosses, of picking out a "head" in the sequence such that the distribution of the remaining coin tosses (centred around the picked coin) is still that of the original sequence. It turns out that the extra head problem can be solved using *transport properties* of point processes.

Before addressing the extra head problem we need to introduce a purely deterministic concept. It is convenient to add the point ∞ to $\mathbb{R}^d$ and to define $\mathbb{R}^d_\infty := \mathbb{R}^d \cup \{\infty\}$. We equip this space with the σ-field generated by $\mathcal{B}^d \cup \{\{\infty\}\}$. We define $\infty + x = \infty - x := \infty$ for all $x \in \mathbb{R}^d$. Recall from Definition 6.6 that $\mathbf{N}_{ls} := \mathbf{N}_{ls}(\mathbb{R}^d)$ is the space of all locally finite simple counting measures. Every $\mu \in \mathbf{N}_{ls}$ is identified with its support $\operatorname{supp} \mu$, defined at (8.4).

Definition 10.1 An *allocation* is a measurable mapping $\tau\colon \mathbb{R}^d \times \mathbf{N}_{ls} \to \mathbb{R}^d_\infty$ such that

$$\tau(x, \mu) \in \mu \cup \{\infty\}, \quad x \in \mathbb{R}^d,\ \mu \in \mathbf{N}_{ls},$$

and such that τ is *covariant* under shifts, i.e.

$$\tau(x - y, \theta_y \mu) = \tau(x, \mu) - y, \quad x, y \in \mathbb{R}^d,\ \mu \in \mathbf{N}_{ls}, \tag{10.3}$$

where the shift operator θ_y was defined at (8.1).

Given an allocation τ, define

$$C^\tau(x, \mu) = \{y \in \mathbb{R}^d : \tau(y, \mu) = x\}, \quad x \in \mathbb{R}^d,\ \mu \in \mathbf{N}_{ls}. \tag{10.4}$$

Since $\tau(\cdot, \mu)$ is measurable for each $\mu \in \mathbf{N}_{ls}$, the set $C^\tau(x, \mu)$ is Borel for each $x \in \mu$. Note that $C^\tau(x, \mu) = \emptyset$ whenever $x \notin \mu$ and that $\{C^\tau(x, \mu) : x \in \mu\}$ forms a partition of $\{x \in \mathbb{R}^d : \tau(x, \mu) \neq \infty\}$. The covariance property (10.3) implies that

$$C^\tau(x - y, \theta_y \mu) = C^\tau(x, \mu) - y, \quad x, y \in \mathbb{R}^d,\ \mu \in \mathbf{N}_{ls}. \tag{10.5}$$

An example is the Voronoi tessellation discussed in Chapter 9. In general we do not assume that $x \in C^\tau(x, \mu)$ or even that $C^\tau(x, \mu) \neq \emptyset$ for $x \in \mu$.

Turning our attention back to the point process η, we may consider η as a random element of $\mathbf{N}_{ls}$. We can then assume that the Palm version η^0 of η is a random element of $\mathbf{N}_{ls}$.

Theorem 10.2 *Let τ be an allocation and let $f, g \in \mathbb{R}_+(\mathbf{N}_{ls})$. Then*

$$\mathbb{E}[\mathbf{1}\{\tau(0,\eta) \neq \infty\} f(\eta) g(\theta_{\tau(0,\eta)}\eta)] = \gamma \mathbb{E}\Big[g(\eta^0) \int_{C^\tau(0,\eta^0)} f(\theta_x \eta^0)\, dx\Big]. \tag{10.6}$$

Proof The proof is similar to that of Theorem 9.6. Apply (9.15) (a direct consequence of the refined Campbell theorem) to the function $(x, \mu) \mapsto f(\mu) g(\theta_x \mu) \mathbf{1}\{\tau(0, \mu) = x\}$. □

Definition 10.3 Let $\alpha > 0$. An allocation τ is said to be *α-balanced for η* if

$$\mathbb{P}(\lambda_d(C^\tau(x,\eta)) = \alpha \text{ for all } x \in \eta) = 1. \tag{10.7}$$

Lemma 10.4 *An allocation τ is α-balanced for η if and only if*

$$\mathbb{P}(\lambda_d(C^\tau(0,\eta^0)) = \alpha) = 1. \tag{10.8}$$

Proof Because $(x, y, \mu) \mapsto \mathbf{1}\{\tau(y, \mu) = x\}$ is a measurable mapping on $\mathbb{R}^d \times \mathbb{R}^d \times \mathbf{N}_{ls}$,

$$\lambda_d(C^\tau(x,\mu)) = \int \mathbf{1}\{\tau(y,\mu) = x\}\, dy$$

depends measurably on (x, μ). Therefore Proposition 2.7 shows that the integral $\int \mathbf{1}\{\lambda_d(C^\tau(x,\eta)) \neq \alpha\}\, \eta(dx)$ is a random variable. By (10.5) we have $C^\tau(0, \theta_x \eta) = C^\tau(x, \eta) - x$ for all $x \in \mathbb{R}^d$. Therefore the refined Campbell theorem (Theorem 9.1) shows that

$$\begin{aligned}\mathbb{E}\Big[\int \mathbf{1}\{\lambda_d(C^\tau(x,\eta)) \neq \alpha\}\, \eta(dx)\Big] &= \mathbb{E}\Big[\int \mathbf{1}\{\lambda_d(C^\tau(0,\theta_x\eta)) \neq \alpha\}\, \eta(dx)\Big] \\ &= \gamma \int \mathbb{P}(\lambda_d(C^\tau(0,\eta^0)) \neq \alpha)\, dx,\end{aligned}$$

and the result follows. □

The following theorem clarifies the relevance of balanced allocations for the extra head problem.

Theorem 10.5 *Let $\alpha > 0$. Let τ be an allocation and put $T := \tau(0, \eta)$. Then τ is α-balanced for η if and only if*

$$\mathbb{P}(T \neq \infty) = \alpha\gamma \tag{10.9}$$

and

$$\mathbb{P}(\theta_T \eta \in \cdot \mid T \neq \infty) = \mathbb{P}^0_\eta. \tag{10.10}$$

Proof If τ is α-balanced for η, then (10.6) with $f \equiv 1$ yields

$$\mathbb{E}[\mathbf{1}\{T \neq \infty\}g(\theta_T\eta)] = \gamma\alpha\,\mathbb{E}[g(\eta^0)].$$

This yields both (10.9) and (10.10). Assume, conversely, that (10.9) and (10.10) hold. Let $C_0 := C^\tau(0,\eta^0)$. Using (10.6) with $f \equiv 1$ followed by a multiplication by $\mathbb{P}(T < \infty)^{-1}$ gives us

$$\mathbb{E}[g(\eta^0)] = \mathbb{P}(T \neq \infty)^{-1}\gamma\,\mathbb{E}[g(\eta^0)\lambda_d(C_0)] = \alpha^{-1}\mathbb{E}[g(\eta^0)\lambda_d(C_0)].$$

In particular, $\mathbb{E}[\lambda_d(C_0)] = \alpha$. Choosing $g(\mu) = \lambda_d(C^\tau(0,\mu))$ yields

$$\mathbb{E}[\lambda_d(C_0)^2] = \alpha\,\mathbb{E}[\lambda_d(C_0)] = \alpha^2.$$

Since this implies (10.8), τ is α-balanced for η. □

Theorem 10.5 shows that for $\alpha = \gamma^{-1}$ a solution to the extra head problem (10.1) may be obtained from an α-balanced allocation if one exists.

10.2 The Point-Optimal Gale–Shapley Algorithm

The identity (10.9) shows that α-balanced allocations can exist only if $\alpha \leq \gamma^{-1}$. The extra head problem arises in the case $\alpha = \gamma^{-1}$.

We now describe one way to construct α-balanced allocations. Suppose that each point of $\mu \in \mathbf{N}_{ls}$ starts growing at time 0 at unit speed, trying to capture a region of volume α. In the absence of any interaction with other growing points, for $t \leq \alpha^{1/d}\kappa_d^{-1/d}$ a point $x \in \mu$ grows to a ball $B(x,t)$ by time t, where

$$B(x,t) := \{y \in \mathbb{R}^d : \|y - x\| \leq t\} \tag{10.11}$$

and where we recall that $\kappa_d = \lambda_d(B(x,1))$. However, a growing point can only capture sites that have not been claimed by some other point before. Once a region reaches volume α, it stops growing. This idea is formalised as follows.

Algorithm 10.6 Let $\alpha > 0$ and $\mu \in \mathbf{N}_{ls}$. For $n \in \mathbb{N}$, $x \in \mu$ and $z \in \mathbb{R}^d$, define the sets $C_n(x) \subset \mathbb{R}^d$ (in words, the set of sites *claimed* by x at stage n), $R_n(x) \subset \mathbb{R}^d$ (the set of sites *rejecting* x during the first n stages) and $A_n(z) \subset \mu$ (the set of points of μ claiming site z in the first n stages) via the following recursion. Define $R_0(x) := \emptyset$ for all $x \in \mu$ and for $n \in \mathbb{N}$:

(i) For $x \in \mu$, define

$$r_n(x) := \inf\{r \geq 0 : \lambda_d(B(x,r) \setminus R_{n-1}(x)) \geq \alpha\},$$
$$C_n(x) := B(x, r_n(x)).$$

(ii) For $z \in \mathbb{R}^d$, define

$$A_n(z) := \{x \in \mu : z \in C_n(x)\}.$$

If $A_n(z) \neq \emptyset$ then define

$$\tau_n(z) := l(\{x \in A_n(z) : \|z - x\| = d(z, A_n(z))\})$$

as the point *shortlisted* by site z at stage n, where $l(B)$ denotes the lexicographic minimum of a finite non-empty set $B \subset \mathbb{R}^d$ and where $d(\cdot, \cdot)$ is defined by (A.21). If $A_n(z) = \emptyset$ then define $\tau_n(z) := \infty$.

(iii) For $x \in \mu$, define

$$R_n(x) := \{z \in C_n(x) : \tau_n(z) \neq x\}.$$

The *point-optimal Gale–Shapley allocation* with *appetite* α is denoted $\tau^{\alpha,\mathrm{p}}$ and defined as follows. (The superscript p stands for "point-optimal".) Consider Algorithm 10.6 for $\mu \in \mathbf{N}_{ls}$ and let $z \in \mathbb{R}^d$. If $\tau_n(z) = \infty$ (that is, $A_n(z) = \emptyset$) for all $n \in \mathbb{N}$ we put $\tau^{\alpha,\mathrm{p}}(\mu, z) := \infty$. Otherwise, set $\tau^{\alpha,\mathrm{p}}(\mu, z) := \lim_{n\to\infty} \tau_n(z)$. We argue as follows that this limit exists. Defining $r_0(x) := 0$ for all $x \in \mu$, we assert that for all $n \in \mathbb{N}$ the following holds:

$$r_n(x) \geq r_{n-1}(x), \quad x \in \mu, \tag{10.12}$$

$$A_n(z) \supset A_{n-1}(z), \quad z \in \mathbb{R}^d, \tag{10.13}$$

$$R_n(x) \supset R_{n-1}(x), \quad x \in \mu. \tag{10.14}$$

This is proved by induction; clearly (10.12) implies (10.13) and (10.13) implies (10.14), while (10.14) implies that (10.12) holds for the next value of n. By (10.13), $\|\tau_n(z) - z\|$ is decreasing in n, and hence, since μ is locally finite, there exist $x \in \mu$ and $n_0 \in \mathbb{N}$ such that $\tau_n(z) = x$ for all $n \geq n_0$. In this case we define $\tau^{\alpha,\mathrm{p}}(z, \mu) := x$.

We have used the lexicographic minimum in (ii) to break ties in a shift-covariant way. An alternative is to leave $\tau_n(z)$ undefined whenever z has the same distance from two different points of μ. We shall prove that $\tau^{\alpha,\mathrm{p}}$ has the following properties.

Definition 10.7 Let $\alpha \in (0, \infty)$. An allocation τ is said to have *appetite* $\alpha > 0$ if both

$$\lambda_d(C^\tau(x, \mu)) \leq \alpha, \quad x \in \mu, \ \mu \in \mathbf{N}_{ls}, \tag{10.15}$$

and there is no $\mu \in \mathbf{N}_{ls}$ satisfying

$$\{z \in \mathbb{R}^d : \tau(z, \mu) = \infty\} \neq \emptyset \quad \text{and} \quad \{x \in \mu : \lambda_d(C^\tau(x, \mu)) < \alpha\} \neq \emptyset. \tag{10.16}$$

Lemma 10.8 *The point-optimal Gale–Shapley allocation $\tau^{\alpha,\mathrm{p}}$ is an allocation with appetite α.*

Proof It follows by induction over the stages of Algorithm 10.6 that the mappings τ_n are measurable as a function of both z and μ. (The proof of this fact is left to the reader.) Hence $\tau^{\alpha,\mathrm{p}}$ is measurable. Moreover it is clear that $\tau^{\alpha,\mathrm{p}}$ has the covariance property (10.3). Upon defining $\tau^{\alpha,\mathrm{p}}$ we noted that for each $z \in \mathbb{R}^d$, either $\tau^{\alpha,\mathrm{p}}(z,\mu) = \infty$ or $\tau_n(z) = x$ for some $x \in \mu$ and all sufficiently large $n \in \mathbb{N}$. Therefore

$$\mathbf{1}\{\tau^{\alpha,\mathrm{p}}(z,\mu) = x\} = \lim_{n\to\infty} \mathbf{1}\{z \in C_n(x) \setminus R_{n-1}(x)\}, \quad z \in \mathbb{R}^d. \tag{10.17}$$

On the other hand, by Algorithm 10.6(i) we have $\lambda_d(C_n(x) \setminus R_{n-1}(x)) \le \alpha$, so that (10.15) follows from Fatou's lemma (Lemma A.7).

Assume the strict inequality $\lambda_d(C^{\tau^{\alpha,\mathrm{p}}}(x,\mu)) < \alpha$ for some $x \in \mu$. We assert that the radii $r_n(x)$ defined in step (i) of the algorithm diverge. To see this, suppose on the contrary that $r(x) := \lim_{n\to\infty} r_n(x) < \infty$. By (10.17) there exist $n_0 \in \mathbb{N}$ and $\alpha_1 < \alpha$ such that $\lambda_d(B(x, r_n(x)) \setminus R_{n-1}(x)) \le \alpha_1$ for $n \ge n_0$. Hence there exists $\alpha_2 \in (\alpha_1, \alpha)$ such that $\lambda_d(B(x, r(x)) \setminus R_{n-1}(x)) \le \alpha_2$ for $n \ge n_0$, implying the contradiction $r_{n+1}(x) > r(x)$ for $n \ge n_0$. Now taking $z \in \mathbb{R}^d$, we hence have $z \in C_n(x)$ for some $n \ge 1$, so that z shortlists either x or some closer point of μ. In either case, $\tau(\mu, z) \ne \infty$. □

10.3 Existence of Balanced Allocations

We now return to the stationary point process η. Under an additional hypothesis on η we shall prove that any allocation with appetite $\alpha \le \gamma^{-1}$ (and in particular the Gale–Shapley allocation) is α-balanced for η. To formulate this condition, recall the definitions (8.13) and (8.15) of the invariant σ-fields $\mathcal{I}$ and $\mathcal{I}_\eta$. We say that η is *pseudo-ergodic* if

$$\mathbb{E}[\eta([0,1]^d) \mid \mathcal{I}_\eta] = \gamma, \quad \mathbb{P}\text{-a.s.} \tag{10.18}$$

Every ergodic point process is pseudo-ergodic. Exercise 10.10 shows that the converse of this statement does not hold. If (10.18) holds then Exercise 8.11 shows that

$$\mathbb{E}[\eta(B) \mid \mathcal{I}_\eta] = \gamma\lambda_d(B), \quad \mathbb{P}\text{-a.s.}, \ B \in \mathcal{B}^d. \tag{10.19}$$

Theorem 10.9 *Assume that η is pseudo-ergodic and let τ be an allocation with appetite $\alpha \in (0, \gamma^{-1}]$. Then τ is α-balanced for η.*

Proof Let A be the set of all $\mu \in \mathbf{N}_{ls}$ with $\{x \in \mu : \lambda_d(C^\tau(x,\mu)) < \alpha\} \ne \emptyset$.

Then $A \in \mathcal{I}$ by (10.5). In view of (10.16) we obtain from Theorem 10.2 that

$$\mathbb{P}(\eta \in A) = \mathbb{P}(\tau(0,\eta) \neq \infty, \eta \in A) = \gamma\, \mathbb{E}[\mathbf{1}\{\eta^0 \in A\}\lambda_d(C^\tau(0,\eta^0))].$$

Therefore by (9.5), for all $B \in \mathcal{B}^d$ with $0 < \lambda_d(B) < \infty$, we have

$$\begin{aligned}\mathbb{P}(\eta \in A) &= \lambda_d(B)^{-1}\mathbb{E}\Big[\int_B \mathbf{1}\{\theta_x\eta \in A\}\lambda_d(C^\tau(0,\theta_x\eta))\,\eta(dx)\Big] \\ &= \lambda_d(B)^{-1}\mathbb{E}\Big[\mathbf{1}\{\eta \in A\}\int_B \lambda_d(C^\tau(x,\eta))\,\eta(dx)\Big], \qquad (10.20)\end{aligned}$$

where we have used the invariance of A under translations and (10.5). Using (10.15), this yields

$$\begin{aligned}\mathbb{P}(\eta \in A) &\leq \lambda_d(B)^{-1}\alpha\, \mathbb{E}[\mathbf{1}\{\eta \in A\}\eta(B)] \qquad (10.21)\\ &= \lambda_d(B)^{-1}\alpha\, \mathbb{E}[\mathbf{1}\{\eta \in A\}\mathbb{E}[\eta(B) \mid \mathcal{I}_\eta]] \\ &= \alpha\gamma\, \mathbb{P}(\eta \in A) \leq \mathbb{P}(\eta \in A),\end{aligned}$$

where we have used (10.19) (a consequence of assumption (10.18)) and the assumption that $\alpha \leq \gamma^{-1}$. Therefore inequality (10.21) is in fact an equality, so that, by (10.21) and (10.20),

$$\mathbb{E}\Big[\mathbf{1}\{\eta \in A\}\int_B (\alpha - \lambda_d(C^\tau(x,\eta)))\,\eta(dx)\Big] = 0.$$

Taking $B \uparrow \mathbb{R}^d$, this yields

$$\mathbf{1}\{\eta \in A\}\int (\alpha - \lambda_d(C^\tau(x,\eta)))\,\eta(dx) = 0, \qquad \mathbb{P}\text{-a.s.}$$

Hence $\lambda_d(C^\tau(x,\eta(\omega))) = \alpha$ for all $x \in \eta(\omega)$ for $\mathbb{P}$-a.e. $\omega \in \{\eta \in A\}$. By definition of A this is possible only if $\mathbb{P}(\eta \in A) = 0$. Hence τ is α-balanced for η. □

Corollary 10.10 *There is an allocation that is γ^{-1}-balanced for η if and only if η is pseudo-ergodic. In this case the point-optimal Gale–Shapley allocation $\tau^{\gamma^{-1},p}$ is one possible choice.*

Proof If (10.18) holds, then by Theorem 10.9 and Lemma 10.8 the allocation $\tau^{\alpha^{-1},p}$ is γ^{-1}-balanced. For the converse implication, suppose that τ is an allocation that is γ^{-1}-balanced for η. Then for each $B \in \mathcal{B}^d$ we have almost surely that

$$\gamma \int \mathbf{1}\{\tau(z,\eta) \in B\}\,dz = \gamma \sum_{x\in\eta}\int \mathbf{1}\{\tau(z,\eta) = x, x \in B\}\,dz = \eta(B).$$

Taking $A \in \mathcal{I}$ (so that $\{\theta_z\eta \in A\} = \{\eta \in A\}$) and using the shift-covariance property (10.3), we obtain

$$\mathbb{E}[\mathbf{1}\{\eta \in A\}\eta(B)] = \gamma\,\mathbb{E}\Big[\int \mathbf{1}\{\theta_z\eta \in A, \tau(0,\theta_z\eta) + z \in B\}\,dz\Big]$$
$$= \gamma\,\mathbb{E}\Big[\int \mathbf{1}\{\eta \in A, \tau(0,\eta) + z \in B\}\,dz\Big] = \gamma\,\mathbb{P}(A)\lambda_d(B),$$

where we have used Fubini's theorem and stationarity to get the last two identities. This proves (10.18). □

By Proposition 8.13 the next corollary applies in particular to a stationary Poisson process.

Corollary 10.11 *Suppose that the point process η is ergodic. Then the point-optimal Gale–Shapley allocation $\tau^{\gamma^{-1},p}$ provides a solution of the general extra head problem* (10.1).

Proof Condition (10.18) follows by ergodicity. Then by Lemma 10.8 and Theorem 10.9 the allocation $\tau^{\gamma^{-1},p}$ is γ^{-1}-balanced, and then by Theorem 10.5 we have (10.1) for $T = \tau(0,\eta)$. □

10.4 Allocations with Large Appetite

Let the point process η and the constant $\gamma \in (0,\infty)$ be as before. If τ is an allocation with appetite $\alpha = \gamma^{-1}$ then $\mathbb{P}(\tau(0,\eta) \neq \infty) = 1$ by Theorem 10.9 and (10.9). The following result shows that this remains true for $\alpha > \gamma^{-1}$.

Proposition 10.12 *Assume that η is pseudo-ergodic and let τ be an allocation with appetite $\alpha \geq \gamma^{-1}$. Then $\mathbb{P}(\tau(0,\eta) \neq \infty) = 1$ and*

$$\mathbb{E}[\lambda_d(C^\tau(0,\eta^0))] = \gamma^{-1}. \tag{10.22}$$

If $\alpha > \gamma^{-1}$, then $\mathbb{P}$-a.s. there are infinitely many points $x \in \eta$ with the property $\lambda_d(C^\tau(x,\eta)) < \alpha$.

Proof Let $A' \in \mathcal{I}$. Applying (10.6) with $f \equiv 1$ and $g = \mathbf{1}_{A'}$ yields

$$\mathbb{P}(\eta \in A', \tau(0,\eta) \neq \infty) = \gamma\,\mathbb{E}[\mathbf{1}\{\eta^0 \in A'\}\lambda_d(C^\tau(0,\eta^0))]. \tag{10.23}$$

Among other things, this implies (10.22) once we have proved the first assertion. Since A' is translation invariant it follows from the definition (9.3) of $\mathbb{P}^0_\eta$ and the definition (8.15) of $\mathcal{I}_\eta$ that

$$\mathbb{P}(\eta^0 \in A') = \gamma^{-1}\mathbb{E}[\mathbf{1}\{\eta \in A'\}\eta([0,1]^d)]$$
$$= \gamma^{-1}\mathbb{E}[\mathbf{1}\{\eta \in A'\}\mathbb{E}[\eta([0,1]^d) \mid \mathcal{I}_\eta]] = \mathbb{P}(\eta \in A'), \tag{10.24}$$

where we have used assumption (10.18). The identity (10.23) can hence be written as

$$\gamma\,\mathbb{E}[\mathbf{1}\{\eta^0 \in A'\}(\alpha - \lambda_d(C^\tau(0,\eta^0)))] - \mathbb{P}(\eta \in A', \tau(0,\eta) = \infty)$$
$$= (\gamma\alpha - 1)\,\mathbb{P}(\eta \in A'). \quad (10.25)$$

We now choose A' as the set of all $\mu \in \mathbf{N}_{ls}$ with $\int \mathbf{1}\{\tau(x,\mu) = \infty\}\,dx > 0$. It is easy to check that $A' \in \mathcal{I}$. By Fubini's theorem, invariance of A', stationarity and the covariance property (10.3),

$$\mathbb{E}\Big[\mathbf{1}\{\eta \in A'\} \int \mathbf{1}\{\tau(x,\eta) = \infty\}\,dx\Big] = \int \mathbb{P}(\theta_x\eta \in A', \tau(x,\eta) = \infty)\,dx$$
$$= \int \mathbb{P}(\eta \in A', \tau(0,\eta) = \infty)\,dx.$$

Assuming $\mathbb{P}(\eta \in A') > 0$, this yields $\mathbb{P}(\eta \in A', \tau(0,\eta) = \infty) > 0$. Let $A \in \mathcal{I}$ be defined as in the proof of Theorem 10.9. Definition 10.7 shows that $A' \subset \mathbf{N}_{ls} \setminus A$, so that the first term on the left-hand side of (10.25) vanishes. Since this contradicts our assumption $\gamma\alpha - 1 \geq 0$, we must have $\mathbb{P}(\eta \in A') = 0$, and hence

$$0 = \mathbb{E}\Big[\int \mathbf{1}\{\tau(x,\eta) = \infty\}\,dx\Big] = \int \mathbb{P}(\tau(x,\eta) = \infty)\,dx$$
$$= \int \mathbb{P}(\tau(0,\eta) = \infty)\,dx,$$

where the last line comes from the covariance property and stationarity. Hence we have the first assertion of the proposition.

Assume, finally, that $\alpha > \gamma^{-1}$ and consider the point process

$$\eta' := \int \mathbf{1}\{x \in \cdot, \lambda_d(C^\tau(x,\eta)) < \alpha\}\,\eta(dx).$$

Since λ_d is translation invariant it follows that η' is stationary. In view of Proposition 8.4 we need to show that $\mathbb{P}(\eta' = 0) = 0$. Let A' be the set of all $\mu \in \mathbf{N}_{ls}$ such that $\mu(\{x \in \mathbb{R}^d : \lambda_d(C^\tau(x,\mu)) < \alpha\}) = 0$. Then $A' \in \mathcal{I}$ and $\{\eta' = 0\} = \{\eta \in A'\}$. Since τ has appetite α, we have

$$\mathbb{E}[\mathbf{1}\{\eta^0 \in A'\}\lambda_d(C^\tau(0,\eta^0))] = \alpha\,\mathbb{P}(\eta^0 \in A') = \alpha\,\mathbb{P}(\eta \in A'),$$

where we have used (10.24). Hence we obtain from (10.23) and the first assertion that $\mathbb{P}(\eta \in A') = \gamma\alpha\,\mathbb{P}(\eta \in A')$. This shows that $\mathbb{P}(\eta \in A') = 0$. □

10.5 The Modified Palm Distribution

Corollary 10.10 and Theorem 10.5 show that the extra head problem (10.1) can only be solved under the assumption (10.18). To indicate what happens without this assumption we recall the definition (8.21) of the sample intensity $\hat{\eta}$ of η. We assume that $\mathbb{P}(0 < \hat{\eta} < \infty) = 1$ and say that an allocation τ is *balanced for* η if

$$\mathbb{P}(\lambda_d(C^\tau(x,\eta)) = \hat{\eta}^{-1} \text{ for all } x \in \eta) = 1. \tag{10.26}$$

It is possible to generalise the proof of Theorem 10.9 so as to show that balanced allocations exist without further assumptions on η. (The idea is to use allocations with a random appetite $\hat{\eta}^{-1}$.) If τ is such a balanced allocation and $T := \tau(\eta, 0)$, then one can show that

$$\mathbb{P}(\theta_T \in \cdot) = \mathbb{P}^*_\eta, \tag{10.27}$$

where

$$\mathbb{P}^*_\eta := \mathbb{E}\Big[\hat{\eta}^{-1} \int \mathbf{1}\{x \in [0,1]^d, \theta_x\eta \in \cdot\}\,\eta(dx)\Big] \tag{10.28}$$

is the *modified Palm distribution* of η. This distribution describes the statistical behaviour of η as seen from a *randomly chosen* point of η. For a pseudo-ergodic point process it coincides with the Palm distribution $\mathbb{P}^0_\eta$. We do not give further details.

10.6 Exercises

Exercise 10.1 Let $(\mathbb{Y}, \mathcal{Y}, \mathbb{Q})$ be an s-finite measure space and suppose that ξ is a Poisson process on $\mathbb{R}^d \times \mathbb{Y}$ with intensity measure $\gamma\lambda_d \otimes \mathbb{Q}$. Show that $\vartheta_x\xi \overset{d}{=} \xi$ for all $x \in \mathbb{R}^d$, where $\vartheta_x \colon \mathbf{N}(\mathbb{R}^d \times \mathbb{Y}) \to \mathbf{N}(\mathbb{R}^d \times \mathbb{Y})$ is the measurable mapping (shift) defined by

$$\vartheta_x\mu := \int \mathbf{1}\{(x' - x, y) \in \cdot\}\,\mu(d(x', y)).$$

Assume in addition that $\mathbb{Y}$ is a CSMS and that $\mathbb{Q}$ is locally finite. Show that ξ has the mixing property

$$\lim_{\|x\|\to\infty} \mathbb{P}(\xi \in A, \vartheta_x\xi \in A') = \mathbb{P}(\xi \in A)\mathbb{P}(\xi \in A')$$

for all $A, A' \in \mathcal{N}(\mathbb{R}^d \times \mathbb{Y})$. (Hint: The proof of Proposition 8.13 applies with a more general version of Proposition 8.12.)

Exercise 10.2 Let η be a stationary Poisson process on $\mathbb{R}$ and let $X \in \eta$ be the point of η closest to the origin. Show that $\theta_X \eta \setminus \delta_0$ is not a Poisson process. (Hint: Use Exercise 7.9 and Theorem 7.2 (the interval theorem).)

Exercise 10.3 Extend the assertion of Exercise 10.2 to arbitrary dimensions.

Exercise 10.4 Prove the analogue of Theorem 9.6 for a general allocation, defining C_0 as the set of sites allocated to 0.

Exercise 10.5 Formulate and prove a generalisation of Proposition 9.7 and its consequence (9.21) to an arbitrary allocation τ. (Hint: Use Exercise 10.4.)

Exercise 10.6 Assume that η is pseudo-ergodic and let τ be an allocation with appetite $\alpha \leq \gamma^{-1}$. Let $g \in \mathbb{R}_+(\mathbf{N}_{ls})$. Show that

$$\mathbb{E}[\mathbf{1}\{\tau(0,\eta) \neq \infty\} g(\theta_{\tau(0,\eta)}\eta)] = \gamma\alpha\, \mathbb{E}[g(\eta^0)].$$

(Hint: Use Theorems 10.5 and 10.9.)

Exercise 10.7 Assume that η is pseudo-ergodic and let τ be an allocation with appetite $\alpha < \gamma^{-1}$. Show that $\mathbb{P}(\tau(0,\eta) = \infty) > 0$. (Hint: Take in Exercise 10.6 the function g as the indicator function of all $\mu \in \mathbf{N}_{ls}$ satisfying $\int \mathbf{1}\{\tau(x,\mu) = \infty\}\, dx = 0$ and use (10.24).)

Exercise 10.8 Let η be a stationary locally finite point process on $\mathbb{R}^d$ such that $\mathbb{P}(0 < \hat{\eta} < \infty) = 1$, where $\hat{\eta}$ is given by (8.21). Define the modified Palm distribution $\mathbb{P}^*_\eta$ of η by (10.28). Show that $\mathbb{P}_\eta$ and $\mathbb{P}^*_\eta$ coincide on the invariant σ-field $\mathcal{I}$, defined by (8.13).

Exercise 10.9 Let the point process η be as in Exercise 10.8. Show that there exists $g \in \mathbb{R}_+(\mathbf{N}_l)$ satisfying $g(\theta_x\mu) = g(\mu)$ for all $(x,\mu) \in \mathbb{R}^d \times \mathbf{N}_l$ and such that $\hat{\eta}^{-1} = g(\eta)$ almost surely. Show then that $\mathbb{P}^*_\eta$ is absolutely continuous with respect to the Palm distribution $\mathbb{P}^0_\eta$ of η with density γg. (Hint: Apply the refined Campbell theorem to prove the second assertion.)

Exercise 10.10 For $t \geq 0$ let η_t be a stationary Poisson process on $\mathbb{R}^d$ with intensity t. Let $y \in \mathbb{R}^d \setminus \{0\}$ and show that $\eta' := \eta_1 + \theta_y\eta_1$ is stationary. Let X be a $\{0,1\}$-valued random variable and assume that η_1, η_2, X are independent. Show that $\eta := \mathbf{1}\{X = 0\}\eta' + \mathbf{1}\{X = 1\}\eta_2$ is stationary and pseudo-ergodic. Show also that η is not ergodic unless X is deterministic.

11

Stable Allocations

In the Gale–Shapley allocation introduced in Chapter 10, the idea is that points and sites both prefer to be allocated as close as possible. As a result there is no point and no site that prefer each other over their current partners. This property is called stability. Stable allocations are essentially unique. To prove this, it is useful to introduce a site-optimal version of the Gale–Shapley algorithm.

11.1 Stability

Let $d \in \mathbb{N}$. Recall that $\mathbf{N}_{ls} = \mathbf{N}_{ls}(\mathbb{R}^d)$ is the space of all locally finite simple counting measures on $\mathbb{R}^d$ and that we identify each $\mu \in \mathbf{N}_{ls}$ with its support.

We start by defining the key concept of this chapter.

Definition 11.1 Let τ be an allocation with appetite $\alpha > 0$. Let $\mu \in \mathbf{N}_{ls}$, $x \in \mu$ and $z \in \mathbb{R}^d$. We say the site z *desires* x if $\|z - x\| < \|z - \tau(z, \mu)\|$, where $\|\infty\| := \infty$. We say the point x *covets* z if

$$\|x - z\| < \|x - z'\| \text{ for some } z' \in C^\tau(x, \mu), \text{ or } \lambda_d(C^\tau(x, \mu)) < \alpha.$$

The pair (z, x) is said to be *unstable* (for μ and with respect to τ) if z desires x and x covets z. The allocation τ is said to be *stable* if there is no μ with an unstable pair.

Lemma 11.2 *The point-optimal Gale–Shapley allocation with appetite $\alpha > 0$ is stable.*

Proof Take $\mu \in \mathbf{N}_{ls}$, $x \in \mu$ and $z \in \mathbb{R}^d$. Consider Algorithm 10.6. If z desires x, then $z \notin C_n(x)$ for all $n \geq 1$. But if x covets z, then $z \in C_n(x)$ for some $n \geq 1$. (Note that if $\lambda_d(C^{\tau^{\alpha,p}}(x, \mu)) < \alpha$ then the radii $r_n(x)$ diverge, as explained in the proof of Lemma 10.8.) Therefore (z, x) cannot be an unstable pair. □

11.2 The Site-Optimal Gale–Shapley Allocation

Our goal is to prove that stable allocations are essentially uniquely determined. To achieve this goal, it is helpful to introduce the *site-optimal* Gale–Shapley allocation.

Algorithm 11.3 Let $\alpha > 0$ and $\mu \in \mathbf{N}_{ls}$. For $n \in \mathbb{N}$, $x \in \mu$ and $z \in \mathbb{R}^d$, define the point $\tau_n(x) \in \mu \cup \{\infty\}$ (in words, the point *claimed* by z at stage n) and define the sets $R_n(z) \subset \mu$ (the set of points *rejecting* x during the first n stages), $A_n(x) \subset \mathbb{R}^d$ (the set of sites claiming point x at stage n) and $S_n(x) \subset \mathbb{R}^d$ (the set of sites *shortlisted* by x at stage n) via the following recursion. Define $R_0(z) := \emptyset$ for all $z \in \mathbb{R}^d$ and take $n \in \mathbb{N}$.

(i) For $z \in \mathbb{R}^d$, define

$$\tau_n(z) := \begin{cases} l(\{x \in \mu : \|z - x\| = d(z, \mu \setminus R_{n-1}(z))\}), & \text{if } \mu \setminus R_{n-1}(z) \neq \emptyset, \\ \infty, & \text{otherwise.} \end{cases}$$

(ii) For $x \in \mu$, define (recall the convention $\inf \emptyset := \infty$)

$$A_n(x) := \{z \in \mathbb{R}^d : \tau_n(z) = x\},$$
$$r_n(x) := \inf\{r \geq 0 : \lambda_d(B(x, r) \cap A_n(x)) \geq \alpha\},$$
$$S_n(x) := A_n(x) \cap B(x, r_n(x)).$$

(iii) For $z \in \mathbb{R}^d$ let $R_n(z) := R_{n-1}(z) \cup \{x\}$ if $\tau_n(z) = x \in \mu$ and $z \notin B(x, r_n(x))$. Otherwise define $R_n(z) := R_{n-1}(z)$.

Given $\alpha > 0$, the *site-optimal Gale–Shapley allocation* $\tau^{\alpha,s}$ with appetite α is defined as follows. (The superscript s stands for "site-optimal".) Let $\mu \in \mathbf{N}_{ls}$ and consider Algorithm 11.3. For $z \in \mathbb{R}^d$ there are two cases. In the first case z is rejected by every point, that is $\mu = \cup_{n=1}^{\infty} R_n(z)$. Then we define $\tau^{\alpha,s}(z) := \infty$. In the second case there exist $x \in \mu$ and $n_0 \in \mathbb{N}$ such that $z \in S_n(x)$ for all $n \geq n_0$. In this case we define $\tau^{\alpha,s}(z, \mu) := x$.

The following lemma shows that $\tau^{\alpha,s}$ has properties similar to those of the point-optimal version. The proof can be given as before and is left to the reader.

Lemma 11.4 *The site-optimal Gale–Shapley allocation $\tau^{\alpha,s}$ is a stable allocation with appetite α.*

11.3 Optimality of the Gale–Shapley Algorithms

In this section we prove some optimality properties of the Gale–Shapley allocations. They are key to the forthcoming proof of the uniqueness of stable allocations. We start with two lemmas. We say that a site $x \in \mathbb{R}^d$ is *normal* for $\mu \in \mathbf{N}_{ls}$ if the distances from x to the points of μ are all distinct. Note that λ_d-almost all sites have this property.

Lemma 11.5 *Let τ be a stable allocation with appetite $\alpha > 0$ and let $\mu \in \mathbf{N}_{ls}$. Then for λ_d-a.e. z with $\tau(z,\mu) \neq \infty$, the point $\tau(z,\mu)$ does not reject z in the site-optimal Gale–Shapley algorithm for μ. In particular,*

$$\|\tau^{\alpha,s}(z,\mu) - z\| \leq \|\tau(z,\mu) - z\|, \quad \lambda_d\text{-a.e. } z. \tag{11.1}$$

Proof For each $n \in \mathbb{N}$ we need to show that for λ_d-a.e. z with $\tau(z,\mu) \neq \infty$, the point $\tau(z,\mu)$ does not reject z in the first n stages of the site-optimal Gale–Shapley algorithm. We do this by induction on n. For $n = 1$ let us assume the opposite. Then there exist $x \in \mu$ and a measurable set

$$R_1 \subset C^\tau(x,\mu) \cap (A_1(x) \setminus B(x, r_1(x)))$$

with $\lambda_d(R_1) > 0$. Then $S_1 := S_1(x) = A_1(x) \cap B(x, r_1(x))$ satisfies $\lambda_d(S_1) = \alpha$. Since $R_1 \cap S_1 = \emptyset$ we obtain for the set $T_1 := S_1 \setminus C^\tau(x,\mu)$ that

$$\begin{aligned}\lambda_d(T_1) &= \lambda_d(S_1 \setminus (C^\tau(x,\mu) \setminus R_1)) \geq \lambda_d(S_1) - \lambda_d(C^\tau(x,\mu) \setminus R_1)\\ &= \alpha - \lambda_d(C^\tau(x,\mu)) + \lambda_d(R_1) > 0,\end{aligned}$$

where we have used that $\lambda_d(C^\tau(x,\mu)) \leq \alpha$. Therefore we can pick normal $z \in T_1$ and $z' \in R_1$. Then $\tau(z,\mu) \neq x$, but x is the closest point of μ to z (since $z \in S_1$) so z desires x. Also $\|z - x\| \leq r_1(x) < \|z' - x\|$ so x covets z. This contradicts the assumed stability of τ.

For the induction step we let $n \in \mathbb{N}$ and assume, for λ_d-a.e. z with $\tau(z,\mu) \neq \infty$, that the point $\tau(z,\mu)$ does not reject z in the first n stages of the site-optimal Gale–Shapley algorithm. To show that this is also true for $n+1$ in place of n we assume the opposite. Then there exist $x \in \mu$ and a measurable set

$$R \subset C^\tau(x,\mu) \cap (A_{n+1}(x) \setminus B(x, r_{n+1}(x)))$$

with $\lambda_d(R) > 0$. Then $S := S_{n+1}(x) = A_{n+1}(x) \cap B(x, r_{n+1}(x))$ satisfies $\lambda_d(S) = \alpha$ and every site in S is closer to x than every site in R is to x. As before we obtain the fact that $T := S \setminus C^\tau(x,\mu)$ satisfies $\lambda_d(T) > 0$. A site $z \in S \supset T$ claims x in stage $n+1$ of the algorithm. Therefore z must have been rejected by all closer points of μ in one of the first n stages. Combining this with $\tau(z,\mu) \neq x$ for $z \in T$ and with the induction hypothesis shows that $\|\tau(z,\mu) - z\| > \|x - z\|$ for λ_d-a.e. $z \in T$. As we have already seen this contradicts the stability of τ.

The final assertion follows upon noting that a normal site z is allocated in $\tau^{\alpha,s}$ to the closest point in μ which does not reject it. □

The proof of the following lemma is similar to the preceding one and is left as an exercise; see Exercise 11.5.

Lemma 11.6 *Let τ be a stable allocation with appetite $\alpha > 0$. Let $\mu \in \mathbf{N}_{ls}$. Then for λ_d-a.e. z with $\tau(z,\mu) \neq \infty$, the site z never rejects $\tau(z,\mu)$ in the point-optimal Gale–Shapley algorithm for μ.*

Given an allocation τ, we define functions $g_\tau, h_\tau \colon \mathbb{R}^d \times \mathbf{N}_{ls} \times \overline{\mathbb{R}}_+ \to \mathbb{R}_+$ by

$$g_\tau(z,\mu,r) := \mathbf{1}\{\|\tau(z,\mu) - z\| \le r\},$$
$$h_\tau(x,\mu,r) := \lambda_d(C^\tau(x,\mu) \cap B(x,r)),$$

where $B(x,\infty) := \mathbb{R}^d$ and $\|\infty\| := \infty$. In a sense these functions describe the quality of the allocation for a site $z \in \mathbb{R}^d$ or a point $x \in \mu$, respectively.

Proposition 11.7 *Let τ be a stable allocation with appetite $\alpha > 0$. Let $\mu \in \mathbf{N}_{ls}$ and $r \in \overline{\mathbb{R}}_+$. Then*

$$g_{\tau^{\alpha,s}}(z,\mu,r) \ge g_\tau(z,\mu,r) \ge g_{\tau^{\alpha,p}}(z,\mu,r), \quad \lambda_d\text{-a.e. } z, \tag{11.2}$$
$$h_{\tau^{\alpha,p}}(x,\mu,r) \ge h_\tau(x,\mu,r) \ge h_{\tau^{\alpha,s}}(x,\mu,r), \quad x \in \mu. \tag{11.3}$$

Proof By (11.1) the first inequality of (11.2) holds for λ_d-a.e. $z \in \mathbb{R}^d$.

Now we prove the first inequality in (11.3). Assume the contrary, so that there exists $x \in \mu$ such that $h_{\tau^{\alpha,p}}(x,\mu,r) < h_\tau(x,\mu,r)$. Then

$$T := B(x,r) \cap (C^\tau(x,\mu) \setminus C^{\tau^{\alpha,p}}(x,\mu))$$

is a set with positive Lebesgue measure. Moreover, since $\tau^{\alpha,p}$ and τ both have appetite α, either $\lambda_d(C^{\tau^{\alpha,p}}(x,\mu)) < \alpha$ or $C^{\tau^{\alpha,p}}(x,\mu) \setminus B(x,r) \neq \emptyset$. In the first case x has claimed every site and must therefore have been rejected by the sites in T. In the second case x has claimed all sites in $B(x,r)$. Again it must have been rejected by all sites in T. This contradicts Lemma 11.6.

Next we take $x \in \mu$ and $r \ge 0$ and prove the second inequality in (11.3). Assume on the contrary that

$$\lambda_d(C^\tau(x,\mu) \cap B(x,r)) < \lambda_d(C^{\tau^{\alpha,s}}(x,\mu) \cap B(x,r)). \tag{11.4}$$

Then either $\lambda_d(C^\tau(x,\mu)) < \alpha$ or

$$C^\tau(x,\mu) \cap (\mathbb{R}^d \setminus B(x,r)) \neq \emptyset. \tag{11.5}$$

(If (11.5) fails, then $\lambda_d(C^\tau(x,\mu)) < \lambda_d(C^{\tau^{\alpha,s}}(x,\mu)) \le \alpha$.) Further we obtain from (11.4) that $\lambda_d(T) > 0$, where

$$T := B(x,r) \cap (C^{\tau^{\alpha,s}}(x,\mu) \setminus C^\tau(x,\mu)).$$

For $z \in T$ we have that $\tau^{\alpha,s}(z,\mu) = x$ and $\tau(z,\mu)$ are distinct points of μ so that the first inequality in (11.2) implies $\|z - x\| < \|z - \tau(z,\mu)\|$ for λ_d-a.e. $z \in T$. In particular, there is a site $z \in T$ that desires x. On the other hand, we obtain from (11.5) and $z \in B(x,r)$ that x covets z. Hence (z,x) is an unstable pair with respect to τ, contradicting our assumption.

The second inequality in (11.2) can be proved with the help of the first inequality in (11.3). Since it will not be used in what follows, we leave the proof to the reader. □

11.4 Uniqueness of Stable Allocations

In this section we let η be a stationary simple point process on $\mathbb{R}^d$ with intensity $\gamma \in (0,\infty)$ and with $\mathbb{P}(\eta = 0) = 0$. We shall need the following consequence of Theorem 10.2.

Lemma 11.8 *Let τ be an allocation. Then*

$$\mathbb{E}[g_\tau(0,\eta,r)] = \gamma\,\mathbb{E}[h_\tau(0,\eta^0,r)], \quad r \in [0,\infty].$$

Proof Let $r \in \overline{\mathbb{R}}_+$ and define $f \in \mathbb{R}_+(\mathbf{N}_{ls})$ by $f(\mu) := \mathbf{1}\{\|\tau(0,\mu)\| \le r\}$, $\mu \in \mathbf{N}_{ls}$. Using Theorem 10.2 with $g \equiv 1$ gives

$$\begin{aligned}\mathbb{E}[g_\tau(0,\eta,r)] = \mathbb{E}[f(\eta)] &= \gamma\,\mathbb{E}\Big[\int_{C^\tau(0,\eta^0)} \mathbf{1}\{\|\tau(0,\theta_z\eta^0)\| \le r\}\,dz\Big]\\ &= \gamma\,\mathbb{E}\Big[\int_{C^\tau(0,\eta^0)} \mathbf{1}\{\|\tau(z,\eta^0) - z\| \le r\}\,dz\Big],\end{aligned}$$

where we have used (10.3) to get the last identity. But if $z \in C^\tau(0,\eta^0)$ then $\tau(z,\eta^0) = 0$, provided that $0 \in \eta^0$, an event of probability 1. The result follows. □

We are now able to prove the following uniqueness result.

Theorem 11.9 *Let τ be a stable allocation with appetite $\alpha > 0$. Then*

$$\mathbb{P}(\tau(0,\eta) = \tau^{\alpha,s}(0,\eta)) = 1. \tag{11.6}$$

Proof Let $r \in [0,\infty]$. By (10.3) and stationarity we have for all $z \in \mathbb{R}^d$

that $\mathbb{E}[g_\tau(z,\eta,r)] = \mathbb{E}[g_\tau(0,\eta,r)]$. Applying Proposition 11.7 and Lemma 11.8 yields

$$\begin{aligned}\mathbb{E}[g_{\tau^{\alpha,s}}(0,\eta,r)] \geq \mathbb{E}[g_\tau(0,\eta,r)] &= \gamma\,\mathbb{E}[h_\tau(0,\eta^0,r)]\\ &\geq \gamma\,\mathbb{E}[h_{\tau^{\alpha,s}}(0,\eta^0,r)] = \mathbb{E}[g_{\tau^{\alpha,s}}(0,\eta^0,r)].\end{aligned}$$

Therefore the above inequalities are all equalities and

$$\mathbb{E}[g_{\tau^{\alpha,s}}(z,\eta,r)] = \mathbb{E}[g_\tau(z,\eta,r)], \quad z \in \mathbb{R}^d.$$

By Proposition 11.7,

$$g_{\tau^{\alpha,s}}(z,\eta,r) \geq g_\tau(z,\eta,r), \quad \lambda_d\text{-a.e. } z,$$

so that

$$g_{\tau^{\alpha,s}}(z,\eta,r) = g_\tau(z,\eta,r), \quad \mathbb{P}\text{-a.s, } \lambda_d\text{-a.e. } z. \tag{11.7}$$

Now we take $\mu \in \mathbf{N}_{ls}$ and $z \in \mathbb{R}^d$ such that

$$\mathbf{1}\{\|\tau^{\alpha,s}(z,\mu) - z\| \leq r\} = \mathbf{1}\{\|\tau(z,\mu) - z\| \leq r\}, \quad r \in D,$$

where $D \subset \mathbb{R}_+$ is countable and dense. Then $\|\tau^{\alpha,s}(z,\mu) - z\| = \|\tau(z,\mu) - z\|$ and hence $\tau^{\alpha,s}(z,\mu) = \tau(z,\mu)$ for all normal sites z. Therefore (11.7) implies

$$\lambda_d(\{z \in \mathbb{R}^d : \tau(z,\eta) \neq \tau^{\alpha,s}(z,\eta)\}) = 0, \quad \mathbb{P}\text{-a.s.} \tag{11.8}$$

By the covariance property (10.3) we have $\tau(z,\eta) \neq \tau^{\alpha,s}(z,\eta)$ if and only if $\tau(0,\theta_z\eta) \neq \tau^{\alpha,s}(0,\theta_z\eta)$. Therefore (11.6) follows by Exercise 8.13. □

11.5 Moment Properties

Finally in this chapter we show that stable allocations with appetite α have poor moment properties. In view of Theorem 10.5 we consider only the case $\alpha = \gamma^{-1}$. First we need the following generalisation of Lemma 11.5.

Lemma 11.10 *Let τ be a stable allocation with appetite $\alpha > 0$ and let $\mu_0, \mu \in \mathbf{N}_{ls}$ such that $\mu_0 \leq \mu$. Then for λ_d-a.e. z with $\tau(z,\mu_0) \neq \infty$, the point $\tau(z,\mu_0)$ does not reject z in the site-optimal Gale–Shapley algorithm for μ. In particular,*

$$\|\tau^{\alpha,s}(z,\mu) - z\| \leq \|\tau^{\alpha,s}(z,\mu_0) - z\|, \quad \lambda_d\text{-a.e. } z. \tag{11.9}$$

Proof The proof of Lemma 11.5 extends with obvious changes. □

Theorem 11.11 *Let η be a stationary Poisson process on $\mathbb{R}^d$ with intensity $\gamma > 0$ and let τ be a stable allocation with appetite γ^{-1}. Then $\mathbb{E}[\|\tau(0,\eta)\|^d] = \infty$.*

Proof Let $\alpha := \gamma^{-1}$. By Theorem 10.5 we have $\mathbb{P}(\tau(0,\eta) \neq \infty) = 1$. By Theorem 11.9 there is (essentially) only one stable allocation with appetite α. By Lemma 11.4 we can therefore assume that $\tau = \tau^{\alpha,s}$. We first prove that

$$\lambda_d(\{z \in \mathbb{R}^d : \|\tau(z,\eta) - z\| \geq \|z\| - 1\}) = \infty, \quad \mathbb{P}\text{-a.s.} \tag{11.10}$$

Let $m \in \mathbb{N}$ and let $U_1, \dots, U_m$ be independent and uniformly distributed on the unit ball B^d, independent of η. Define the point process

$$\eta' := \eta + \sum_{i=1}^{m} \delta_{U_i}.$$

We can assume that η and η' are random elements of $\mathbf{N}_{ls}$. Let

$$A := \{\mu \in \mathbf{N}_{ls} : \lambda_d(C^\tau(x,\mu)) = \alpha \text{ for all } x \in \mu\}.$$

Theorem 10.9 and Lemma 11.4 show that $\mathbb{P}(\eta \in A) = 1$. By Exercise 4.10, $\mathbb{P}(\eta' \in A) = 1$. Therefore

$$\lambda_d(\{z \in \mathbb{R}^d : \tau(z,\eta') \in B^d\}) \geq m\alpha, \quad \mathbb{P}\text{-a.s.}$$

But Lemma 11.10 shows, for λ_d-a.e. z with $\tau(z,\eta') \in B^d$, that

$$\|\tau(z,\eta) - z\| \geq \|\tau(z,\eta') - z\| \geq \|z\| - 1.$$

Since m is arbitrary, (11.10) follows.

By Fubini's theorem, (10.3) and stationarity,

$$\begin{aligned}
\mathbb{E}\Big[\int \mathbf{1}\{\|\tau(z,\eta) - z\| \geq \|z\| - 1\}\,dz\Big] &= \int \mathbb{P}(\|\tau(z,\eta) - z\| \geq \|z\| - 1)\,dz \\
&= \int \mathbb{P}(\|\tau(0,\eta)\| \geq \|z\| - 1)\,dz = \mathbb{E}\Big[\int \mathbf{1}\{\|\tau(0,\eta)\| \geq \|z\| - 1\}\,dz\Big] \\
&= \kappa_d\,\mathbb{E}[(\|\tau(0,\eta)\| + 1)^d].
\end{aligned}$$

The relationship (11.10) implies $\mathbb{E}[(\|\tau(0,\eta)\| + 1)^d] = \infty$ and hence the assertion. □

11.6 Exercises

Exercise 11.1 Let $d = 1$ and $\mu := \delta_0 + \delta_1$. Compute the point-optimal Gale–Shapley allocations $\tau^{\alpha,p}(\cdot,\mu)$ for $\alpha = 1$ and for $\alpha = 2$. Do they coincide with the site-optimal allocations $\tau^{\alpha,s}(\cdot,\mu)$?

Exercise 11.2 Let $\mu \in \mathbf{N}_{<\infty}(\mathbb{R}^d)$ and let τ be a stable allocation with appetite $\alpha > 0$. Show that

$$\tau(x, \mu) = \tau^{\alpha, \mathrm{p}}(x, \mu), \quad \lambda_d\text{-a.e. } x.$$

(Hint: There is an $n \in \mathbb{N}$ and a cube $C \subset \mathbb{R}^d$ with side length n such that $\mu(\mathbb{R}^d \setminus C) = 0$. Apply Theorem 11.9 to the stationary point process η defined as in Exercise 8.3.)

Exercise 11.3 Give an example of an allocation with appetite α that is not stable. (Hint: Use a version of the point-optimal Gale–Shapley Algorithm 11.3 with *impatient* sites.)

Exercise 11.4 Prove Lemma 11.4.

Exercise 11.5 Prove Lemma 11.6. (Hint: Proceed similarly to the proof of Lemma 11.5).

Exercise 11.6 A point process η on $\mathbb{R}^d$ is said to be *insertion tolerant* if, for each Borel set $B \subset \mathbb{R}^d$ with $0 < \lambda_d(B) < \infty$ and each random vector X that is uniformly distributed on B and independent of η, the distribution of $\eta + \delta_U$ is absolutely continuous with respect to $\mathbb{P}(\eta \in \cdot)$. Show that Theorem 11.11 remains valid for a stationary insertion tolerant point process η.

Exercise 11.7 Suppose that η_1 and η_2 are independent stationary point processes. Assume moreover, that η_2 is a Poisson process with positive intensity. Show that $\eta := \eta_1 + \eta_2$ is insertion tolerant. (Hint: Use the Mecke equation for η_2.)

Exercise 11.8 Let χ be a stationary Poisson cluster process as in Exercise 8.2. Show that χ is the sum of a stationary Poisson cluster process and an independent stationary Poisson process with intensity $\gamma\,\mathbb{Q}(\{\delta_0\})$. Deduce from Exercise 11.7 that χ is insertion tolerant, provided that $\gamma\,\mathbb{Q}(\{\delta_0\}) > 0$.

12

Poisson Integrals

The Wiener–Itô integral is the centred Poisson process integral. By means of a basic isometry equation it can be defined for any function that is square integrable with respect to the intensity measure. Wiener–Itô integrals of higher order are defined in terms of factorial measures of the appropriate order. Joint moments of such integrals can be expressed in terms of combinatorial diagram formulae. This yields moment formulae and central limit theorems for Poisson U-statistics. The theory is illustrated with a Poisson process of hyperplanes.

12.1 The Wiener–Itô Integral

In this chapter we fix an s-finite measure space $(\mathbb{X}, \mathcal{X}, \lambda)$. Let η denote a Poisson process on $\mathbb{X}$ with intensity measure λ. Let $f \in \mathbb{R}(\mathbb{X}^m)$ for some $m \in \mathbb{N}$. Corollary 4.10 shows that if $f \in L^1(\lambda^m)$ then

$$\mathbb{E}\left[\int f \, d\eta^{(m)}\right] = \int f \, d\lambda^m, \tag{12.1}$$

where the factorial measures $\eta^{(m)}$ are defined by Proposition 4.3. Our aim is to compute joint moments of random variables of the type $\int f \, d\eta^{(m)}$.

We start with a necessary and sufficient condition on $f \in \mathbb{R}_+(\mathbb{X})$ for the integral $\eta(f) = \int f \, d\eta$ to be almost surely finite.

Proposition 12.1 *Let $f \in \mathbb{R}_+(\mathbb{X})$. If*

$$\int (f \wedge 1) \, d\lambda < \infty, \tag{12.2}$$

then $\mathbb{P}(\eta(f) < \infty) = 1$. If (12.2) fails, then $\mathbb{P}(\eta(f) = \infty) = 1$.

Proof Assume that (12.2) holds and without loss of generality that η is proper. Then $\lambda(\{f \geq 1\}) < \infty$ and therefore $\eta(\{f \geq 1\}) < \infty$ a.s. Hence

$\eta(\mathbf{1}\{f \geq 1\}f) < \infty$ a.s. Furthermore we have from Proposition 2.7 that

$$\mathbb{E}[\eta(\mathbf{1}\{f < 1\}f)] = \lambda(\mathbf{1}\{f < 1\}f) < \infty,$$

so that $\eta(\mathbf{1}\{f < 1\}f) < \infty$ almost surely.

Assume, conversely, that (12.2) fails. By Theorem 3.9,

$$\mathbb{E}[e^{-\eta(f)}] = \exp[-\lambda(1 - e^{-f})]. \tag{12.3}$$

The inequality $(1 - e^{-t}) \geq (1 - e^{-1})(t \wedge 1)$, $t \geq 0$, implies $\lambda(1 - e^{-f}) = \infty$ and hence $\mathbb{E}[e^{-\eta(f)}] = 0$. Therefore $\mathbb{P}(\eta(f) = \infty) = 1$. □

Recall that $L^p(\lambda) = \{f \in \mathbb{R}(\mathbb{X}) : \lambda(|f|^p) < \infty\}$; see Section A.1. For $f \in L^1(\lambda)$ the *compensated integral* of f with respect to η is defined by

$$I(f) := \eta(f) - \lambda(f). \tag{12.4}$$

It follows from Campbell's formula (Proposition 2.7) that

$$\mathbb{E}[I(f)] = 0. \tag{12.5}$$

The random variable $I(f)$ is also denoted $\int f \, d(\eta - \lambda)$ or $\int f \, d\hat{\eta}$, where $\hat{\eta} := \eta - \lambda$. However, the reader should keep in mind that $\hat{\eta}$ is not defined on on all of $\mathcal{X}$ but only on $\{B \in \mathcal{X} : \lambda(B) < \infty\}$. Let $L^{1,2}(\lambda) := L^1(\lambda) \cap L^2(\lambda)$. The compensated integral has the following useful *isometry* property.

Lemma 12.2 *Suppose $f, g \in L^{1,2}(\lambda)$. Then*

$$\mathbb{E}[I(f)I(g)] = \int fg \, d\lambda. \tag{12.6}$$

Proof Equation (4.26) shows that $\mathbb{E}[\eta(f)^2] < \infty$. By (12.5), $\mathbb{E}[I(f)I(g)]$ is just the covariance between $\eta(f)$ and $\eta(g)$. A simple calculation gives

$$\mathbb{E}[I(f)I(g)] = \mathbb{E}[\eta(f)\eta(g)] - \lambda(f)\lambda(g).$$

Applying (4.26) yields (12.6). □

The next lemma shows that $L^{1,2}(\lambda)$ is a dense subset of $L^2(\lambda)$.

Lemma 12.3 *Let $f \in L^2(\lambda)$ and $n \in \mathbb{N}$. Then $f_n := \mathbf{1}\{|f| \geq 1/n\}f \in L^1(\mathbb{P})$. Moreover, $f_n \to f$ in $L^2(\mathbb{P})$ as $n \to \infty$.*

Proof For each $c > 0$ we have that

$$\int f^2 \, d\lambda \geq \int \mathbf{1}\{f^2 \geq c^2\} f^2 \, d\lambda \geq c^2 \lambda(|f| \geq c).$$

Therefore,

$$\int \mathbf{1}\{|f| \geq c\}|f|\,d\lambda = \int \mathbf{1}\{1 \geq |f| \geq c\}|f|\,d\lambda + \int \mathbf{1}\{|f| > 1\}|f|\,d\lambda$$
$$\leq \lambda(|f| \geq c) + \int f^2\,d\lambda < \infty,$$

so that $\mathbf{1}\{|f| \geq c\}f \in L^1(\lambda)$. Since $\lim_{n\to\infty} f_n(x) = f(x)$ for each $x \in \mathbb{X}$ and $|f - f_n| \leq 2|f|$, dominated convergence shows that $\lambda((f - f_n)^2) \to 0$. □

Lemma 12.3 can be used to extend I to a mapping from $L^2(\lambda)$ to $L^2(\mathbb{P})$ as follows.

Proposition 12.4 *The mapping* $I\colon L^{1,2}(\lambda) \to L^2(\mathbb{P})$ *defined by* (12.4) *can be uniquely extended to a linear mapping* $I\colon L^2(\lambda) \to L^2(\mathbb{P})$ *such that* (12.5) *and* (12.6) *hold for all* $f \in L^2(\lambda)$.

Proof The proof is based on basic Hilbert space arguments. For $f \in L^2(\lambda)$ we define (f_n) as in Lemma 12.3 and then obtain from (12.6) that

$$\mathbb{E}[(I(f_m) - I(f_n))^2] = \mathbb{E}[(I(f_m - f_n))^2] = \lambda((f_m - f_n)^2),$$

which tends to 0 as $m, n \to \infty$. Since $L^2(\mathbb{P})$ is complete, the sequence $(I(f_n))$ converges in $L^2(\mathbb{P})$ to some element of $L^2(\mathbb{P})$; we define $I(f)$ to be this limit. If in addition $f \in L^1(\lambda)$, then, by dominated convergence, $\lambda(f_n) \to \lambda(f)$, while dominated convergence and Proposition 12.1 show that $\eta(f_n) \to \eta(f)$ almost surely. Hence our new definition is consistent with (12.4). Since $\mathbb{E}[I(f_n)] = 0$ for all $n \in \mathbb{N}$, the L^2-convergence yields $\mathbb{E}[I(f)] = 0$. Furthermore,

$$\mathbb{E}[I(f)^2] = \lim_{n\to\infty} \mathbb{E}[I(f_n)^2] = \lim_{n\to\infty} \lambda(f_n^2) = \lambda(f^2),$$

where we have used Lemma 12.2 and, for the final identity, dominated convergence. The linearity

$$I(af + bg) = aI(f) + bI(g), \quad \mathbb{P}\text{-a.s.},\ f, g \in L^2(\lambda),\ a, b \in \mathbb{R}, \tag{12.7}$$

follows from the linearity of I on $L^1(\lambda)$.

If $f, g \in L^2(\lambda)$ coincide λ-a.e., then (12.6) implies that

$$\mathbb{E}[(I(f) - I(g))^2] = \mathbb{E}[(I(f - g))^2] = \lambda((f - g)^2) = 0,$$

so that $I(f) = I(g)$ a.s. Hence $I\colon L^2(\lambda) \to L^2(\mathbb{P})$ is a well-defined mapping.

If I' is another extension with the same properties as I then we can use the Minkowski inequality and $I(f_n) = I'(f_n)$ to conclude that

$$(\mathbb{E}[(I(f) - I'(f))^2])^{1/2} \leq (\mathbb{E}[(I(f) - I(f_n))^2])^{1/2} + (\mathbb{E}[(I'(f) - I'(f_n))^2])^{1/2}$$

for all $n \in \mathbb{N}$. By the isometry (12.6), both terms on the right-hand side of the preceding equation tend to 0 as $n \to \infty$. □

Definition 12.5 For $f \in L^2(\lambda)$ the random variable $I(f) \in L^2(\mathbb{P})$ is called the (stochastic) *Wiener–Itô integral* of f.

Let $f \in \mathbb{R}(\mathbb{X})$ and define (f_n) as in Lemma 12.3. If $f \in L^1(\lambda)$, then as shown in the preceding proof the sequence $I(f_n)$ converges almost surely towards $I(f)$, defined pathwise (that is, for every $\omega \in \Omega$) by (12.4). If, however, $f \in L^2(\lambda) \setminus L^1(\lambda)$, then $I(f_n)$ converges to $I(f)$ in $L^2(\mathbb{P})$ and hence only in probability.

Note that $L^1(\lambda)$ is not contained in $L^2(\lambda)$, and, unless $\lambda(\mathbb{X}) < \infty$, neither is $L^2(\lambda)$ contained in $L^1(\lambda)$. Exercise 12.3 shows that it is possible to extend I to the set of all $f \in \mathbb{R}(\mathbb{X})$ satisfying

$$\int |f| \wedge f^2 \, d\lambda < \infty. \tag{12.8}$$

12.2 Higher Order Wiener–Itô Integrals

In this section we turn to Poisson integrals of higher order. For $m \in \mathbb{N}$ and $f \in L^1(\lambda^m)$ define

$$I_m(f) := \sum_{J \subset [m]} (-1)^{m-|J|} \iint f(x_1, \dots, x_m) \, \eta^{(|J|)}(dx_J) \, \lambda^{m-|J|}(dx_{J^c}), \tag{12.9}$$

where $[m] := \{1, \dots, m\}$, $J^c := [m] \setminus J$, $x_J := (x_j)_{j \in J}$ and where $|J| :=$ card J denotes the number of elements of J. The inner integral in (12.9) is interpreted as $f(x_1, \dots, x_m)$ when $J = \emptyset$. This means that we set $\eta^{(0)}(c) := c$ for all $c \in \mathbb{R}$. Similarly, when $J = [m]$ the outer integration is performed according to the rule $\lambda^0(c) := c$ for all $c \in \mathbb{R}$. For each $J \subset [m]$ it follows from (12.1) and Fubini's theorem that

$$\mathbb{E}\left[\iint |f(x_1, \dots, x_m)| \, \eta^{(|J|)}(dx_J) \, \lambda^{m-|J|}(dx_{J^c})\right] = \int |f| \, d\lambda^m < \infty.$$

Therefore, $I_m(f)$ is an almost surely finite random variable and

$$\mathbb{E}[I_m(f)] = 0. \tag{12.10}$$

A function $f \colon \mathbb{X}^m \to \mathbb{R}$ is said to be *symmetric* if

$$f(x_1, \dots, x_m) = f(x_{\pi(1)}, \dots, x_{\pi(m)}), \quad (x_1, \dots, x_m) \in \mathbb{X}^m, \pi \in \Sigma_m, \tag{12.11}$$

where Σ_m is the set of permutations of $[m]$, that is the set of all bijective mappings from $[m]$ to $[m]$. If $f \in L^1(\lambda^m)$ is symmetric, then the symmetry of $\eta^{(m)}$ (see (A.17)) and λ^m implies that

$$I_m(f) = \sum_{k=0}^{m} (-1)^{m-k} \binom{m}{k} \eta^{(k)} \otimes \lambda^{m-k}(f), \tag{12.12}$$

where $\eta^{(k)} \otimes \lambda^{m-k}(f)$ is the integral of f with respect to the product measure $\eta^{(k)} \otimes \lambda^{m-k}$. In accordance with our convention $\eta^{(0)}(c) = \lambda^0(c) = c$, we have set $\eta^{(0)} \otimes \lambda^m := \lambda^m$ and $\eta^{(m)} \otimes \lambda^0 := \eta^{(m)}$.

For $m = 1$ the definition (12.9) reduces to (12.4). For $m = 2$ we have

$$\begin{aligned} I_2(f) = &\int f(x_1, x_2)\, \eta^{(2)}(d(x_1, x_2)) - \iint f(x_1, x_2)\, \eta(dx_1)\, \lambda(dx_2) \\ &- \iint f(x_1, x_2)\, \lambda(dx_1)\, \eta(dx_2) + \int f(x_1, x_2)\, \lambda^2(d(x_1, x_2)). \end{aligned}$$

In general, $I_m(f)$ is a linear combination of $\eta^{(m)}(f)$ and integrals of the type $\eta^{(k)}(f_k)$ for $k \le m - 1$, where f_k is obtained from f by integrating $m - k$ variables with respect to λ^{m-k}. In view of the forthcoming orthogonality relations (12.19) there are some advantages in dealing with $I_m(f)$ rather than with $\eta^{(m)}(f)$.

We shall prove formulae for the mixed moments $\mathbb{E}[\prod_{i=1}^{\ell} \eta^{(n_i)}(f_i)]$ and $\mathbb{E}[\prod_{i=1}^{\ell} I_{n_i}(f_i)]$, where $\ell, n_1, \dots, n_\ell \in \mathbb{N}$ and $f_i \in L^1(\lambda^{n_i})$ for $i \in [\ell]$. To do so we shall employ combinatorial arguments.

Let $n \in \mathbb{N}$. A *subpartition* of $[n]$ is a family of disjoint non-empty subsets of $[n]$, which we call *blocks*. A *partition* of $[n]$ is a subpartition σ of $[n]$ such that $\cup_{J \in \sigma} J = [n]$. We denote by Π_n (resp. Π_n^*) the system of all partitions (resp. subpartitions) of $[n]$. The cardinality of $\sigma \in \Pi_n^*$ (i.e., the number of blocks of σ) is denoted by $|\sigma|$, while the cardinality of $\cup_{J \in \sigma} J$ is denoted by $\|\sigma\|$. If σ is a partition, then $\|\sigma\| = n$.

Let $\ell, n_1, \dots, n_\ell \in \mathbb{N}$. Define $n := n_1 + \dots + n_\ell$ and

$$J_i := \{j \in \mathbb{N} : n_1 + \dots + n_{i-1} < j \le n_1 + \dots + n_i\}, \quad i = 1, \dots, \ell. \tag{12.13}$$

Let $\pi := \{J_i : 1 \le i \le \ell\}$ and let $\Pi(n_1, \dots, n_\ell) \subset \Pi_n$ (resp. $\Pi^*(n_1, \dots, n_\ell) \subset \Pi_n^*$) denote the set of all $\sigma \in \Pi_n$ (resp. $\sigma \in \Pi_n^*$) with $|J \cap J'| \le 1$ for all $J \in \sigma$ and for all $J' \in \pi$. Let $\Pi_{\ge 2}(n_1, \dots, n_\ell)$ (resp. $\Pi_{=2}(n_1, \dots, n_\ell)$) denote the set of all $\sigma \in \Pi(n_1, \dots, n_\ell)$ with $|J| \ge 2$ (resp. $|J| = 2$) for all $J \in \sigma$. Let $\Pi^*_{\ge 2}(n_1, \dots, n_\ell)$ denote the set of all $\sigma \in \Pi^*(n_1, \dots, n_\ell)$ with $|J| \ge 2$ for all $J \in \sigma$.

Let $\sigma \in \Pi(n_1, \dots, n_\ell)$. It is helpful to visualise the pair (π, σ) as a *diagram* with rows $J_1, \dots, J_\ell$, where the elements in each block $J \in \sigma$ are encircled by a closed curve. Since the blocks of σ are not allowed to contain more than one entry from each row, one might say that the diagram (π, σ) is *non-flat*.

The *tensor product* $\otimes_{i=1}^{\ell} f_i$ (also written $f_1 \otimes \cdots \otimes f_\ell$) of functions $f_i : \mathbb{X}^{n_i} \to \mathbb{R}$, $i \in \{1, \ldots, \ell\}$, is the function from $\mathbb{X}^n$ to $\mathbb{R}$ which maps each $(x_1, \ldots, x_n)$ to $\prod_{i=1}^{\ell} f_i(x_{J_i})$. In the case that $n_1 = \cdots = n_\ell$ and $f_1 = \cdots = f_\ell = f$ for some f, we write $f^{\otimes \ell}$ instead of $\otimes_{i=1}^{\ell} f_i$.

For any function $f : \mathbb{X}^n \to \mathbb{R}$ and $\sigma \in \Pi_n^*$ we define $f_\sigma : \mathbb{X}^{n+|\sigma|-\|\sigma\|} \to \mathbb{R}$ by identifying those arguments which belong to the same block of σ. (The arguments $x_1, \ldots, x_{n+|\sigma|-\|\sigma\|}$ are inserted in the order of first occurrence.) For example, if $n = 4$ and $\sigma = \{\{1, 3\}, \{4\}\}$, then $f_\sigma(x_1, x_2, x_3) = f(x_1, x_2, x_1, x_3)$. The partition $\{\{2\}, \{1, 3\}, \{4\}\}$ and the subpartition $\{\{1, 3\}\}$ lead to the same function.

We now give a formula for the expectation of a product of integrals with respect to factorial measures.

Proposition 12.6 *Let $f_i \in L^1(\lambda^{n_i})$, $i \in \{1, \ldots, \ell\}$, where $\ell, n_1, \ldots, n_\ell \in \mathbb{N}$. Let $n := n_1 + \cdots + n_\ell$ and assume that*

$$\int (|f_1| \otimes \cdots \otimes |f_\ell|)_\sigma \, d\lambda^{|\sigma|} < \infty, \quad \sigma \in \Pi(n_1, \ldots, n_\ell). \tag{12.14}$$

Then

$$\mathbb{E}\Big[\prod_{i=1}^{\ell} \eta^{(n_i)}(f_i)\Big] = \sum_{\sigma \in \Pi_{\geq 2}^*(n_1, \ldots, n_\ell)} \int (f_i \otimes \cdots \otimes f_\ell)_\sigma \, d\lambda^{n+|\sigma|-\|\sigma\|}. \tag{12.15}$$

Proof Since we consider a distributional property, we can assume that η is proper and given by (2.4). Using (4.4) for each $\eta^{(n_i)}$, we obtain

$$\prod_{i=1}^{\ell} \eta^{(n_i)}(f_i) = \sum_{i_1, \ldots, i_n \leq \kappa}^{*} (f_1 \otimes \cdots \otimes f_\ell)(X_{i_1}, \ldots, X_{i_n}), \tag{12.16}$$

where $\sum^*$ means that in the sum the i_j and i_k must be distinct whenever $j \neq k$ and j, k lie in the same block of π, where π was defined after (12.13). (The convergence of this possibly infinite series will be justified at the end of the proof.) Each multi-index $(i_1, \ldots, i_n)$ induces a subpartition in $\Pi_{\geq 2}^*(n_1, \ldots, n_\ell)$ by taking $j \neq k$ to be in the same block whenever $i_j = i_k$. Then the right-hand side of (12.16) equals

$$\sum_{\sigma \in \Pi_{\geq 2}^*(n_1, \ldots, n_\ell)} \sum_{i_1, \ldots, i_{n+|\sigma|-\|\sigma\|} \leq \kappa}^{\neq} (f_1 \otimes \cdots \otimes f_\ell)_\sigma (X_{i_1}, \ldots, X_{i_{n+|\sigma|-\|\sigma\|}}).$$

Corollary 4.10 yields the asserted result (12.15). The same computation also shows that the series on the right-hand side of (12.16) has a finite expectation upon replacing f_i by $|f_i|$ for each $i \in [\ell]$ and using assumption (12.14). In particular this series converges absolutely, almost surely. □

The following result is consistent with (12.5), (12.6) and (12.10).

Theorem 12.7 *Let $f_1, \dots, f_\ell$ be as in Proposition 12.6 and assume that* (12.14) *holds. Then*

$$\mathbb{E}\Big[\prod_{i=1}^{\ell} I_{n_i}(f_i)\Big] = \sum_{\sigma \in \Pi_{\geq 2}(n_1,\dots,n_\ell)} \int (f_1 \otimes \cdots \otimes f_\ell)_\sigma \, d\lambda^{|\sigma|}. \tag{12.17}$$

Proof By the definition (12.9) and Fubini's theorem,

$$\prod_{i=1}^{\ell} I_{n_i}(f_i) = \sum_{I \subset [n]} (-1)^{n-|I|} \int \cdots \int f_1 \otimes \cdots \otimes f_\ell$$
$$\times \eta^{(|I \cap J_1|)}(dx_{I \cap J_1}) \cdots \eta^{(|I \cap J_\ell|)}(dx_{I \cap J_\ell}) \, \lambda^{n-|I|}(dx_{I^c}), \tag{12.18}$$

where $I^c := [n] \setminus I$ and where we use definition (12.13) of J_i. By Proposition 12.6,

$$\mathbb{E}\Big[\prod_{i=1}^{\ell} I_{n_i}(f_i)\Big] = \sum_{I \subset [n]} (-1)^{n-|I|} \sum_{\sigma \in \Pi^*_{\geq 2}(n_1,\dots,n_\ell): \sigma \subset I} \int (\otimes_{i=1}^{\ell} f_i)_\sigma \, d\lambda^{n+|\sigma|-\|\sigma\|},$$

where $\sigma \subset I$ means that every block of σ is contained in I. Interchanging the order of the above summations, and noting that for any given σ the sum $\sum_{I:\sigma \subset I} (-1)^{n-|I|}$ comes to zero except when σ is a partition, in which case it comes to one, gives us (12.17). □

Given $n \geq 1$, let $L^2_s(\lambda^n)$ denote the set of all $f \in L^2(\lambda^n)$ that are symmetric. Let $L_{0,s}(\lambda^n) \subset L^2(\lambda^n)$ denote the set of all symmetric $f \in \mathbb{R}(\mathbb{X}^n)$ such that f is bounded and $\lambda^n(\{f \neq 0\}) < \infty$. The following result generalises Lemma 12.2.

Corollary 12.8 *Let $m, n \in \mathbb{N}$, $f \in L_{0,s}(\lambda^m)$ and $g \in L_{0,s}(\lambda^n)$. Then*

$$\mathbb{E}[I_m(f) I_n(g)] = \mathbf{1}\{m = n\} m! \int f g \, d\lambda^m. \tag{12.19}$$

Proof The assumptions allow us to apply Theorem 12.7 with $\ell = 2$, $f_1 = f$ and $g_2 = g$. First assume $m = n$. Then each element of $\Pi_{\geq 2}(m, n)$ is a partition of $[2m]$ with m blocks each having two elements, one from $\{1, \dots, m\}$ and one from $\{m+1, \dots, 2m\}$. We identify each element of

$\Pi_{\geq 2}(m,m)$ with a permutation of $[m]$ in the obvious manner. With Σ_m denoting the set of all such permutations, (12.17) gives us

$$\mathbb{E}[I_m(f)I_n(g)] = \sum_{\sigma\in\Sigma_m} \int f(x_1,\dots,x_m)g(x_{\sigma(1)},\dots,x_{\sigma(m)})\,\lambda^m(d(x_1,\dots,x_m)).$$

By symmetry, each term in the expression on the right-hand side equals $\int fg\,d\lambda^m$. Hence (12.19) holds for $m = n$. If $m \neq n$ then $\Pi_{\geq 2}(m,n) = \emptyset$, so that (12.19) holds in this case too. □

The following result can be proved in the same manner as Proposition 12.4; see Exercise 12.4.

Proposition 12.9 *The mappings $I_m\colon L_{0,s}(\lambda^m) \to L^2(\mathbb{P})$, $m \in \mathbb{N}$, can be uniquely extended to linear mappings $I_m\colon L_s^2(\lambda^m) \to L^2(\mathbb{P})$ so that* (12.10) *and* (12.19) *hold for all $f \in L_s^2(\lambda^m)$ and $g \in L_s^2(\lambda^n)$. If $f \in L^1(\lambda^m) \cap L_s^2(\lambda^m)$, then $I_m(f)$ is almost surely given by* (12.9).

Definition 12.10 For $m \in \mathbb{N}$ and $f \in L_s^2(\lambda^m)$ the random variable $I_m(f) \in L^2(\mathbb{P})$ is called the (stochastic, m-th order) *Wiener–Itô integral* of f.

12.3 Poisson U-Statistics

In the rest of this chapter we apply the preceding results to *U-statistics*. Let $m \in \mathbb{N}$ and $h \in L_s^1(\lambda^m)$ and set

$$U := \int h(x_1,\dots,x_m)\,\eta^{(m)}(d(x_1,\dots,x_m)). \tag{12.20}$$

Then U is known as a *Poisson U-statistic* with *kernel function* h. For $n \in \{0,\dots,m\}$, define $h_n \in L_s^1(\lambda^n)$ by

$$h_n(x_1,\dots,x_n) := \binom{m}{n} \int h(x_1,\dots,x_n,y_1,\dots,y_{m-n})\,\lambda^{m-n}(d(y_1,\dots,y_{m-n})), \tag{12.21}$$

where $h_0 := \lambda^m(h)$. With our earlier convention $\lambda^0(c) = c$, $c \in \mathbb{R}$, this means that $h_m = h$. A Poisson U-statistic can be decomposed into mutually orthogonal terms as follows.

Proposition 12.11 *Let U be the Poisson U-statistic given by* (12.20) *and define the functions h_n by* (12.21). *Then*

$$U = \mathbb{E}[U] + \sum_{n=1}^{m} I_n(h_n), \quad \mathbb{P}\text{-a.s.} \tag{12.22}$$

Proof We note that $\mathbb{E}[U] = \lambda^m(h) = h_0$ and set $\eta^{(0)} \otimes \lambda^0(h_0) := h_0$. Then we obtain from (12.12) that

$$\begin{aligned}\mathbb{E}[U] + \sum_{n=1}^{m} I_n(h_n) &= \sum_{n=0}^{m}\sum_{k=0}^{n}(-1)^{n-k}\binom{n}{k}\eta^{(k)} \otimes \lambda^{n-k}(h_n)\\ &= \sum_{k=0}^{m}\sum_{n=k}^{m}(-1)^{n-k}\binom{n}{k}\binom{m}{n}\eta^{(k)} \otimes \lambda^{m-k}(h)\\ &= \sum_{k=0}^{m}\binom{m}{k}\eta^{(k)} \otimes \lambda^{m-k}(h)\sum_{r=0}^{m-k}(-1)^r\binom{m-k}{r}, \end{aligned} \tag{12.23}$$

where we have used the substitution $r := n - k$ and the combinatorial identity

$$\binom{k+r}{k}\binom{m}{k+r} = \binom{m}{k}\binom{m-k}{r}.$$

The inner sum at (12.23) vanishes for $m > k$ and equals 1 otherwise. The result follows. □

Together with Proposition 12.9 the preceding proposition yields the following result.

Proposition 12.12 *Let the Poisson U-statistic U be given by* (12.20) *and assume that the functions h_n defined by* (12.21) *are square integrable with respect to λ^n for all $n \in [m]$. Then U is square integrable with variance*

$$\mathbb{V}\mathrm{ar}[U] = \sum_{n=1}^{m} n! \int h_n^2 \, d\lambda^n. \tag{12.24}$$

Proof Proposition 12.11, the assumption $h_n \in L^2(\lambda^n)$ and Proposition 12.9 imply that U is square integrable. Moreover, the isometry relation (12.19) (see Proposition 12.9) gives us (12.24). □

If, in the situation of Proposition 12.12, $\lambda(h_1^2) = 0$ then we say that U is *degenerate*. This happens if and only if

$$\lambda\Big\{x_1 : \int h(x_1, \dots, x_m)\, \lambda^{m-1}(d(x_2, \dots, x_m)) \neq 0\Big\} = 0. \tag{12.25}$$

Therefore $\lambda^m(h) \neq 0$ is sufficient for U not to be degenerate.

Next we generalise Proposition 12.12. For $\ell \in \mathbb{N}$ and $\sigma \in \Pi(m, \dots, m)$ (m occurs ℓ times) let

$$[\sigma] := \{i \in [\ell] : \text{there exists } J \in \sigma, J \cap \{m(i-1)+1, \dots, mi\} \neq \emptyset\},$$

i.e. $[\sigma]$ is the set of rows of the diagram of $(m,\dots,m)$ that are visited by σ. Let $\Pi^{**}_{\geq 2}(m,\dots,m)$ be the set of subpartitions in $\Pi^{*}_{\geq 2}(m,\dots,m)$ that satisfy $[\sigma]=[\ell]$, i.e. that visit every row of the diagram.

Proposition 12.13 *Let the Poisson U-statistic U be given by* (12.20). *Let $\ell \geq 2$ be an integer such that $\int (|h|^{\otimes \ell})_\sigma \, d\lambda^{|\sigma|} < \infty$ for all $\sigma \in \Pi(m,\dots,m)$ (with ℓ occurring m times). Then*

$$\mathbb{E}[(U-\mathbb{E}[U])^\ell] = \sum_{\sigma \in \Pi^{**}_{\geq 2}(m,\dots,m)} \int (h^{\otimes \ell})_\sigma \, d\lambda^{m\ell + |\sigma| - \|\sigma\|}. \tag{12.26}$$

Proof We have

$$\mathbb{E}[(U-\mathbb{E}[U])^\ell] = \sum_{I \subset [\ell]} \mathbb{E}[U^{|I|}](-\mathbb{E}[U])^{\ell - |I|},$$

and, using (12.15) along with the fact that $\mathbb{E}[U]=\lambda^m(h)$, we have that this equals

$$\sum_{I \subset [\ell]} (-1)^{\ell-|I|} \sum_{\sigma \in \Pi^{**}_{\geq 2}(m,\dots,m):[\sigma]\subset I} \int (h \otimes \cdots \otimes h)_\sigma \, d\lambda^{m\ell+|\sigma|-\|\sigma\|}.$$

Interchanging the summations and arguing as in the proof of Theorem 12.7 gives the result. □

In the remainder of this chapter we extend our setting by considering for $t>0$ a Poisson process η_t with intensity measure $\lambda_t := t\lambda$. We study Poisson U-statistics of the form

$$U_t := b(t) \int h(x_1,\dots,x_m)\, \eta_t^{(m)}(d(x_1,\dots,x_m)), \tag{12.27}$$

where $m \in \mathbb{N}$, $b(t)>0$ for all $t>0$ and the kernel function $h \in L^1_s(\lambda^m)$ does not depend on t. We recall the definition (B.6) of the double factorial $(\ell-1)!!$ for even integer $l \geq 2$. This is the number of (perfect) *matchings* π of $[\ell]$. Such a matching is a partition of $[\ell]$ whose blocks are all of size 2. If ℓ is odd, then no such matching exists.

Theorem 12.14 *Let $m \in \mathbb{N}$ and $h \in L^1_s(\lambda^m)$. Let U_t be the Poisson U-statistic given by* (12.27) *and let $\ell \geq 2$ be an integer. Assume that $\int (|h|^{\otimes \ell})_\sigma \, d\lambda^{|\sigma|} < \infty$ for all $\sigma \in \Pi(m,\dots,m)$. Assume also that $\lambda(h_1^2)>0$, where h_1 is given by* (12.21) *for $n=1$. Then*

$$\lim_{t\to\infty} \frac{\mathbb{E}[(U_t - \mathbb{E}U_t)^\ell]}{(\mathbb{V}\mathrm{ar}[U_t])^{\ell/2}} = \begin{cases} (\ell-1)!!, & \text{if } \ell \text{ is even,} \\ 0, & \text{if } \ell \text{ is odd.} \end{cases} \tag{12.28}$$

Proof Let $n \in \{1, \dots, m\}$. Our assumptions on h imply that the mapping that sends the vector $(x_1, \dots, x_n, y_1, \dots, y_{m-n}, z_1, \dots, z_{m-n})$ to the product of $|h(x_1, \dots, x_n, y_1, \dots, y_{m-n})|$ and $|h(x_1, \dots, x_n, z_1, \dots, z_{m-n})|$ is integrable with respect to λ^{2m-n}. (To see this we might take a σ containing the pairs $\{1, m+1\}, \dots, \{n, m+n\}$ while the other blocks are all singletons.) Therefore by Fubini's theorem $h_n \in L^2(\lambda^n)$, where h_n is given by (12.21). For $t > 0$, define the function $h_{t,n}$ by (12.21) with (h, λ) replaced by $(b(t)h, \lambda_t)$. Then $h_{t,n} = b(t)t^{m-n}h_n$, so that Proposition 12.12 yields that U_t is square integrable and

$$\mathbb{V}\mathrm{ar}[U_t] = b(t)^2 \sum_{n=1}^{m} n! t^{2m-n} \int h_n^2 \, d\lambda^n. \tag{12.29}$$

In the remainder of the proof we assume without loss of generality that $b(t) = 1$ for all $t > 0$. Since we assume $\lambda(h_1^2) > 0$, we obtain from (12.29) that

$$(\mathbb{V}\mathrm{ar}[U_t])^{\ell/2} = (\lambda(h_1^2))^{\ell/2} t^{\ell(m-1/2)} + p(t), \tag{12.30}$$

where the remainder term $p(\cdot)$ is a polynomial of degree strictly less than $\ell(m-1/2)$.

In the present setting we can rewrite (12.26) as

$$\mathbb{E}[(U_t - \mathbb{E}[U_t])^\ell] = \sum_{\sigma \in \Pi^{**}_{\geq 2}(m,\dots,m)} t^{m\ell + |\sigma| - \|\sigma\|} \int (h^{\otimes \ell})_\sigma \, d\lambda^{m\ell + |\sigma| - \|\sigma\|}. \tag{12.31}$$

Note that for any $\sigma \in \Pi^{**}_{\geq 2}(m, \dots, m)$ we have $\|\sigma\| \geq \ell$ and $|\sigma| \leq \|\sigma\|/2$. Therefore, if ℓ is even, $m\ell + |\sigma| - \|\sigma\|$ is maximised (over subpartitions $\sigma \in \Pi^{**}_{\geq 2}(m, \dots, m)$) by taking σ such that $|\sigma| = \ell/2$ and each block has size 2 and $[\sigma] = \ell$. The number of such σ is $(\ell - 1)!!m^\ell$ so the leading order term in the expression (12.31) comes to

$$t^{m\ell - \ell/2} \left(\int h_1^2 d\lambda \right)^{\ell/2} (\ell - 1)!!$$

as required. For ℓ odd, the above inequalities show that

$$m\ell + |\sigma| - \|\sigma\| \leq m\ell - (\ell + 1)/2$$

so that $\lim_{t \to \infty} t^{-\ell(m-1/2)} \mathbb{E}[(U_t - \mathbb{E}[U_t])^\ell] = 0$. □

The preceding proof implies the following result on the asymptotic variance of a Poisson U-statistic.

Proposition 12.15 *Let U_t be the Poisson U-statistic given by* (12.27). *Let the functions h_n be given by* (12.21) *and assume that $h_n \in L^2(\lambda^n)$ for each $n \in [m]$. Then*

$$\lim_{t\to\infty} b(t)^{-2} t^{1-2m}\, \mathbb{V}\mathrm{ar}[U_t] = \int h_1^2\, d\lambda. \tag{12.32}$$

Proof It is easy to see (and is explained at the beginning of the proof of Theorem 12.14) that our assumption that $h_n \in L^2(\lambda^n)$ for each $n \in [m]$ is equivalent to the integrability assumption of Theorem 12.14 in the case $\ell = 2$. Hence we have (12.29) and the result follows. □

Theorem 12.14 leads to the following central limit theorem.

Theorem 12.16 (Central limit theorem for Poisson U-statistics) *Suppose that $m \in \mathbb{N}$ and $h \in L_s^1(\lambda^m)$ and that U_t is given by* (12.27). *Assume that $\int (|h|^{\otimes \ell})_\sigma\, d\lambda^{|\sigma|} < \infty$ for all $\ell \in \mathbb{N}$ and all $\sigma \in \Pi(m, \dots, m)$. Assume also that $\lambda(h_1^2) > 0$, where h_1 is given by* (12.21) *for $n = 1$. Then*

$$(\mathbb{V}\mathrm{ar}[U_t])^{-1/2}(U_t - \mathbb{E}U_t) \xrightarrow{d} N \text{ as } t \to \infty, \tag{12.33}$$

where $\xrightarrow{d}$ denotes convergence in distribution and N is a standard normal random variable.

Proof By (B.5) we have $\mathbb{E}[N^\ell] = (\ell - 1)!!$ if ℓ is even and $\mathbb{E}[N^\ell] = 0$ otherwise. Moreover, with the help of Proposition B.4 one can show that the distribution of N is determined by these moments. Theorem 12.14 says that the moments of the random variable $(\mathbb{V}\mathrm{ar}[U_t])^{-1/2}(U_t - \mathbb{E}U_t)$ converge to the corresponding moments of N. The method of moments (see Proposition B.12) gives the asserted convergence in distribution. □

The following lemma is helpful for checking the integrability assumptions of Theorem 12.16.

Lemma 12.17 *Let $\ell \in \mathbb{N}$ such that $h \in L_s^\ell(\lambda^m)$. Assume that $\{h \neq 0\} \subset B^m$, where $B \in \mathcal{X}$ satisfies $\lambda(B) < \infty$. Then we have $\int (|h|^{\otimes \ell})_\sigma\, d\lambda^{|\sigma|} < \infty$ for all $\sigma \in \Pi(m, \dots, m)$.*

Proof Apply Exercise 12.9 in the case $f_1 = \cdots = f_\ell = h$. □

12.4 Poisson Hyperplane Processes

Finally in this chapter we discuss a model from stochastic geometry. Let $d \in \mathbb{N}$ and let $\mathbb{H}_{d-1}$ denote the space of all hyperplanes in $\mathbb{R}^d$. Any such hyperplane H is of the form

$$H_{u,r} := \{y \in \mathbb{R}^d : \langle y, u\rangle = r\}, \tag{12.34}$$

where u is an element of the unit sphere $\mathbb{S}^{d-1}$, $r \geq 0$ and $\langle\cdot,\cdot\rangle$ denotes the Euclidean scalar product. We can make $\mathbb{H}_{d-1}$ a measurable space by introducing the smallest σ-field $\mathcal{H}$ containing the sets

$$[K] := \{H \in \mathbb{H}_{d-1} : H \cap K \neq \emptyset\}, \quad K \in C^d,$$

where C^d denotes the system of all compact subsets of $\mathbb{R}^d$. In fact, $\mathbb{H}_{d-1} \cup \{\emptyset\}$ is a closed subset of the space $\mathcal{F}^d$ of all closed subsets of $\mathbb{R}^d$, equipped with the Fell topology, as defined in Section A.3.

We fix a measure λ on $\mathbb{H}_{d-1}$ satisfying

$$\lambda([K]) < \infty, \quad K \in C^d. \tag{12.35}$$

In particular, λ is σ-finite. As before, for $t > 0$ let η_t be a Poisson process with intensity measure $\lambda_t := t\lambda$. The point process η_t is called a *Poisson hyperplane process*, while the union

$$Z := \bigcup_{\eta\{H\}>0} H$$

of all hyperplanes in η is called a *Poisson hyperplane tessellation*. (It can be shown that the singletons in $\mathbb{H}_{d-1}$ are closed and hence measurable.) The cells of this tessellation are the (closures of the) connected components of the complement $\mathbb{R}^d \setminus Z$.

Recall from Section A.3 that $\mathcal{K}^d$ denotes the space of all convex compact subsets of $\mathbb{R}^d$. Let $W \in \mathcal{K}^d$ and $m \in \mathbb{N}$. Let $\psi\colon \mathcal{K}^d \to \mathbb{R}$ be a measurable function. By Lemma A.30, $(H_1,\dots,H_m) \mapsto \psi(H_1 \cap \cdots \cap H_m \cap W)$ is a measurable mapping from $(\mathbb{H}_{d-1})^m$ to $\mathbb{R}$. We also assume that $\psi(\emptyset) = 0$ and that $\psi(H_1 \cap \cdots \cap H_m \cap W) \leq c_W$ for λ^m-a.e. $(H_1,\dots,H_m)$, where c_W depends on W. (We shall give some examples at the end of this section.) We consider the Poisson U-statistic given by

$$U_t := \frac{1}{m!}\int \psi(H_1 \cap \cdots \cap H_m \cap W)\, \eta_t^{(m)}(d(H_1,\dots,H_m)), \quad t > 0. \tag{12.36}$$

To formulate a corollary to Theorem 12.16 we define, for $n \in [m]$, a function $\psi_n \in \mathbb{R}((\mathbb{H}_{d-1})^n)$ by

$$\psi_n(H_1,\dots,H_n) := \int \psi(H_1 \cap \cdots \cap H_m \cap W)\, \lambda^{m-n}(d(H_{n+1},\dots,H_m)). \tag{12.37}$$

Corollary 12.18 *Let ψ satisfy the assumptions formulated before* (12.36) *and define U_t by* (12.36). *Then*

$$\mathbb{E}[U_t] = \frac{t^m}{m!} \int \psi(H_1 \cap \cdots \cap H_m \cap W)\, \lambda^m(d(H_1, \dots, H_m)), \tag{12.38}$$

$$\mathbb{V}\mathrm{ar}[U_t] = \sum_{n=1}^{m} \frac{1}{n!((m-n)!)^2} t^{2m-n} \int \psi_n^2 \, d\lambda^n, \tag{12.39}$$

and if, moreover,

$$\int \psi(H_1 \cap \cdots \cap H_m \cap W)\, \lambda^m(d(H_1, \dots, H_m)) \neq 0, \tag{12.40}$$

then the central limit theorem (12.33) *holds.*

Proof We apply the results of the preceding section, taking

$$h(H_1, \dots, H_m) := \frac{1}{m!} \psi(H_1 \cap \cdots \cap H_m \cap W).$$

For $n \in [m]$ the function (12.21) is then given by $h_n = \binom{m}{n}(m!)^{-1}\psi_n$. It follows from (12.35) and Lemma 12.17 that U_t satisfies the integrability assumptions of Proposition 12.12 and Theorem 12.16. Hence (12.38) follows from (12.1) and (12.39) follows from (12.29). As noted at (12.25), the inequality (12.40) implies the non-degeneracy assumption of Theorem 12.16 and hence the central limit theorem. □

Under weak assumptions on λ the intersection of m different hyperplanes from η_t is almost surely either empty or an affine space of dimension $d-m$. If we choose $\psi(K) = V_{d-m}(K)$ as the $(d-m)$-th intrinsic volume of $K \in \mathcal{K}^d$, then U_t is the total volume of the intersections of the $(d-m)$-dimensional faces of the hyperplane tessellation with W. This ψ satisfies the preceding assumptions. Another possible choice is $\psi(K) = \mathbf{1}\{K \neq \emptyset\}$. In that case U_t is the number of $(d-m)$-dimensional faces intersecting W.

12.5 Exercises

Exercise 12.1 Let η be a Poisson process on $(0, \infty)$ with an intensity measure ν satisfying $\int x \wedge 1\, \nu(dx) < \infty$. Show that $X := \int x\, \eta(dx)$ is a finite random variable with Laplace transform

$$\mathbb{E}[\exp(-tX)] = \exp\left[-\int (1 - e^{-tx})\, \nu(dx)\right], \quad t \geq 0.$$

Show also that for each $m \in \mathbb{N}$ there are independent random variables $X_1, \dots, X_m$ with equal distribution such that $X \overset{d}{=} X_1 + \cdots + X_m$. (Such an X is said to be *infinitely divisible*.)

Exercise 12.2 Suppose that ν and ν' are measures on $(0, \infty)$ satisfying $\int x \wedge 1\,(\nu + \nu')(dx) < \infty$. Assume that $\int (1 - e^{-tx})\,\nu(dx) = \int (1 - e^{-tx})\,\nu'(dx)$, $t \geq 0$. Show that $\nu = \nu'$. (Hint: Take derivatives and apply Proposition B.4 to suitable transforms of ν and ν'.)

Exercise 12.3 Let $f \in \mathbb{R}(\mathbb{X})$ satisfy (12.8); define $g := \mathbf{1}\{|f| \leq 1\}f$ and $h := \mathbf{1}\{|f| > 1\}f$. Show that

$$\int g^2\,d\lambda + \int |h|\,d\lambda = \int |f| \wedge |f|^2\,d\lambda.$$

This result justifies the definition $I(f) := I(g) + I(h)$, where $I(g)$ is given by Definition 12.5 and $I(h) := \eta(h) - \lambda(h)$. Let $f_n \in \mathbb{R}(\mathbb{X})$, $n \in \mathbb{N}$, be bounded such that $\lambda(\{f_n \neq 0\}) < \infty$, $|f_n| \leq |f|$ and $f_n \to f$. Show that $I(f_n) \to I(f)$ in probability.

Exercise 12.4 Prove Proposition 12.9.

Exercise 12.5 Let $f, g \in L^{1,2}(\lambda)$ and assume moreover that the functions fg^2, f^2g and f^2g^2 are all in $L^1(\lambda)$. Show that

$$\mathbb{E}[I(f)^2 I(g)^2] = \lambda(f^2)\lambda(g^2) + 2[\lambda(fg)]^2 + \lambda(f^2g^2).$$

In particular, $\mathbb{E}[I(f)^4] = 3[\lambda(f^2)]^2 + \lambda(f^4)$ provided that $f \in L^1(\lambda) \cap L^4(\lambda)$.

Exercise 12.6 Let $f \in L^1_s(\lambda^2)$ and $g \in L^1(\lambda)$. Show that

$$\mathbb{E}[I_2(f)^2 I_1(g)] = 4 \int f(x_1, x_2)^2 g(x_1)\,\lambda^2(d(x_1, x_2))$$

holds under suitable integrability assumptions on f and g.

Exercise 12.7 Let $f, g \in L^2(\lambda)$ be such that $fg \in L^2(\lambda)$. Show that $I_1(f)I_1(g) = I_2(f \otimes g) + I_1(fg) + \lambda(fg)$.

Exercise 12.8 Let $m, n \in \mathbb{N}$. Let $f \in L^1(\lambda^m)$ and $g \in L^1(\lambda^n)$ be such that $f \otimes g = 0$ on the generalised diagonal of $\mathbb{X}^{m+n}$. Show that $I_m(f)I_n(g) = I_{m+n}(f \otimes g)$.

Exercise 12.9 Let $f_i \in L^1_s(\lambda^{n_i})$, $i = 1, \dots, \ell$, where $\ell, n_1, \dots, n_\ell \in \mathbb{N}$. Assume for each $i \in [\ell]$ that $f_i \in L^\ell(\lambda^{n_i})$ and $\{f_i \neq 0\} \subset B^{n_i}$, where $B \in \mathcal{X}$

satisfies $\lambda(B) < \infty$. Show for $\sigma \in \Pi(n_1, \dots, n_\ell)$ that

$$\left(\int (\otimes_{i=1}^{\ell} |f_i|)_\sigma \, d\lambda^{|\sigma|}\right)^{\ell} \le \lambda(B)^{|\sigma|-n_1} \int |f_1|^\ell \, d\lambda^{n_1} \cdots \lambda(B)^{|\sigma|-n_\ell} \int |f_\ell|^\ell \, d\lambda^{n_\ell}.$$

Note that this implies (12.14). (Hint: Apply the multivariate Hölder inequality (A.3) in the case $m = \ell$ and $p_1 = \cdots = p_m = m$.)

Exercise 12.10 (Moment and factorial moment measures) Suppose that η is a proper point process with factorial moment measures α_j, $j \in \mathbb{N}$. Let $m \in \mathbb{N}$ and $f \in \mathbb{R}(\mathbb{X}^m)$ such that $\mathbb{E}[\int |f| \, d\eta^m] < \infty$. Show that

$$\mathbb{E}\left[\int f \, d\eta^m\right] = \sum_{\sigma \in \Pi_m} \int f_\sigma \, d\alpha_{|\sigma|}.$$

Exercise 12.11 Suppose that η is a Poisson process on $\mathbb{X}$ with s-finite intensity measure λ. Let $m \in \mathbb{N}$ and $f \in \mathbb{R}_+(\mathbb{X}^m \times \mathbf{N})$. Show that

$$\mathbb{E}\left[\int f(x_1, \dots, x_m, \eta) \, \eta^m(d(x_1, \dots, x_m))\right]$$
$$= \sum_{\sigma \in \Pi_m} \int \mathbb{E}[f_\sigma(x_1, \dots, x_{|\sigma|}, \eta + \delta_{x_1} + \cdots + \delta_{x_{|\sigma|}})] \lambda^{|\sigma|}(d(x_1, \dots, x_{|\sigma|})),$$

where $f_\sigma(\cdot, \mu) := (f(\cdot, \mu))_\sigma$ for each $\mu \in \mathbf{N}$.

Exercise 12.12 Show that the mapping $(u, r) \mapsto H_{u,r}$ from $\mathbb{S}^{d-1} \times \mathbb{R}$ to $\mathbb{H}_{d-1}$ is measurable; see (12.34). (Hint: For any compact set $K \subset \mathbb{R}^d$ the set $\{(u, r) : H(u, r) \cap K = \emptyset\}$ is open in $\mathbb{S}^{d-1} \times \mathbb{R}$.)

Exercise 12.13 Let the measure λ_1 on $\mathbb{H}_{d-1}$ be given by

$$\lambda_1(\cdot) := \int_{\mathbb{S}^{d-1}} \int_{\mathbb{R}} \mathbf{1}\{H(u, r) \in \cdot\} \, dr \, \mathbb{Q}(du), \tag{12.41}$$

where $\gamma > 0$ and the *directional distribution* $\mathbb{Q}$ is a probability measure on the unit sphere. For $t > 0$ let η_t be a Poisson process with intensity measure $\lambda_t := t\lambda$. Show that λ_t satisfies (12.35). Show also that η_t is *stationary* in the sense that the distribution of $\theta_x \eta_t$ does not depend on $x \in \mathbb{R}^d$. Here, for any point $x \in \mathbb{R}^d$ and any measure μ on $\mathbb{H}_{d-1}$, $\theta_x \mu$ denotes the measure $\mu(\{H : H - x \in \cdot\})$.

13

Random Measures and Cox Processes

A Cox process is a Poisson process with a random intensity measure and hence the result of a doubly stochastic procedure. The study of Cox processes requires the concept of a random measure, a natural and important generalisation of a point process. The distribution of a Cox process determines that of its random intensity measure. Mecke's characterisation of the Poisson process via a functional integral equation extends to Cox processes.

13.1 Random Measures

A Cox process (here denoted η) can be interpreted as the result of a *doubly stochastic* procedure, which generates first a random measure ξ and then a Poisson process with intensity measure ξ. Before making this idea precise we need to introduce the concept of a random measure. Fortunately, the basic definitions and results are natural extensions of what we have seen before.

Let $(\mathbb{X}, \mathcal{X})$ be a measurable space and let $\mathbf{M}(\mathbb{X}) \equiv \mathbf{M}$ denote the set of all s-finite measures μ on $\mathbb{X}$. Let $\mathcal{M}(\mathbb{X}) \equiv \mathcal{M}$ denote the σ-field generated by all sets of the form

$$\{\mu \in \mathbf{M} : \mu(B) \leq t\}, \quad B \in \mathcal{X}, t \in \mathbb{R}_+.$$

This is the smallest σ-field of subsets of $\mathbf{M}$ such that $\mu \mapsto \mu(B)$ is a measurable mapping for all $B \in \mathcal{X}$.

For the following and later definitions we recall that all random elements are defined on a fixed probability space $(\Omega, \mathcal{F}, \mathbb{P})$.

Definition 13.1 A *random measure* on $\mathbb{X}$ is a random element ξ of the space $(\mathbf{M}, \mathcal{M})$, that is, a measurable mapping $\xi : \Omega \to \mathbf{M}$.

As in the case of point processes, if ξ is a random measure and $B \in \mathcal{X}$, then we denote by $\xi(B)$ the random variable $\omega \mapsto \xi(\omega, B) := \xi(\omega)(B)$. The mapping $(\omega, B) \mapsto \xi(\omega, B)$ is a kernel from Ω to $\mathbb{X}$ with the additional property that the measure $\xi(\omega, \cdot)$ is s-finite for each $\omega \in \Omega$.

The distribution of a random measure ξ on $\mathbb{X}$ is the probability measure $\mathbb{P}_\xi$ on $(\mathbf{M}, \mathcal{M})$, given by $A \mapsto \mathbb{P}(\xi \in A)$. As in the point process setting this distribution is determined by the family of random vectors $(\xi(B_1), \dots, \xi(B_m))$ for pairwise disjoint $B_1, \dots, B_m \in \mathcal{X}$ and $m \in \mathbb{N}$. It is also determined by the *Laplace functional* $L_\xi \colon \mathbb{R}_+(\mathbb{X}) \to [0, 1]$ defined by

$$L_\xi(u) := \mathbb{E}\Big[\exp\Big(-\int u(x)\,\xi(dx)\Big)\Big], \quad u \in \mathbb{R}_+(\mathbb{X}).$$

For ease of reference we summarise these facts with the following Proposition.

Proposition 13.2 *Let η and η' be random measures on $\mathbb{X}$. Then the assertions (i)–(iv) of Proposition 2.10 are equivalent.*

Proof The proof is essentially the same as that of Proposition 2.10. □

The *intensity measure* of a random measure ξ is the measure ν on $\mathbb{X}$ defined by $\nu(B) := \mathbb{E}[\xi(B)]$, $B \in \mathcal{X}$. It satisfies *Campbell's formula*

$$\mathbb{E}\Big[\int \upsilon(x)\,\xi(dx)\Big] = \int \upsilon(x)\,\nu(dx), \quad \upsilon \in \mathbb{R}_+(\mathbb{X}), \tag{13.1}$$

which can be proved in the same manner as Proposition 2.7. For $m \in \mathbb{N}$ we can form the m-th power ξ^m of ξ, that is $\xi^m := \xi(\cdot)^m$. This is a random measure on $\mathbb{X}^m$; see Exercise 13.3. The m-th *moment measure* of a random measure ξ is the measure β_m on $\mathbb{X}^m$ defined by

$$\beta_m(B) := \mathbb{E}[\xi^m(B)], \quad B \in \mathcal{X}^m. \tag{13.2}$$

By definition, any point process is a random measure, even though, in the present generality, we cannot prove that $\mathbf{N}$ is a measurable subset of $\mathbf{M}$. Here is another class of random measures. Later in the book we shall encounter further examples.

Example 13.3 Let $(Y(x))_{x \in \mathbb{X}}$ be a non-negative *random field* on $\mathbb{X}$, that is a family of $\mathbb{R}_+$-valued variables $\omega \mapsto Y(\omega, x)$. Assume that the random field is *measurable*, meaning that $(\omega, x) \mapsto Y(\omega, x)$ is a measurable mapping on $\Omega \times \mathbb{X}$. Let ν be a σ-finite measure on $\mathbb{X}$. Then

$$\xi(B) := \int \mathbf{1}_B(x) Y(x)\,\nu(dx), \quad B \in \mathcal{X},$$

defines a random measure ξ. Indeed, that $\xi(\omega, \cdot)$ is s-finite is a general fact from measure theory and easy to prove. The same is true for the measurablity of $\omega \mapsto \xi(\omega, B)$; see also the paragraph preceding Fubini's theorem (Theorem A.13). The intensity measure of ξ is the measure with density $\mathbb{E}[Y(x)]$ with respect to ν.

13.2 Cox Processes

Given $\lambda \in \mathbf{M}(\mathbb{X})$, let Π_λ denote the distribution of a Poisson process with intensity measure λ. The existence of Π_λ is guaranteed by Theorem 3.6.

Lemma 13.4 *Let $f \in \overline{\mathbb{R}}_+(\mathbf{N})$. Then $\lambda \mapsto \Pi_\lambda(f) = \int f \, d\Pi_\lambda$ is a measurable mapping from $\mathbf{M}$ to $\overline{\mathbb{R}}_+$.*

Proof Assume first that $f(\mu) = \mathbf{1}\{\mu(B_1) = k_1, \ldots, \mu(B_m) = k_m\}$ for some $m \in \mathbb{N}$, $B_1, \ldots, B_m \in \mathcal{X}$ and $k_1, \ldots, k_m \in \mathbb{N}_0$. Let $C_1, \ldots, C_n$ be the atoms of the field generated by $B_1, \ldots, B_m$; see Section A.1. Then $\Pi_\lambda(f)$ is a linear combination of the probabilities

$$\Pi_\lambda(\{\mu \in \mathbf{N} : \mu(C_1) = \ell_1, \ldots, \mu(C_n) = \ell_n\}) = \prod_{i=1}^{n} \frac{\lambda(C_i)^{\ell_i}}{\ell_i!} \exp[-\lambda(C_i)],$$

where $\ell_1, \ldots, \ell_n \in \mathbb{N}_0$. These products are clearly measurable functions of λ. By the monotone class theorem (Theorem A.1) the measurability property extends to $f = \mathbf{1}_A$ for arbitrary $A \in \mathcal{N}$. The general case follows from monotone convergence. □

Definition 13.5 Let ξ be a random measure on $\mathbb{X}$. A point process η on $\mathbb{X}$ is called a *Cox process directed by* ξ if

$$\mathbb{P}(\eta \in A \mid \xi) = \Pi_\xi(A), \quad \mathbb{P}\text{-a.s.}, \; A \in \mathcal{N}. \tag{13.3}$$

Then ξ is called a *directing random measure* of η.

Let ξ be a random measure on $\mathbb{X}$. By Lemma 13.4 the right-hand side of (13.3) is a random variable for any fixed $A \in \mathcal{N}$. The left-hand side is a conditional probability as defined in Section B.4. Equation (13.3) is equivalent to

$$\mathbb{E}[h(\xi)\mathbf{1}\{\eta \in A\}] = \mathbb{E}[h(\xi)\Pi_\xi(A)], \quad A \in \mathcal{N}, \; h \in \mathbb{R}_+(\mathbf{M}). \tag{13.4}$$

By the monotone class theorem and monotone convergence, this equation can be extended to

$$\mathbb{E}[g(\xi, \eta)] = \mathbb{E}\Big[\int g(\xi, \mu) \, \Pi_\xi(d\mu)\Big], \quad g \in \mathbb{R}_+(\mathbf{M} \times \mathbf{N}). \tag{13.5}$$

Given any probability measure $\mathbb{Q}$ on $(\mathbf{M}, \mathcal{M})$, it is always the case that a Cox process directed by a random measure with distribution $\mathbb{Q}$ exists. For instance, $(\Omega, \mathcal{F}, \mathbb{P})$ can be taken as $(\Omega, \mathcal{F}) = (\mathbf{M} \times \mathbf{N}, \mathcal{M} \otimes \mathcal{N})$ and

$$\mathbb{P}(\cdot) := \iint \mathbf{1}\{(\lambda, \mu) \in \cdot\} \, \Pi_\lambda(d\mu) \, \mathbb{Q}(d\lambda).$$

Taking ξ and η as the first and second projections from Ω to $\mathbf{M}$ and $\mathbf{N}$, respectively, it is easy to check that $\mathbb{P}_\xi = \mathbb{Q}$ and that (13.3) holds.

Suppose η is a Cox process on $\mathbb{X}$ directed by a random measure ξ. Taking in (13.5) a function g of product form yields

$$\mathbb{E}[f(\eta) \mid \xi] = \int f(\mu) \, \Pi_\xi(d\mu), \quad \mathbb{P}\text{-a.s.}, \ f \in \mathbb{R}_+(\mathbf{N}). \tag{13.6}$$

As a first consequence we obtain for all $B \in \mathcal{X}$ that

$$\mathbb{E}[\eta(B)] = \mathbb{E}[\mathbb{E}[\eta(B) \mid \xi]] = \mathbb{E}\Big[\int \mu(B) \, \Pi_\xi(d\mu)\Big] = \mathbb{E}[\xi(B)], \tag{13.7}$$

where we have used that $\int \mu(B) \, \Pi_\lambda(d\mu) = \lambda(B)$ for all $\lambda \in \mathbf{M}$. Hence ξ and η have the same intensity measure. The next result deals with the second moment measure.

Proposition 13.6 *Let η be a Cox process on $\mathbb{X}$ with directing random measure ξ. Let $v \in L^1(\nu)$, where ν is the intensity measure of ξ. Then $\mathbb{E}[\eta(v)^2] < \infty$ if and only if $\mathbb{E}[\xi(v)^2] < \infty$ and $v \in L^2(\nu)$. In this case*

$$\mathbb{V}\mathrm{ar}[\eta(v)] = \nu(v^2) + \mathbb{V}\mathrm{ar}[\xi(v)]. \tag{13.8}$$

Proof Let $\lambda \in \mathbf{M}$ and assume that $v \in L^1(\lambda)$ with $v \geq 0$. By (4.26)

$$\int (\mu(v))^2 \, \Pi_\lambda(d\mu) = \lambda(v^2) + (\lambda(v))^2, \tag{13.9}$$

first under the additional assumption $v \in L^2(\lambda)$ but then, allowing for the value ∞ on both sides of (13.9), for general $v \in L^1(\lambda)$. From equation (13.7) and Campbell's formula (13.1) we have $\mathbb{E}[\eta(v)] = \mathbb{E}[\xi(v)] = \nu(v) < \infty$, so that we can apply (13.9) for $\mathbb{P}_\xi$-a.e. λ. It follows that

$$\mathbb{E}[(\eta(v))^2] = \mathbb{E}[\mathbb{E}[(\eta(v))^2 \mid \xi]] = \mathbb{E}[\xi(v^2)] + \mathbb{E}[(\xi(v))^2] = \nu(v^2) + \mathbb{E}[(\xi(v))^2].$$

This shows the asserted equivalence for $v \geq 0$. The general case follows by taking positive and negative parts of v. The formula (13.8) for the variance follows upon subtracting $(\mathbb{E}[\eta(v)])^2 = (\mathbb{E}[\xi(v)])^2$ from both sides. □

If η' is a Poisson process with intensity measure ν, then $\nu(\upsilon^2)$ is the variance of $\eta'(\upsilon)$ by (4.26). If η is a Cox process with intensity measure ν, then (13.8) shows that the variance of $\eta(\upsilon)$ is at least the variance of $\eta'(\upsilon)$. A Cox process is Poisson only if the directing measure is deterministic.

Corollary 4.10 and the law of total expectation show that the factorial moment measures of a Cox process η directed by ξ are given by

$$\mathbb{E}[\eta^{(m)}(\cdot)] = \mathbb{E}[\xi^m(\cdot)], \quad m \in \mathbb{N}, \tag{13.10}$$

where $\mathbb{E}[\xi^m(\cdot)]$ is the m-th moment measure of ξ; see (13.2).

The next result shows that the distribution of a Cox process determines that of the directing random measure.

Theorem 13.7 *Let η and η' be Cox processes on $\mathbb{X}$ with directing random measures ξ and ξ', respectively. Then $\eta \overset{d}{=} \eta'$ if and only if $\xi \overset{d}{=} \xi'$.*

Proof Suppose $\xi \overset{d}{=} \xi'$. By (13.3) we have for each $A \in \mathcal{N}$ that

$$\mathbb{P}(\eta \in A) = \mathbb{E}[\Pi_\xi(A)] = \mathbb{E}[\Pi_{\xi'}(A)] = \mathbb{P}(\eta' \in A),$$

and hence $\eta \overset{d}{=} \eta'$.

Assume, conversely, that $\eta \overset{d}{=} \eta'$. By (13.6) and Theorem 3.9,

$$\mathbb{E}[\exp[-\eta(u)]] = \mathbb{E}\Big[\exp\Big(-\int (1 - e^{-u(x)})\,\xi(dx)\Big)\Big], \quad u \in \mathbb{R}_+(\mathbb{X}). \tag{13.11}$$

A similar equation holds for the pair (η', ξ'). Let $\upsilon\colon \mathbb{X} \to [0, 1)$ be measurable. Taking $u := -\log(1 - \upsilon)$ in (13.11) shows that

$$\mathbb{E}[\exp[-\xi(\upsilon)]] = \mathbb{E}[\exp[-\xi'(\upsilon)]].$$

For any such υ we then also have $\mathbb{E}[\exp[-t\xi(\upsilon)]] = \mathbb{E}[\exp[-t\xi'(\upsilon)]]$ for each $t \in [0, 1]$. Proposition B.5 shows that $\xi(\upsilon) \overset{d}{=} \xi'(\upsilon)$. The latter identity can be extended to arbitrary bounded υ and then to any $\upsilon \in \mathbb{R}_+(\mathbb{X})$ using monotone convergence. An application of Proposition 13.2 concludes the proof. □

13.3 The Mecke Equation for Cox Processes

In this section we extend Theorem 4.1.

Theorem 13.8 *Let ξ be a random measure on $\mathbb{X}$ and let η be a point process on $\mathbb{X}$. Then η is a Cox process directed by ξ if and only if we have for every $f \in \mathbb{R}_+(\mathbb{X} \times \mathbf{N} \times \mathbf{M})$ that*

$$\mathbb{E}\Big[\int f(x, \eta, \xi)\,\eta(dx)\Big] = \mathbb{E}\Big[\int f(x, \eta + \delta_x, \xi)\,\xi(dx)\Big]. \tag{13.12}$$

Proof Suppose that η is a Cox process directed by the random measure ξ and let $f \in \mathbb{R}_+(\mathbb{X}\times\mathbf{N}\times\mathbf{M})$. By taking $g(\xi,\eta) = \int f(x,\eta,\xi)\,\eta(dx)$ in (13.5), the left-hand side of (13.12) equals

$$\mathbb{E}\Big[\iint f(x,\mu,\xi)\,\mu(dx)\,\Pi_\xi(d\mu)\Big] = \mathbb{E}\Big[\iint f(x,\mu+\delta_x,\xi)\,\xi(dx)\,\Pi_\xi(d\mu)\Big],$$

where we have used Theorem 4.1 to get the equality. Applying (13.5) again yields (13.12).

Assume, conversely, that

$$\mathbb{E}\Big[h(\xi)\int g(x,\eta)\,\eta(dx)\Big] = \mathbb{E}\Big[\int h(\xi)g(x,\eta+\delta_x)\,\xi(dx)\Big]$$

for all $g \in \mathbb{R}_+(\mathbb{X}\times\mathbf{N})$ and $h \in \mathbb{R}_+(\mathbf{M})$. This implies that

$$\mathbb{E}\Big[\int g(x,\eta)\,\eta(dx)\;\Big|\;\xi\Big] = \mathbb{E}\Big[\int g(x,\eta+\delta_x)\,\xi(dx)\;\Big|\;\xi\Big], \quad \mathbb{P}\text{-a.s.} \tag{13.13}$$

If there were a regular conditional probability distribution of η given ξ, we could again appeal to Theorem 4.1 to conclude that η is a Cox process. As this cannot be guaranteed in the present generality, we have to resort to the proof of Theorem 4.1 and take disjoint sets $A_1,\dots,A_m$ in $\mathcal{X}$ and $k_1,\dots,k_m \in \overline{\mathbb{N}}_0$ with $k_1 > 0$. Then (13.13) implies, as in the proof of Theorem 4.1, that

$$\begin{aligned} &k_1\mathbb{P}(\eta(A_1)=k_1,\dots,\eta(A_m)=k_m \mid \xi)\\ &\quad = \xi(A_1)\mathbb{P}(\eta(A_1)=k_1-1,\eta(A_2)=k_2,\dots,\eta(A_m)=k_m \mid \xi), \quad \mathbb{P}\text{-a.s.,} \end{aligned}$$

with the measure theory convention $\infty\cdot 0 := 0$. This implies that

$$\mathbb{P}(\eta(A_1)=k_1,\dots,\eta(A_m)=k_m \mid \xi) = \prod_{j=1}^m \frac{\xi(A_j)^{k_j}}{k_j!}\exp[-\xi(A_j)], \quad \mathbb{P}\text{-a.s.,}$$

for all $k_1,\dots,k_m \in \mathbb{N}_0$, where we recall that $(\infty^k)\exp[-\infty] := 0$ for all $k \in \mathbb{N}_0$. This is enough to imply (13.3). □

13.4 Cox Processes on Metric Spaces

In this section we assume that $\mathbb{X}$ is a complete separable metric space. Let $\mathbf{M}_l$ denote the set of all locally finite measures on $\mathbb{X}$. Also let $\mathbf{M}_{ld}$ denote the set of all locally finite measures on $\mathbb{X}$ that are also diffuse.

Lemma 13.9 *The sets $\mathbf{M}_l$ and $\mathbf{M}_{ld}$ are measurable subsets of $\mathbf{M}$.*

Proof Just as in the case of $\mathbf{N}_l$, the measurability of $\mathbf{M}_l$ follows from the fact that a measure μ on $\mathbb{X}$ is locally finite if and only if $\mu(B_n) < \infty$ for all $n \in \mathbb{N}$, where B_n denotes the ball $B(x_0, n)$ for some fixed $x_0 \in \mathbb{X}$.

To prove the second assertion, we note that a measure $\mu \in \mathbf{M}$ is in $\mathbf{M}_{ld}$ if and only if $\mu_{B_n} \in \mathbf{M}_{<\infty} \cap \mathbf{M}_{ld}$ for each $n \in \mathbb{N}$, where $\mathbf{M}_{<\infty}$ is the set of all finite measures on $\mathbb{X}$. Hence it suffices to prove that $\mathbf{M}_{<\infty} \cap \mathbf{M}_{ld}$ is measurable. This follows from Exercise 13.10. Indeed, a measure $\mu \in \mathbf{M}_{<\infty}$ is diffuse if and only if $\tau_n(\mu) = 0$ for each $n \in \{1, \ldots, k(\mu)\}$. □

Definition 13.10 A random measure ξ on a metric space $\mathbb{X}$ is said to be *locally finite* if $\mathbb{P}(\xi(B) < \infty) = 1$ for each bounded $B \in \mathcal{X}$. A locally finite random measure ξ on $\mathbb{X}$ is said to be *diffuse* if $\mathbb{P}(\xi \in \mathbf{M}_{ld}) = 1$.

The next result says that the one-dimensional marginals of a diffuse random measure determine its distribution. Recall from Definition 2.11 that $\mathcal{X}_b$ denotes the system of bounded sets in $\mathcal{X}$.

Theorem 13.11 *Let ξ and ξ' be locally finite random measures on $\mathbb{X}$ and assume that ξ is diffuse. If $\xi(B) \overset{d}{=} \xi'(B)$ for all $B \in \mathcal{X}_b$, then $\xi \overset{d}{=} \xi'$.*

Proof Let η and η' be Cox processes directed by ξ and ξ', respectively. Then η and η' are locally finite. Moreover, by Proposition 6.9 and Lemma 13.9 the point process η is simple. Assuming $\xi(B) \overset{d}{=} \xi'(B)$ for all $B \in \mathcal{X}_b$ we have for all such B that $\mathbb{P}(\eta(B) = 0) = \mathbb{P}(\eta'(B) = 0)$. Let

$$\eta^* := \int \eta'\{x\}^{\oplus} \mathbf{1}\{x \in \cdot\}\, \eta'(dx)$$

be the simple point process with the same support as η', where we recall from Exercise 8.7 that $a^{\oplus} := \mathbf{1}\{a \neq 0\} a^{-1}$ is the generalised inverse of $a \in \mathbb{R}$. Then $\eta \overset{d}{=} \eta^*$ by Theorem 6.11. Since $\mathbb{E}[\eta(B)] = \mathbb{E}[\eta'(B)]$ we have in particular that $\mathbb{E}[\eta'(B)] = \mathbb{E}[\eta^*(B)]$, and since $\eta'(B) \geq \eta^*(B)$ we obtain the relation $\mathbb{P}(\eta'(B) = \eta^*(B)) = 1$. By the separability of $\mathbb{X}$, there is an at most countably infinite π-system $\mathcal{H} \subset \mathcal{X}$ containing only bounded sets and generating $\mathcal{X}$. We have just proved that η and η' almost surely coincide on $\mathcal{H}$. Theorem A.5 shows that this extends to $\mathcal{X}$ and hence that $\eta \overset{d}{=} \eta'$. Now we can conclude the proof using Theorem 13.7. □

13.5 Exercises

Exercise 13.1 Let $f \in \mathbb{R}_+(\mathbb{X} \times \Omega)$ and let ξ be a random measure on $\mathbb{X}$. For $\omega \in \Omega$ and $B \in \mathcal{X}$ define $\xi'(\omega, B) := \int \mathbf{1}\{x \in B\} f(x, \omega)\, \xi(\omega, dx)$. Show that $\omega \mapsto \xi'(\omega, \cdot)$ is a random measure on $\mathbb{X}$.

Exercise 13.2 Let η be a Cox process directed by ξ and let $f \in \mathbb{R}_+(\mathbb{X}\times\mathbf{N})$. Show that

$$\mathbb{E}\Big[\int f(x,\eta)\,\xi(dx)\,\Big|\,\xi\Big] = \int \mathbb{E}[f(x,\eta)\mid\xi]\,\xi(dx), \quad \mathbb{P}\text{-a.s.} \tag{13.14}$$

(Hint: Use the monotone class theorem.)

Exercise 13.3 Let ξ be a random measure on $\mathbb{X}$ and let $m \in \mathbb{N}$. Show that ξ^m is a random measure on $\mathbb{X}^m$.

Exercise 13.4 Let η be a Cox process directed by a random measure of the form $Y\rho$, where $Y \geq 0$ is a random variable and ρ is an s-finite measure on $\mathbb{X}$. Then η is called a *mixed Poisson process*. Assume now, in particular, that Y has a *Gamma distribution* with shape parameter a and scale parameter b, where $a, b > 0$; see (1.27). Let $B \in \mathcal{X}$ with $0 < \rho(B) < \infty$; show that

$$\mathbb{P}(\eta(B) = n) = \frac{\Gamma(n+a)}{\Gamma(n+1)\Gamma(a)}\left[\frac{\rho(B)}{b+\rho(B)}\right]^n\left[\frac{b}{b+\rho(B)}\right]^a, \quad n \in \mathbb{N}_0.$$

This is a negative binomial distribution with parameters $p = \frac{b}{b+\rho(B)}$ and a; see (1.22).

Exercise 13.5 Let η be a Cox process directed by ξ and let $B \in \mathcal{X}$. Show that η_B is a Cox process directed by ξ_B.

Exercise 13.6 Let ξ be a random measure on $\mathbb{X}$ with $\mathbb{P}(0 < \xi(B) < \infty) = 1$ for some $B \in \mathcal{X}$. For $m \in \mathbb{N}$, define a probability measure $\mathbb{Q}_m$ on $\mathbf{M} \times \mathbb{X}^m$ by

$$\iint \mathbf{1}\{(\lambda, x_1, \dots, x_m) \in \cdot\}\,(\lambda(B)^m)^{\oplus}(\lambda_B)^m(d(x_1,\dots,x_m))\,\mathbb{V}(d\lambda),$$

where $\mathbb{V}$ is the distribution of ξ. Show that there is a unique probability measure $\mathbb{Q}$ on $\mathbf{M}\times\mathbb{X}^\infty$ satisfying $\mathbb{Q}(A \times B \times \mathbb{X}^\infty) = \mathbb{Q}_m(A \times B)$ for all $m \in \mathbb{N}$, $A \in \mathcal{M}$ and $B \in \mathcal{X}^m$. (Hint: Use Theorem B.2.)

Exercise 13.7 Let ξ be a random measure on $\mathbb{X}$ such that $\mathbb{P}(\xi(B_n) < \infty) = 1$ for some measurable partition $\{B_n : n \in \mathbb{N}\}$ of $\mathbb{X}$. Construct a suitable probability space supporting a proper Cox process directed by ξ.

Exercise 13.8 Let η be a Cox process directed by a random measure ξ satisfying the assumption of Exercise 13.7. Assume η to be proper and let χ be a K-marking of η, where K is a probability kernel from $\mathbb{X}$ to some measurable space $(\mathbb{Y}, \mathcal{Y})$. Show that χ has the distribution of a Cox process directed by the random measure $\tilde{\xi} := \iint \mathbf{1}\{(x,y) \in \cdot\}\,K(x,dy)\,\xi(dx)$. Show in particular that, for any measurable $p\colon \mathbb{X} \to [0,1]$, a p-thinning of η has the distribution of a Cox process directed by the random measure $p(x)\xi(dx)$. (Hint: Use Proposition 5.4 and (13.11).)

Exercise 13.9 Suppose that the assumptions of Exercise 13.8 are satisfied. Assume in addition that the sequence $(Y_n)_{n\geq 1}$ used in Definition 5.3 to define the K-marking of η is conditionally independent of ξ given $(\kappa, (X_n)_{n\leq\kappa})$. Show that χ is a Cox process directed by $\tilde{\xi}$.

Exercise 13.10 Let $\mathbb{X}$ be a Borel space and let $\mathbf{M}_{<\infty}$ denote the space of all finite measures on $\mathbb{X}$. Show that there are measurable mappings $\tau_n\colon \mathbf{M}_{<\infty} \to (0,\infty)$ and $\pi_n\colon \mathbf{M}_{<\infty} \to \mathbb{X}$, $n \in \mathbb{N}$, along with measurable mappings $k\colon \mathbf{M}_{<\infty} \to \mathbb{N}_0$ and $D\colon \mathbf{M}_{<\infty} \to \mathbf{M}_{<\infty}$ such that $\sum_{n=1}^{k(\mu)} \delta_{\pi_n(\mu)}$ is simple, $D(\mu)$ is diffuse for each $\mu \in \mathbf{M}_{<\infty}$ and

$$\mu = D(\mu) + \sum_{n=1}^{k(\mu)} \tau_n(\mu)\delta_{\pi_n(\mu)}, \qquad \mu \in \mathbf{M}_{<\infty}. \tag{13.15}$$

(Hint: Extend the method used in the proof of Proposition 6.2.)

Exercise 13.11 Assume that $\mathbb{X}$ is a CSMS. Show that $\mathbf{N}_l$ is a measurable subset of $\mathbf{M}_l$. (Hint: Use that $\mathcal{X}$ has a countable generator).

Exercise 13.12 Let $d \in \mathbb{N}$. Given $x \in \mathbb{R}^d$ and $\mu \in \mathbf{M}(\mathbb{R}^d)$, define $\theta_x\mu \in \mathbf{M}(\mathbb{R}^d)$ as in (8.1). A random measure ξ on $\mathbb{R}^d$ is said to be *stationary* if $\theta_x\xi \overset{d}{=} \xi$ for each $x \in \mathbb{R}^d$. In this case the number $\mathbb{E}[\xi[0,1]^d]$ is called the *intensity* of ξ. Show that the intensity measure of a stationary random measure with finite intensity is a multiple of Lebesgue measure.

Exercise 13.13 Let $d \in \mathbb{N}$ and let η be a stationary random measure on $\mathbb{R}^d$ with finite intensity. Show that Theorem 8.14 remains valid with an appropriately defined invariant σ-field $\mathcal{I}_\eta$.

Exercise 13.14 Let $p \in (0,1)$. Suppose that η_p (resp. η'_p) is a p-thinning of a proper point process η (resp. η'). Assume that $\eta_p \overset{d}{=} \eta'_p$ and show that $\eta \overset{d}{=} \eta'$ (Hint: Use Exercise 5.4 and the proof of Theorem 13.7.)

Exercise 13.15 Let $p \in (0,1)$ and let η be a proper point process with s-finite intensity measure. Suppose that η_p is a p-thinning of η and that η_p and $\eta - \eta_p$ are independent. Show that η is a Poisson process. (Hint: Use Exercise 5.9 to show that η_p satisfies the Mecke equation. Then use Exercise 13.14.)

14

Permanental Processes

For $\alpha > 0$, an α-permanental point process is defined by explicit algebraic formulae for the densities of its factorial moment measures. These densities are the α-permanents arising from a given non-negative definite kernel function K and determine the distribution. If 2α is an integer, then an α-permanental process can be constructed as a Cox process, whose directing random measure is determined by independent Gaussian random fields with covariance function K. The proof of this fact is based on moment formulae for Gaussian random variables. The Janossy measures of a permanental Cox process are given as the α-permanent of a suitably modified kernel function. The number of points in a bounded region is a sum of independent geometric random variables.

14.1 Definition and Uniqueness

In this chapter the state space $(\mathbb{X}, \mathcal{X})$ is assumed to be a locally compact separable metric space equipped with its Borel σ-field; see Section A.2. A set $B \subset \mathbb{X}$ is said to be *relatively compact* if its closure is compact. Let $\mathcal{X}_{rc}$ denote the system of all relatively compact $B \in \mathcal{X}$. We fix a measure ν on $\mathbb{X}$ such that $\nu(B) < \infty$ for every $B \in \mathcal{X}_{rc}$. Two important examples are $\mathbb{X} = \mathbb{R}^d$ with ν being Lebesgue measure and $\mathbb{X} = \mathbb{N}$ with ν being counting measure.

Let $m \in \mathbb{N}$ and let $\sigma \in \Sigma_m$ be a permutation of $[m]$. A *cycle* of σ is a k-tuple $(i_1 \dots i_k) \in [m]^k$ with distinct entries (written without commas), where $k \in [m]$, $\sigma(i_j) = i_{j+1}$ for $j \in [k-1]$ and $\sigma(i_k) = i_1$. In this case $(i_2 \dots i_k\, i_1)$ denotes the same cycle, that is cyclic permutations of a cycle are identified. The number k is called the *length* of the cycle. Let $\#\sigma$ denote the number of cycles of $\sigma \in \Sigma_m$. For $r \in [m]$ let $\Sigma_m^{(r)}$ be the set of permutations of $[m]$ with exactly r cycles.

Definition 14.1 Let $m \in \mathbb{N}$. Let $A = (a_{i,j})_{i,j\in[m]}$ be an $(m \times m)$-matrix of real numbers. For $r \in [m]$ let

$$\mathrm{per}^{(r)}(A) := \sum_{\sigma \in \Sigma_m^{(r)}} \prod_{i=1}^{m} a_{i,\sigma(i)}.$$

For $\alpha \in \mathbb{R}$ the *α-permanent* of A is defined by

$$\mathrm{per}_\alpha(A) := \sum_{\sigma \in \Sigma_m} \alpha^{\#\sigma} \prod_{i=1}^{m} a_{i,\sigma(i)} = \sum_{r=1}^{m} \alpha^r \, \mathrm{per}^{(r)}(A).$$

The number $\mathrm{per}_1(A)$ is called the *permanent* of A.

We note that $\mathrm{per}_{-1}(-A)$ is the determinant of A.

Let $\mathbb{X}^*$ denote the support of ν. In this chapter we fix a symmetric jointly continuous function (sometimes called a *kernel*) $K \colon \mathbb{X}^* \times \mathbb{X}^* \to \mathbb{R}$. We assume that K is non-negative definite; see (B.11). We extend K to $\mathbb{X} \times \mathbb{X}$ by setting $K(x,y) := 0$ for $(x,y) \notin \mathbb{X}^* \times \mathbb{X}^*$. For $m \in \mathbb{N}$ and $x_1, \dots, x_m \in \mathbb{X}$ we define $[K](x_1, \dots, x_m)$ to be the $(m \times m)$-matrix with entries $K(x_i, x_j)$. Since K is continuous it is bounded on compact subsets of $\mathbb{X}^*$, so that

$$\sup\{|K(x,y)| : x, y \in B\} < \infty, \quad B \in \mathcal{X}_{rc}. \tag{14.1}$$

Definition 14.2 Let $\alpha > 0$. A point process η on $\mathbb{X}$ is said to be an *α-permanental* process with kernel K (with respect to ν) if for every $m \in \mathbb{N}$ the m-th factorial moment measure of η is given by

$$\alpha_m(d(x_1, \dots, x_m)) = \mathrm{per}_\alpha([K](x_1, \dots, x_m))\, \nu^m(d(x_1, \dots, x_m)). \tag{14.2}$$

If η is α-permanental with kernel K, then the case $m = 1$ of (14.2) implies that the intensity measure of η is given by

$$\mathbb{E}[\eta(B)] = \int_B \alpha K(x,x)\, \nu(dx), \quad B \in \mathcal{X}. \tag{14.3}$$

If $B \in \mathcal{X}_{rc}$, then the relation (14.1) and our assumption $\nu(B) < \infty$ imply that $\mathbb{E}[\eta(B)] < \infty$ and in particular $\mathbb{P}(\eta(B) < \infty) = 1$.

Proving existence of an α-permanental point process with a given kernel is a non-trivial task and is one of the themes of the rest of this chapter. We note first that the distribution of a permanental process is uniquely determined by its kernel.

Proposition 14.3 *Let $\alpha > 0$. Suppose that η and η' are α-permanental processes with kernel K. Then $\eta \overset{d}{=} \eta'$.*

Proof Let $B \subset \mathbb{X}$ be compact. It follows from (14.1) and $\nu(B) < \infty$ that

$$\int_{B^m} |\operatorname{per}_\alpha([K](x_1, \dots, x_m))| \, \nu^m(d(x_1, \dots, x_m)) \le m!(\max\{\alpha, 1\})^m c^m \nu(B)^m$$

for some $c > 0$. Since $\mathbb{X}$ is σ-compact (by Lemma A.20), the assertion follows from Proposition 4.12. □

We continue with a simple example.

Example 14.4 Suppose that $\mathbb{X} = \{1\}$ is a singleton and that $\nu\{1\} = 1$. A point process η on $\mathbb{X}$ can be identified with the random variable $\eta\{1\}$. Set $\gamma := K(1, 1) \ge 0$. Let $\alpha > 0$ and $m \in \mathbb{N}$. For $(x_1, \dots, x_m) := (1, \dots, 1)$ and with E_m denoting the $(m \times m)$-matrix with all entries equal to 1, we have that

$$\begin{aligned} \operatorname{per}_\alpha([K](x_1, \dots, x_m)) &= \gamma^m \operatorname{per}_\alpha(E_m) \\ &= \gamma^m \alpha(\alpha + 1) \cdots (\alpha + m - 1), \end{aligned} \tag{14.4}$$

where the second identity follows from Exercise 14.1. By Definition 14.2, (4.7) and (14.4), a point process η on $\mathbb{X}$ is α-permanental with kernel K if

$$\mathbb{E}[(\eta(\mathbb{X}))_m] = \gamma^m \alpha(\alpha + 1) \cdots (\alpha + m - 1), \quad m \ge 1. \tag{14.5}$$

Exercise 14.2 shows that these are the factorial moments of a negative binomial distribution with parameters α and $1/(1+\gamma)$. Hence an α-permanental process with kernel K exists. Proposition 4.12 shows that its distribution is uniquely determined.

14.2 The Stationary Case

In this section we briefly discuss stationary permanental processes, assuming that $\mathbb{X} = \mathbb{R}^d$, $\nu = \lambda_d$ and the translation invariance

$$K(x, y) = K(0, y - x), \quad x, y \in \mathbb{R}^d. \tag{14.6}$$

For a given $\alpha > 0$ let η be an α-permanental process with kernel K (with respect to λ_d). Let $x \in \mathbb{R}^d$ and $n \in \mathbb{N}$. Since

$$(\theta_x \eta)^{(n)} = \int \mathbf{1}\{(x_1 - x, \dots, x_n - x) \in \cdot\} \, \eta^{(n)}(d(x_1, \dots, x_n)),$$

it follows from Definition 14.2 and (14.6) (and a change of variables) that $\theta_x \eta$ is also α-permanental with the same kernel K. Therefore Proposition

14.3 shows that η is stationary. It follows from (14.3) that η has intensity $\alpha K(0,0)$. Since

$$\operatorname{per}_\alpha([K](x,y)) = \alpha^2 K(x,x)K(y,y) + \alpha K(x,y)K(y,x),$$

we can use (14.2) for $m = 2$ and a change of variables in (8.7), to see that the second reduced factorial moment measure $\alpha_2^!$ of η is given by

$$\alpha_2^!(B) = \int_B (\alpha^2 K(0,0)^2 + \alpha K(0,x)^2)\,dx, \quad B \in \mathcal{B}^d.$$

By Definition 8.9 the pair correlation function ρ_2 can be chosen as

$$\rho_2(x) = 1 + \frac{K(0,x)^2}{\alpha K(0,0)^2}, \quad x \in \mathbb{R}^d. \tag{14.7}$$

Hence permanental processes are attractive; see (8.10).

14.3 Moments of Gaussian Random Variables

We now establish a combinatorial formula for mixed moments of normal random variables; later, we shall use this to show the existence in general of permanental processes. For $\ell \in \mathbb{N}$ let $M(\ell)$ denote the set of matchings of $[\ell]$ and note that $M(\ell)$ is empty if ℓ is odd. Recall that Π_ℓ denotes the system of all partitions of $[\ell]$. For any $\pi \in \Pi_\ell$ (in particular for π a matching) we denote the blocks of π (in some arbitrary order) as $J_1(\pi), \dots, J_{|\pi|}(\pi)$. In the case that π is a matching we write $J_r(\pi) = \{k_r(\pi), k'_r(\pi)\}$.

Lemma 14.5 (Wick formula) *Let $\ell \in \mathbb{N}$ and let $f_1, \dots, f_\ell$ be functions on $\mathbb{N}$ such that $\sum_{m=1}^\infty f_i(m)^2 < \infty$ for all $i \in \{1, \dots, \ell\}$. Let $Y_1, Y_2, \dots$ be independent standard normal random variables and define*

$$X_i := \sum_{m=1}^\infty Y_m f_i(m), \quad i = 1, \dots, \ell. \tag{14.8}$$

Then

$$\mathbb{E}\Big[\prod_{i=1}^\ell X_i\Big] = \sum_{\pi \in M(\ell)} \prod_{i=1}^{\ell/2} \Big(\sum_{m=1}^\infty f_{k_i(\pi)}(m) f_{k'_i(\pi)}(m)\Big). \tag{14.9}$$

Proof Let λ_0 denote the counting measure on $\mathbb{N}$. We first show that both sides of (14.9) depend on each of the individual functions $f_1, \dots, f_\ell$ in an $L^2(\lambda_0)$-continuous way. To see this we let $f_1^{(n)} \in L^2(\lambda_0)$ be such that $f_1^{(n)} \to f_1$ in $L^2(\lambda_0)$ as $n \to \infty$. For each $n \in \mathbb{N}$ define $X_1^{(n)}$ by (14.8) with $i = 1$ and $f_1^{(n)}$ in place of f_1. It is easy to check that $\mathbb{E}[(X_1 - X_1^{(n)})^2] \to 0$ as

$n \to \infty$. Since normal random variables have moments of all orders, so does $\prod_{i=2}^{\ell} X_i$. By the Cauchy–Schwarz inequality,

$$\lim_{n\to\infty} \mathbb{E}\Big[|X - X_1^{(n)}| \prod_{i=2}^{\ell} X_i\Big] = 0,$$

so that $\mathbb{E}[X_1^{(n)} \prod_{i=2}^{\ell} X_i] \to \mathbb{E}[\prod_{i=1}^{\ell} X_i]$ as $n \to \infty$. By another application of the Cauchy–Schwarz inequality, the right-hand side of (14.9) also depends on f_1 in an $L^2(\lambda_0)$-continuous manner.

To prove (14.9) we can now assume that $f_i(m) = 0$ for all $i \in [\ell]$ and all sufficiently large m. We then obtain

$$\mathbb{E}\Big[\prod_{i=1}^{\ell} X_i\Big] = \sum_{m_1,\dots,m_\ell=1}^{\infty} f_1(m_1)\cdots f_\ell(m_\ell)\,\mathbb{E}[Y_{m_1}\cdots Y_{m_\ell}].$$

Each $(m_1,\dots,m_\ell)$ in the sum determines a partition $\sigma \in \Pi_\ell$ by letting $i,k \in [\ell]$ be in the same block of σ if and only if $m_i = m_k$. Writing the distinct values of $m_1,\dots,m_\ell$ as $n_1,\dots,n_{|\sigma|}$ and using (B.5) and (B.7), we may deduce that

$$\mathbb{E}\Big[\prod_{i=1}^{\ell} X_i\Big] = \sum_{\sigma\in\Pi_\ell} \sum_{n_1,\dots,n_{|\sigma|}\in\mathbb{N}}^{\neq} \prod_{r=1}^{|\sigma|} \Big(|M(|J_r(\sigma)|)| \prod_{i\in J_r(\sigma)} f_i(n_r)\Big)$$

$$= \sum_{\sigma\in\Pi_\ell} c_\sigma \sum_{n_1,\dots,n_{|\sigma|}\in\mathbb{N}}^{\neq} \prod_{r=1}^{|\sigma|} \prod_{i\in J_r(\sigma)} f_i(n_r), \tag{14.10}$$

where $c_\sigma := \prod_{r=1}^{|\sigma|} |M(|J_r(\sigma)|)|$ is the number of matchings $\pi \in M(\ell)$ such that each block of π is contained in a block of σ (that is, π is a refinement of σ).

Now consider the right-hand side of (14.9). By a similar partitioning argument to the above, this equals

$$\sum_{\pi\in M(\ell)} \sum_{m_1,\dots,m_{\ell/2}=1}^{\infty} \prod_{r=1}^{\ell/2} \prod_{i\in J_r(\pi)} f_i(m_r) = \sum_{\pi\in M(\ell)} \sum_{\sigma\in\Pi_{\ell/2}} \sum_{n_1,\dots,n_{|\sigma|}\in\mathbb{N}}^{\neq} \prod_{r=1}^{|\sigma|} \prod_{i\in J_r(\sigma,\pi)} f_i(n_r),$$

where $J_r(\sigma,\pi) := \cup_{j\in J_r(\sigma)} J_j(\pi)$. Each pair (π,σ) in the sum determines a partition $\pi' \in \Pi_\ell$ by $\pi' := \{J_r(\pi,\sigma) : 1 \le r \le |\sigma|\}$, and each $\pi' \in \Pi_\ell$ is obtained from $c_{\pi'}$ such pairs. Hence, the last display equals the expression (14.10) so the result is proved. □

14.4 Construction of Permanental Processes

In this section we construct α-permanental processes in the case $2\alpha \in \mathbb{N}$. In fact, under certain assumptions on K such a process exists for other values of α (as already shown by Example 14.4) but proving this is beyond the scope of this volume. By Theorem B.17 and Proposition B.19 there exists a measurable centred *Gaussian random field* $Z = (Z(x))_{x\in\mathbb{X}}$ with *covariance function* $K/2$, that is $(Z(x_1), \dots, Z(x_m))$ has a multivariate normal distribution for all $m \in \mathbb{N}$ and $(x_1, \dots, x_m) \in \mathbb{X}^m$, $\mathbb{E}[Z(x)] = 0$ for all $x \in \mathbb{X}$, and

$$\mathbb{E}[Z(x)Z(y)] = \frac{K(x,y)}{2}, \quad x, y \in \mathbb{X}. \tag{14.11}$$

Theorem 14.6 *Let $k \in \mathbb{N}$ and let $Z_1, \dots, Z_k$ be independent measurable random fields with the same distribution as Z. Define a random measure ξ on $\mathbb{X}$ by*

$$\xi(B) := \int_B (Z_1(x)^2 + \dots + Z_k(x)^2)\, \nu(dx), \quad B \in \mathcal{X}, \tag{14.12}$$

and let η be a Cox process directed by ξ. Then η is $(k/2)$-permanental.

The proof of Theorem 14.6 is based on an explicit representation of the Gaussian random field Z in terms of independent standard normal random variables. For $B \in \mathcal{X}_{rc}$ let $B^* \subset \mathbb{X}$ be the support of ν_B. Then B^* is a closed subset of the closure of B (assumed to be compact) and therefore compact. By Lemma A.23,

$$\nu(B \setminus B^*) = 0. \tag{14.13}$$

Applying Mercer's theorem (Theorem B.18) to the metric space B^*, we see that there exist $\gamma_{B,j} \geq 0$ and $v_{B,j} \in L^2(\nu_{B^*})$, $j \in \mathbb{N}$, such that

$$\int_B v_{B,i}(x) v_{B,j}(x)\, \nu(dx) = \mathbf{1}\{\gamma_{B,i} > 0\}\mathbf{1}\{i = j\}, \quad i, j \in \mathbb{N}, \tag{14.14}$$

and

$$K(x,y) = \sum_{j=1}^{\infty} \gamma_{B,j} v_{B,j}(x) v_{B,j}(y), \quad x, y \in B^*, \tag{14.15}$$

where the convergence is uniform and absolute. In (14.14) (and also later) we interpret $v_{B,j}$ as functions on $B \cup B^*$ by setting $v_{B,j}(x) := 0$ for $x \in B \setminus B^*$. In view of (14.13) this modification has no effect on our subsequent calculations. A consequence of (14.15) is

$$\sum_{j=1}^{\infty} \gamma_{B,j} v_{B,j}(x)^2 = K(x,x) < \infty, \quad x \in B^* \cup B. \tag{14.16}$$

Combining this with (14.14), we obtain from monotone convergence that

$$\sum_{j=1}^{\infty} \gamma_{B,j} = \int_B K(x,x)\, \nu(dx). \tag{14.17}$$

By (14.1) and $\nu(B) < \infty$, this is a finite number.

Now let $k \in \mathbb{N}$ and let $Y_{i,j}$, $i = 1, \dots, k$, $j \in \mathbb{N}$, be a family of independent random variables with the standard normal distribution. Define independent measurable random fields $Z_{B,i} = (Z_{B,i}(x))_{x \in B^* \cup B}$, $i \in [k]$, by

$$Z_{B,i}(x) := \frac{1}{\sqrt{2}} \sum_{j=1}^{\infty} \sqrt{\gamma_{B,j}}\, Y_{i,j} v_{B,j}(x), \quad x \in B^* \cup B, \tag{14.18}$$

making the convention that $Z_{B,i}(x) := 0$ whenever the series diverges. By (14.16) and Proposition B.7, (14.18) converges almost surely and in $L^2(\mathbb{P})$. Since componentwise almost sure convergence of random vectors implies convergence in distribution, it follows that $Z_{B,1}, \dots, Z_{B,k}$ are centred Gaussian random fields. By the $L^2(\mathbb{P})$-convergence of (14.18) and (14.15),

$$\mathbb{E}[Z_{B,i}(x) Z_{B,i}(y)] = \frac{1}{2} \sum_{j=1}^{\infty} \gamma_{B,j} v_{B,j}(x) v_{B,j}(y) = \frac{K(x,y)}{2}, \quad x, y \in B^*.$$

It follows that

$$((Z_{B,1}(x))_{x \in B^*}, \dots, (Z_{B,k}(x))_{x \in B^*}) \overset{d}{=} ((Z_1(x))_{x \in B^*}, \dots, (Z_k(x))_{x \in B^*}). \tag{14.19}$$

Therefore, when dealing with the restriction of $(Z_1, \dots, Z_k)$ to a given set $B \in \mathcal{X}_{rc}$, we can work with the explicit representation (14.18). Later we shall need the following fact.

Lemma 14.7 *Let $B \in \mathcal{X}_{rc}$ and $W_B(x) := Z_{B,1}(x)^2 + \dots + Z_{B,k}(x)^2$, $x \in B$. Then we have $\mathbb{P}$-a.s. that*

$$\int_B W_B(x)\, \nu(dx) = \frac{1}{2} \sum_{i=1}^{k} \sum_{j=1}^{\infty} \gamma_{B,j} Y_{i,j}^2. \tag{14.20}$$

Proof Let $i \in \{1, \dots, k\}$ and $n \in \mathbb{N}$. Then, by (14.18),

$$\mathbb{E}\left[\int_{B^*}\left(Z_{B,i}(x) - \frac{1}{\sqrt{2}}\sum_{j=1}^{n}\sqrt{\gamma_{B,j}}Y_{i,j}v_{B,j}(x)\right)^2 \nu(dx)\right]$$

$$= \frac{1}{2}\int_B \mathbb{E}\left[\left(\sum_{j=n+1}^{\infty}\sqrt{\gamma_{B,j}}Y_{i,j}v_{B,j}(x)\right)^2\right]\nu(dx)$$

$$= \frac{1}{2}\int_B\left(\sum_{j=n+1}^{\infty}\gamma_{B,j}v_{B,j}(x)^2\right)\nu(dx) = \frac{1}{2}\sum_{j=n+1}^{\infty}\gamma_{B,j} \to 0 \quad \text{as } n \to \infty,$$

where we have used (14.17) to get the convergence. By Proposition B.8,

$$\int_B\left(Z_{B,i}(x) - \frac{1}{\sqrt{2}}\sum_{j=1}^{n}\sqrt{\gamma_{B,j}}Y_{i,j}v_{B,j}(x)\right)^2 \nu(dx) \to 0$$

in probability. Hence we obtain from the Minkowski inequality that

$$\int_B\left(\frac{1}{\sqrt{2}}\sum_{j=1}^{n}\sqrt{\gamma_{B,j}}Y_{i,j}v_{B,j}(x)\right)^2 \nu(dx) \to \int_B Z_{B,i}(x)^2\,\nu(dx)$$

in probability. By the orthogonality relation (14.14),

$$\int_B\left(\frac{1}{\sqrt{2}}\sum_{j=1}^{n}\sqrt{\gamma_{B,j}}Y_{i,j}v_{B,j}(x)\right)^2 \nu(dx) = \frac{1}{2}\sum_{j=1}^{n}\gamma_{B,j}Y_{i,j}^2 \to \frac{1}{2}\sum_{j=1}^{\infty}\gamma_{B,j}Y_{i,j}^2,$$

with almost sure convergence. Summing both limits over $i \in \{1, \dots, k\}$ proves the result. □

Proof of Theorem 14.6 Let $\alpha := k/2$. Let

$$W_k(x) := Z_1(x)^2 + \cdots + Z_k(x)^2, \quad x \in \mathbb{X}.$$

In view of (14.12) and (13.10), we have for all $m \in \mathbb{N}$ and $B \in \mathcal{X}^m$ that

$$\mathbb{E}[\eta^{(m)}(B)] = \int_B \mathbb{E}[W_k(x_1)\cdots W_k(x_m)]\,\nu^m(d(x_1, \dots, x_m)).$$

Hence by Definition 14.2 we have to show that

$$\mathbb{E}[W_k(x_1)\cdots W_k(x_m)] = \mathrm{per}_{k/2}([K](x_1, \dots, x_m)), \quad \nu^m\text{-a.e. } (x_1, \dots, x_m). \tag{14.21}$$

For the rest of the proof we fix $m \in \mathbb{N}$ and $x_1, \dots, x_m \in \mathbb{X}^*$. By (14.13) and (14.19) we can assume that $Z_i = Z_{B,i}$ for every $i \in \{1, \dots, k\}$, where $B \in \mathcal{X}_{rc}$ satisfies $\{x_1, \dots, x_m\} \subset B^*$.

We first prove (14.21) for $k = 1$. Set $(z_1, \ldots, z_{2m}) := (x_1, x_1, \ldots, x_m, x_m)$ and for $i \in [2m]$, $n \in \mathbb{N}$, set $f_i(n) = \sqrt{\gamma_{B,n}/2}\,v_{B,n}(z_i)$. Then, by (14.18),

$$W_1(x_1)\cdots W_1(x_m) = Z_{B,1}(z_1)\cdots Z_{B,1}(z_{2m}) = \prod_{r=1}^{2m}\left(\sum_{n=1}^{\infty}\sqrt{\frac{\gamma_{B,n}}{2}}v_{B,n}(z_r)Y_{1,n}\right)$$
$$= \prod_{r=1}^{2m}\left(\sum_{j=1}^{\infty} f_r(j)Y_{1,j}\right).$$

Hence we obtain from (14.9) that

$$\mathbb{E}[W_1(x_1)\cdots W_1(x_m)] = \sum_{\pi\in M(2m)}\prod_{i=1}^{m}\left(\sum_{n=1}^{\infty} f_{k_i(\pi)}(n)f_{k_i'(\pi)}(n)\right)$$
$$= \sum_{\pi\in M(2m)}\prod_{i=1}^{m}\left(\sum_{n=1}^{\infty}\frac{\gamma_{B,n}}{2}v_{B,n}(z_{k_i(\pi)})v_{B,n}(z_{k_i'(\pi)})\right),$$

so, by (14.15),

$$\mathbb{E}[W_1(x_1)\cdots W_1(x_m)] = 2^{-m}\sum_{\pi\in M(2m)}\prod_{i=1}^{m} K(z_{k_i(\pi),k_i'(\pi)}). \tag{14.22}$$

Any matching $\pi \in M(2m)$ defines a permutation σ of $[m]$. The cycles of this permutation are defined as follows. Partition $[2m]$ into m blocks $J_i := \{2i-1, 2i\}$, $i \in [m]$. Let $j_2 \in [2m]$ such that $\{1, j_2\} \in \pi$. Then $j_2 \in J_{i_2}$ for some $i_2 \in [m]$ and we define $\sigma(1) := i_2$. If $i_2 = 1$, then (1) is the first cycle of σ. Otherwise there is a $j_2' \in J_i \setminus \{j_2\}$ and a $j_3 \in [2m]$ such that $\{j_2, j_3\} \in \pi$. Then $j_3 \in J_{i_3}$ for some $i_3 \in [m]$; we let $\sigma(i_2) := i_3$. If $i_3 = 1$, then $(1\, i_2)$ is the first cycle. Otherwise we continue with this procedure. After a finite number of recursions we obtain the first cycle $(i_1 \ldots i_k)$ for some $k \in [m]$, where $i_1 := 1$. To get the second cycle we remove the blocks $J_{i_1}, \ldots, J_{i_k}$ and proceed as before (starting with the first available block). This procedure yields the cycles of σ after a finite number of steps. (In the case $m = 3$, for instance, the matching $\{\{1,2\},\{3,5\},\{4,6\}\}$ gives the permutation $\sigma(1) = 1$, $\sigma(2) = 3$ and $\sigma(3) = 2$. This permutation has two cycles.) The corresponding contribution to the right-hand side of (14.22) is

$$2^{-m}\prod_{i=1}^{m} K(x_i, x_{\sigma(i)})$$

and depends only on the permutation and not on the matching. Since any permutation σ of $[m]$ with r cycles of lengths $k_1, \ldots, k_r$ corresponds to $2^{k_1-1}\cdots 2^{k_r-1} = 2^{m-r}$ matchings, the case $k = 1$ of (14.21) follows.

Finally consider a general $k \in \mathbb{N}$. Let $(x_1, \dots, x_m) \in \mathbb{X}^m$ such that (14.21) holds for $k = 1$. Then for $k \in \mathbb{N}$, since $(Z_i(x_1)^2, \dots, Z_i(x_m)^2)$, $i \in [k]$, are independent vectors,

$$\mathbb{E}[W_k(x_1)\cdots W_k(x_m)] = \mathbb{E}\Bigg[\sum_{i_1=1}^{k}\cdots\sum_{i_m=1}^{k} Z_{i_1}(x_1)^2\cdots Z_{i_m}(x_m)^2\Bigg]$$

$$= \sum_{\pi\in\Pi_m}\ \sum_{j_1,\dots,j_{|\pi|}\in[k]}^{\neq} \mathbb{E}\Bigg[\prod_{s=1}^{|\pi|}\prod_{r\in J_s(\pi)} Z_{j_r}(x_r)^2\Bigg]$$

$$= \sum_{\pi\in\Pi_m} (k)_{|\pi|} \prod_{s=1}^{|\pi|} \mathbb{E}\Bigg[\prod_{r\in J_s(\pi)} Z_1(x_r)^2\Bigg], \qquad (14.23)$$

and hence, by the case of (14.21) already proved,

$$\mathbb{E}[W_k(x_1)\cdots W_k(x_m)] = \sum_{\pi\in\Pi_m} (k)_{|\pi|} \prod_{J\in\pi} \mathrm{per}_{1/2}([K]((x_j)_{j\in J}))$$

$$= \sum_{\pi\in\Pi_m} (k)_{|\pi|} \sum_{\sigma\in\Sigma_m:\pi\geq\sigma} (1/2)^{\#\sigma} \prod_{i=1}^{m} K(x_i, x_{\sigma(i)}),$$

where $\pi \geq \sigma$ here means that for each cycle of σ all entries in the cycle lie in the same block of π. Hence

$$\mathbb{E}[W_k(x_1)\cdots W_k(x_m)] = \sum_{\sigma\in\Sigma_m}\ \sum_{\pi\in\Pi_m:\pi\geq\sigma} (k)_{|\pi|}(1/2)^{\#\sigma} \prod_{i=1}^{m} K(x_i, x_{\sigma(i)}),$$

and therefore the general case of (14.21) follows from the algebraic identity

$$\sum_{\pi\in\Pi_n} (k)_{|\pi|} = k^n, \quad k, n \in \mathbb{N}.$$

This identity may be proved by the same argument as in (14.23) (namely decomposition of multi-indices according to the induced partition), but now with each of the variables $Z_i(x_j)$ replaced by the unit constant. □

14.5 Janossy Measures of Permanental Cox Processes

In this section we provide a more detailed analysis of the probabilistic properties of the $(k/2)$-permanental Cox processes for $k \in \mathbb{N}$.

Let $B \in \mathcal{X}_{rc}$ and let $\gamma_{B,1}, \dots, \gamma_{B,k}$ and $v_{B,1}, \dots, v_{B,k}$ be chosen as in Section 14.4; see (14.15). Define a symmetric function $\tilde{K}_B \colon B \times B \to \mathbb{R}$ by

$$\tilde{K}_B(x, y) = \sum_{j=1}^{\infty} \tilde{\gamma}_{B,j} v_{B,j}(x) v_{B,j}(y), \quad x, y \in B, \qquad (14.24)$$

where

$$\tilde{\gamma}_{B,j} := \frac{\gamma_{B,j}}{1+\gamma_{B,j}}, \quad j \in \mathbb{N}. \tag{14.25}$$

Since $0 \le \tilde{\gamma}_{B,j} \le \gamma_{B,j}$ this series converges absolutely. (For $x \in B \setminus B^*$ or $y \in B \setminus B^*$ we have $\tilde{K}_B(x, y) = 0$.) Then $\tilde{K}_B$ is non-negative definite; see Exercise 14.4. Given $\alpha > 0$ we define

$$\delta_{B,\alpha} := \prod_{j=1}^{\infty} \frac{1}{(1+\gamma_{B,j})^{\alpha}} = \prod_{j=1}^{\infty} (1-\tilde{\gamma}_{B,j})^{\alpha}. \tag{14.26}$$

For $B \in \mathcal{X}$ and $m \in \mathbb{N}$ recall from Definition 4.6 that $J_{\eta,B,m}$ denotes the Janossy measure of order m of a point process η restricted to B.

Theorem 14.8 *Let $k \in \mathbb{N}$ and set $\alpha := k/2$. Let η be an α-permanental process with kernel K and let $B \in \mathcal{X}_{rc}$. Then we have for each $m \in \mathbb{N}$ that*

$$J_{\eta,B,m}(d(x_1,\dots,x_m)) = \frac{\delta_{B,\alpha}}{m!} \operatorname{per}_{\alpha}([\tilde{K}_B](x_1,\dots,x_m))\, \nu^m(d(x_1,\dots,x_m)). \tag{14.27}$$

Proof Let $\eta_{(B)}$ be a Cox process directed by the random measure $\xi_{(B)}$, where $\xi_{(B)}(dx) := W_B(x)\,\nu(dx)$ and $W_B(x)$ is defined as in Lemma 14.7. In the proof of Theorem 14.6 we have shown that $\eta_{(B)}$ is α-permanental with kernel K_B, where K_B is the restriction of K to $B^* \times B^*$. On the other hand, it follows from Definition 14.2 that η_B has the same property, so that Proposition 14.3 shows that $\eta_B \overset{d}{=} \eta_{(B)}$. Hence we can assume that $\eta_B = \eta_{(B)}$.

Let $m \in \mathbb{N}$ and let $C \in \mathcal{X}^m$. By conditioning with respect to $\xi_{(B)}$ we obtain from (4.21) that

$$\begin{aligned} J_{\eta,B,m}(C) = \frac{1}{m!}\mathbb{E}\Big[\exp\Big(-\int W_B(x)\,\nu(dx)\Big) \qquad\qquad\qquad\qquad & (14.28) \\ \times \int_{B^m} \mathbf{1}\{(x_1,\dots,x_m) \in C\} W_B(x_1)\cdots W_B(x_m)\, \nu^m(d(x_1,\dots,x_m))\Big]. & \end{aligned}$$

Hence we have to show that for ν^m-a.e. $(x_1,\dots,x_m) \in (B^*)^m$ we have

$$\mathbb{E}\Big[\exp\Big(-\int_B W_B(x)\,\nu(dx)\Big) W_B(x_1)\cdots W_B(x_m)\Big] = \delta_{B,\alpha} \operatorname{per}_{\alpha}([\tilde{K}_B](x_1,\dots,x_m)). \tag{14.29}$$

Let $Y'_{i,j} := \frac{\sqrt{\gamma_{B,j}}}{\sqrt{2}} Y_{i,j}$. By Lemma 14.7 the left-hand side of (14.29) can be written as

$$\mathbb{E}\Big[f(Y')\prod_{i=1}^{k}\prod_{j=1}^{\infty}\exp\big[-(Y'_{i,j})^2\big]\Big], \tag{14.30}$$

where Y' denotes the double array $(Y'_{i,j})$ and f is a well-defined function on $(\mathbb{R}^\infty)^k$. Now let $Y''_{i,j} := \frac{\sqrt{\tilde{\gamma}_{B,j}}}{\sqrt{2}} Y_{i,j}$. If $\gamma_{B,j} > 0$, the densities φ_1 and φ_2 of $Y'_{i,j}$ and $Y''_{i,j}$ respectively are related by

$$\varphi_2(t) = \sqrt{1+\gamma_{B,j}}e^{-t^2}\varphi_1(t), \quad t \in \mathbb{R}.$$

Therefore (14.30) equals

$$\mathbb{E}\Big[f(Y'')\prod_{i=1}^{k}\prod_{j=1}^{\infty}(1+\gamma_{B,j})^{-1/2}\Big] = \delta_{B,\alpha}\,\mathbb{E}[f(Y'')], \tag{14.31}$$

where $Y'' := (Y''_{i,j})$. By (14.18) we have

$$f(Y'') = \sum_{i=1}^{k}\Big(\sum_{j=1}^{\infty}\frac{\sqrt{\tilde{\gamma}_{B,j}}}{\sqrt{2}}Y_{i,j}v_{B,j}(x_1)\Big)^2 \cdots \sum_{i=1}^{k}\Big(\sum_{j=1}^{\infty}\frac{\sqrt{\tilde{\gamma}_{B,j}}}{\sqrt{2}}Y_{i,j}v_{B,j}(x_m)\Big)^2,$$

so we can apply (14.21) (with $\tilde{K}_B$ in place of K) to obtain (14.29) and hence the assertion (14.27). □

14.6 One-Dimensional Marginals of Permanental Cox Processes

Let $k \in \mathbb{N}$ and let η be a $(k/2)$-permanental point process with kernel K. Let $B \in \mathcal{X}_{rc}$. With the correct interpretation, (14.27) remains true for $m = 0$. Indeed, as in (14.28) we have

$$\mathbb{P}(\eta(B)=0) = \mathbb{E}\Big[\exp\Big(-\int W_B(x)\,\nu(dx)\Big)\Big] = \prod_{i=1}^{k}\prod_{j=1}^{\infty}\mathbb{E}\Big[\exp\Big(-\frac{\gamma_{B,j}}{2}Y_{i,j}^2\Big)\Big],$$

where we have used Lemma 14.7 to obtain the second identity. Using (B.8) we obtain

$$\mathbb{P}(\eta(B)=0) = \prod_{i=1}^{k}\prod_{j=1}^{\infty}(1+\gamma_{B,j})^{-1/2},$$

that is

$$J_{\eta,B,0} \equiv \mathbb{P}(\eta(B)=0) = \delta_{B,\alpha}, \tag{14.32}$$

where $\alpha := k/2$. Combining (14.27) with (4.18) yields

$$\mathbb{P}(\eta(B) = m) = \frac{\delta_{B,\alpha}}{m!} \int_{B^m} \mathrm{per}_\alpha([\tilde{K}_B](x_1, \dots, x_m))\, \nu^m(d(x_1, \dots, x_m)). \tag{14.33}$$

Therefore our next result yields the probability generating function of $\eta(B)$.

Proposition 14.9 *Let $\alpha > 0$ and $B \in \mathcal{X}_{rc}$. Define*

$$c := \max\{\alpha, 1\}\nu(B) \sup\{K(x, y) : x, y \in B\}. \tag{14.34}$$

Then we have for $s \in [0, c^{-1} \wedge 1)$ that

$$1 + \sum_{m=1}^{\infty} \frac{s^m}{m!} \int_{B^m} \mathrm{per}_\alpha([\tilde{K}_B](x_1, \dots, x_m))\nu^m(d(x_1, \dots, x_m)) = \prod_{j=1}^{\infty}(1 - s\tilde{\gamma}_{B,j})^{-\alpha}.$$

Proof We abbreviate $\tilde{K} := \tilde{K}_B$, $\tilde{\gamma}_i := \tilde{\gamma}_{B,i}$ and $v_i := v_{B,i}$, $i \in \mathbb{N}$. Let $m \in \mathbb{N}$ and $x_1, \dots, x_m \in B$. For $r \in [m]$ the number $\mathrm{per}^{(r)}([\tilde{K}](x_1, \dots, x_m))$ is given in Definition 14.2. For $r = 1$ we have

$$\begin{aligned}
&\int_{B^m} \mathrm{per}^{(1)}([\tilde{K}](x_1, \dots, x_m))\, \nu^m(d(x_1, \dots, x_m)) \\
&\quad = \sum_{\sigma \in \Sigma_m^{(1)}} \int_{B^m} \prod_{i=1}^m \tilde{K}(x_i, x_{\sigma(i)})\, \nu^m(d(x_1, \dots, x_m)) \\
&\quad = \sum_{\sigma \in \Sigma_m^{(1)}} \sum_{j_1, \dots, j_m = 1}^{\infty} \int_{B^m} \prod_{i=1}^m \tilde{\gamma}_{j_i} v_{j_i}(x_i) v_{j_i}(x_{\sigma(i)})\, \nu^m(d(x_1, \dots, x_m)),
\end{aligned} \tag{14.35}$$

where we have used dominated convergence in (14.15) to interchange summation and integration. (This is possible since the convergence (14.15) is uniform and K is bounded on B^2.) By the invariance of ν^m under permutations of the coordinates, each $\sigma \in \Sigma_m^{(1)}$ makes the same contribution to the right-hand side of (14.35). Moreover, there are $(m-1)!$ permutations with exactly one cycle. Therefore (14.35) equals

$$(m-1)! \sum_{j_1, \dots, j_m = 1}^{\infty} \int_{B^m} \prod_{i=1}^m \tilde{\gamma}_{j_i} v_{j_i}(x_i) v_{j_i}(x_{i+1})\, \nu^m(d(x_1, \dots, x_m)), \tag{14.36}$$

where x_{m+1} is interpreted as x_1. By (14.14) and Fubini's theorem the integral in (14.36) vanishes unless $j_1 = \cdots = j_m$. Therefore

$$\int_{B^m} \mathrm{per}^{(1)}([\tilde{K}](x_1, \dots, x_m))\, \nu^m(d(x_1, \dots, x_m)) = (m-1)! \sum_{j=1}^{\infty} \tilde{\gamma}_j^m.$$

Using the logarithmic series $-\log(1-x) = x + x^2/2 + x^3/3 + \cdots$, $x \in [0,1)$, and noting that $\tilde{\gamma}_j < 1$, it follows that

$$\sum_{m=1}^{\infty} \frac{s^m}{m!} \int_{B^m} \mathrm{per}^{(1)}([\tilde{K}](x_1,\dots,x_m))\, \nu^m(d(x_1,\dots,x_m)) = \sum_{m=1}^{\infty} \frac{s^m}{m} \sum_{j=1}^{\infty} \tilde{\gamma}_j^m = D_s,$$

where

$$D_s := -\sum_{j=1}^{\infty} \log(1 - s\tilde{\gamma}_j). \tag{14.37}$$

Since $-\log(1-x) = xR(x)$, $x \in [0,1)$, with $R(\cdot)$ bounded on $[0,1/2]$, and since $\sum_{j=1}^{\infty} \tilde{\gamma}_j < \infty$, the series (14.37) converges.

Now let $r \in \mathbb{N}$. Then

$$D_s^r = \sum_{m_1,\dots,m_r=1}^{\infty} \frac{s^{m_1+\cdots+m_r}}{m_1!\cdots m_r!} \prod_{j=1}^{r} \int \mathrm{per}^{(1)}([\tilde{K}](\mathbf{x}))\, \nu^{m_j}(d\mathbf{x}),$$

where we identify ν with its restriction to B. Therefore

$$\frac{D_s^r}{r!} = \sum_{m=r}^{\infty} \frac{s^m}{m!} \sum_{\substack{m_1,\dots,m_r=1\\ m_1+\cdots+m_r=m}}^{m} \frac{m!}{r!\,m_1!\cdots m_r!} \prod_{j=1}^{r} \int \mathrm{per}^{(1)}([\tilde{K}](\mathbf{x}))\, \nu^{m_j}(d\mathbf{x}).$$

We assert for all $m \geq r$ that

$$\frac{1}{r!} \sum_{\substack{m_1,\dots,m_r=1\\ m_1+\cdots+m_r=m}}^{m} \frac{m!}{m_1!\cdots m_r!} \prod_{j=1}^{r} \int \mathrm{per}^{(1)}([\tilde{K}](\mathbf{x}))\, \nu^{m_j}(d\mathbf{x})$$
$$= \int \mathrm{per}^{(r)}([\tilde{K}](\mathbf{x}))\, \nu^m(d\mathbf{x}). \tag{14.38}$$

Indeed, if $\sigma \in \Sigma_m$ has r cycles with lengths $m_1,\dots,m_r$, then

$$\int \prod_{i=1}^{m} \tilde{K}(x_i, x_{\sigma(i)})\, \nu^m(d(x_1,\dots,x_m))$$
$$= \prod_{j=1}^{r} \frac{1}{(m_j-1)!} \int \mathrm{per}^{(1)}([\tilde{K}](\mathbf{x}))\, \nu^{m_j}(d\mathbf{x}).$$

The factor $\frac{m!}{m_1\cdots m_r}$ on the left-hand side of (14.38) is the number of ways to create an ordered sequence of r cycles of lengths $m_1,\dots,m_r$ that partition $[m]$. The factor $1/r!$ reflects the fact that any permutation of cycles leads to the same permutation of $[m]$. Exercise 14.5 asks the reader to give a complete proof of (14.38).

By (14.38),

$$\frac{D_s^r}{r!} = \sum_{m=r}^{\infty} \frac{s^m}{m!} \int \mathrm{per}^{(r)}([\tilde{K}](\mathbf{x}))\, \nu^m(d\mathbf{x}).$$

It follows for $\alpha > 0$ that

$$\begin{aligned} e^{\alpha D_s} &= \sum_{r=0}^{\infty} \frac{\alpha^r D_s^r}{r!} = 1 + \sum_{m=1}^{\infty} \frac{s^m}{m!} \sum_{r=1}^{m} \int \alpha^r \,\mathrm{per}^{(r)}([\tilde{K}](\mathbf{x}))\, \nu^m(d\mathbf{x}) \\ &= 1 + \sum_{m=1}^{\infty} \frac{s^m}{m!} \int \mathrm{per}_\alpha([\tilde{K}](\mathbf{x}))\, \nu^m(d\mathbf{x}). \end{aligned}$$

Since $\int |\mathrm{per}_\alpha([\tilde{K}](\mathbf{x}))|\, \nu^m(d\mathbf{x}) \le m! c^m$ and $sc < 1$, this series converges absolutely. In view of the definition (14.37) of D_s, this finishes the proof. □

The following theorem should be compared with Example 14.4.

Theorem 14.10 *Let the assumptions of Theorem 14.8 be satisfied. Then*

$$\eta(B) \overset{d}{=} \sum_{j=1}^{\infty} \zeta_j, \tag{14.39}$$

where ζ_j, $j \in \mathbb{N}$, are independent, and ζ_j has for each $j \in \mathbb{N}$ a negative binomial distribution with parameters α and $1/(1+\gamma_{B,j})$.

Proof Define c by (14.34) and let $s \in [0, c^{-1} \wedge 1)$. By (14.32) and (14.33),

$$\mathbb{E}[s^{\eta(B)}] = \delta_{B,\alpha} + \delta_{B,\alpha} \sum_{m=1}^{\infty} \frac{s^m}{m!} \int_{B^m} \mathrm{per}_\alpha([\tilde{K}_B](x_1,\dots,x_m))\, \nu^m(d(x_1,\dots,x_m)).$$

Using Proposition 14.9 and the definition (14.26) of $\delta_{B,\alpha}$ gives

$$\mathbb{E}[s^{\eta(B)}] = \prod_{j=1}^{\infty} (1-\tilde{\gamma}_{B,j})^{\alpha}(1-s\tilde{\gamma}_{B,j})^{-\alpha} = \prod_{j=1}^{\infty} \mathbb{E}[s^{\zeta_j}], \tag{14.40}$$

where we have used (1.24) (with $p = 1-\tilde{\gamma}_{B,j} = 1/(1+\gamma_{B,j})$ and $a = \alpha$) to get the second identity. The assertion now follows from Proposition B.5. □

Equation (14.39) implies that

$$\mathbb{E}[\eta(B)] = \sum_{j=1}^{\infty} \mathbb{E}[\zeta_j] = \alpha \sum_{j=1}^{\infty} \gamma_{B,j},$$

where the expectation of a negative binomial random variable can be obtained from Exercise 14.2 or from (1.24). This identity is in accordance with (14.3) and (14.17). Since $\sum_{j=1}^{\infty} \gamma_{B,j} < \infty$ we have $\sum_{j=1}^{\infty} \mathbb{E}[\zeta_j] < \infty$, so that almost surely only finitely many of the ζ_j are not zero (even though all $\gamma_{B,j}$ might be positive). Exercise 14.6 provides the variance of $\zeta(B)$.

14.7 Exercises

Exercise 14.1 For $m \in \mathbb{N}$ and $k \in [m]$ let $s(m,k)$ be the number of permutations in Σ_m having exactly k cycles. (These are called the *Stirling numbers* of the first kind.) Show that

$$s(m+1,k) = ms(m,k) + s(m,k-1), \quad m \in \mathbb{N},\ k \in [m+1],$$

where $s(m,0) := 0$. Use this to prove by induction that

$$\sum_{k=1}^{m} s(m,k)x^k = x(x+1)\cdots(x+m-1), \quad x \in \mathbb{R}.$$

Exercise 14.2 Let Z be a random variable having a negative binomial distribution with parameters $r > 0$ and $p \in (0,1]$. Show that the factorial moments of Z are given by

$$\mathbb{E}[(Z)_m] = (1-p)^m p^{-m} r(r+1)\cdots(r+m-1), \quad m \geq 1.$$

Exercise 14.3 Let $\alpha > 0$ and let ξ be an α-permanental process. Let $B_1, B_2 \in \mathcal{X}_{rc}$. Show that $\mathbb{C}\mathrm{ov}[\xi(B_1), \xi(B_2)] \geq 0$.

Exercise 14.4 Show that $\tilde{K}_B$ defined by (14.24) is non-negative definite.

Exercise 14.5 Give a complete argument for the identity (14.38).

Exercise 14.6 Let $\alpha := k/2$ for some $k \in \mathbb{N}$ and let η be α-permanental with kernel K. Let $B \in \mathcal{X}_{rc}$ and show that $\eta(B)$ has the finite variance

$$\mathbb{V}\mathrm{ar}[\eta(B)] = \alpha \sum_{j=1}^{\infty} \gamma_{B,j}(1 + \gamma_{B,j}),$$

where the $\gamma_{B,j}$ are as in (14.15).

Exercise 14.7 Let $B \in \mathcal{X}_{rc}$ and assume that there exists $\gamma > 0$ such that $\gamma_{B,j} \in \{0, \gamma\}$ for each $j \in \mathbb{N}$, where the $\gamma_{B,j}$ are as in (14.15). Let $k \in \mathbb{N}$ and let η be $(k/2)$-permanental with kernel K. Show that $\eta(B)$ has a negative binomial distribution and identify the parameters.

Exercise 14.8 Let $m \in \mathbb{N}$, $r > 0$ and $p \in (0,1]$. Let $(\zeta_1, \ldots, \zeta_m)$ be a random element of $\mathbb{N}_0^m$ such that $\zeta := \zeta_1 + \cdots + \zeta_m$ has a negative binomial distribution with parameters r and p. Moreover, assume for all $\ell \geq 1$ that the conditional distribution of $(\zeta_1, \dots, \zeta_m)$ given $\zeta = \ell$ is multinomial with parameters ℓ and $q_1, \dots, q_m \geq 0$, where $q_1 + \cdots + q_m = 1$. Show that

$$\mathbb{E}[s_1^{\zeta_1} \cdots s_m^{\zeta_m}] = \left(\frac{1}{p} - \frac{1-p}{p} \sum_{j=1}^{m} s_j q_j\right)^{-r}, \quad s_1, \dots, s_m \in [0, 1].$$

(Hint: At some stage one has to use the identity

$$\sum_{n=0}^{\infty} \frac{\Gamma(n+r)}{\Gamma(n+1)\Gamma(r)} q^n = (1-q)^{-r}, \quad q \in [0, 1),$$

which follows from the fact that (1.22) is a probability distribution.)

Exercise 14.9 Let $k \in \mathbb{N}$ and suppose that $\eta_1, \dots, \eta_k$ are independent $(1/2)$-permanental processes with kernel K. Show that $\eta_1 + \cdots + \eta_k$ is $(k/2)$-permanental. (Hint: Assume that $\eta_1, \dots, \eta_k$ are Cox processes and use the identity (13.11).)

Exercise 14.10 For each $k \in \mathbb{N}$ let η_k be a $(k/2)$-permanental process with kernel $(2/k)K$. Let $B \in \mathcal{X}_{rc}$ and show that $\eta_k(B) \xrightarrow{d} \zeta_B$ as $k \to \infty$, where ζ_B has a Poisson distribution with mean $\int_B K(x, x)\, \nu(dx)$. (Hint: Use (14.40), Proposition B.10 and (14.17).)

15

Compound Poisson Processes

A compound Poisson process ξ is a purely discrete random measure that is given as an integral with respect to a Poisson process η on a product space. The coordinates of of the points of η represent the positions and weights of the atoms of ξ. Every compound Poisson process is completely independent. Of particular interest is the case where the intensity measure of ξ is of product form. The second factor is then known as the Lévy measure of ξ. The central result of this chapter asserts that every completely independent random measure without fixed atoms is the sum of a compound Poisson process and a deterministic diffuse measure. The chapter concludes with a brief discussion of linear functionals of ξ and the shot noise Cox process.

15.1 Definition and Basic Properties

Let $(\mathbb{Y}, \mathcal{Y})$ be a measurable space. A *compound Poisson process* is a random measure ξ on $\mathbb{Y}$ of the form

$$\xi(B) = \int_{B\times(0,\infty)} r\,\eta(d(y,r)), \quad B \in \mathcal{Y}, \tag{15.1}$$

where η is a Poisson process on $\mathbb{Y} \times (0, \infty)$ with s-finite intensity measure λ. We might think of a point of η as being a point in $\mathbb{Y}$ with the second coordinate representing its weight. Then the integral (15.1) sums the weights of the points lying in B. Proposition 12.1 implies for each $B \in \mathcal{Y}$ that $\mathbb{P}(\xi(B) < \infty) = 1$ if and only if

$$\int_{B\times(0,\infty)} (r \wedge 1)\,\lambda(d(y,r)) < \infty. \tag{15.2}$$

We need to check that ξ really is a random measure in the sense of Definition 13.1.

Proposition 15.1 *Let η be a point process on $\mathbb{Y} \times (0,\infty)$. Then ξ given by* (15.1) *is a random measure on $\mathbb{Y}$.*

Proof First we fix $\omega \in \Omega$ and write $\xi(\omega, B)$ to denote the dependence of the right-hand side of (15.1) on ω. The measure property of $\xi(\omega) := \xi(\omega, \cdot)$ follows from monotone convergence. We show next that $\xi(\omega)$ is s-finite. To this end we write $\eta(\omega) = \sum_{i=1}^{\infty} \eta_i(\omega)$, where the $\eta_i(\omega)$ are finite measures on $\mathbb{Y} \times (0, \infty)$. Then $\xi(\omega) = \sum_{i,j=1}^{\infty} \xi_{i,j}(\omega)$, where

$$\xi_{i,j}(\omega)(B) := \int_{B\times(0,\infty)} \mathbf{1}\{j-1 < r \le j\} r\, \eta_i(\omega)(d(y, r)), \quad B \in \mathcal{Y}.$$

Since $\xi_{i,j}(\omega)(\mathbb{Y}) \le j\, \eta_i(\omega)(\mathbb{Y} \times (0, \infty)) < \infty$, this shows that $\xi(\omega) \in \mathbf{M}(\mathbb{Y})$.

Finally, by Proposition 2.7, for all $B \in \mathcal{Y}$ the mapping $\xi(B) : \Omega \to \overline{\mathbb{R}}_+$ is measurable and hence $\xi : \Omega \to \mathbf{M}$ is measurable. □

Compound Poisson processes have the following remarkable property. In the special case of point processes we have already come across this property in Chapter 3.

Definition 15.2 A random measure ξ on a measurable space $(\mathbb{X}, \mathcal{X})$ is said to be *completely independent* if, for all $m \in \mathbb{N}$ and pairwise disjoint $B_1, \dots, B_m \in \mathcal{X}$, the random variables $\xi(B_1), \dots, \xi(B_m)$ are stochastically independent. In this case ξ is also called a *completely random measure* for short.

Proposition 15.3 *Let ξ be a compound Poisson process on $\mathbb{Y}$. Then ξ is completely independent.*

Proof Assume that ξ is given by (15.1). Then $\xi = T(\eta)$, where the mapping $T : \mathbf{N}(\mathbb{Y} \times (0, \infty)) \to \mathbf{M}(\mathbb{Y})$ is given by $T(\mu)(B) := \int_{B\times(0,\infty)} r\, \mu(d(y, r))$ for $\mu \in \mathbf{N}(\mathbb{Y} \times (0, \infty))$ and $B \in \mathcal{Y}$. Proposition 15.1 shows that T is measurable. Furthermore, $\xi_B = T(\eta_{B\times(0,\infty)})$, so that the result follows from Theorem 5.2. □

Proposition 15.4 *Let ξ be the compound Poisson process given by* (15.1). *Then*

$$L_\xi(u) = \exp\left[-\int (1 - e^{-ru(y)})\, \lambda(d(y, r))\right], \quad u \in \mathbb{R}_+(\mathbb{Y}). \tag{15.3}$$

Proof Let $u \in \mathbb{R}_+(\mathbb{Y})$. It follows from (15.1) and monotone convergence that

$$\exp[-\xi(u)] = \exp\left(-\int ru(y)\, \eta(d(y, r))\right).$$

Then the assertion follows from Theorem 3.9. □

Let ξ be given by (15.1). Then we can apply (15.3) with $u = t\mathbf{1}_B$ for $t \geq 0$ and $B \in \mathcal{Y}$. Since $1 - e^{-tr\mathbf{1}_B(y)} = 0$ for $y \notin B$, we obtain

$$\mathbb{E}[\exp[-t\xi(B)]] = \exp\left[-\int (1 - e^{-tr})\,\lambda(B, dr)\right], \quad t \geq 0, \tag{15.4}$$

where $\lambda(B, \cdot) := \lambda(B \times \cdot)$. If $\mathbb{P}(\xi(B) < \infty) = 1$ then (15.2) and Exercise 12.2 show that the distribution of $\xi(B)$ determines $\lambda(B, \cdot)$.

Of particular interest is the case where $\lambda = \rho_0 \otimes \nu$, where ν is a measure on $(0, \infty)$ satisfying

$$\int (r \wedge 1)\,\nu(dr) < \infty \tag{15.5}$$

and ρ_0 is a σ-finite measure on $\mathbb{Y}$. Then (15.4) simplifies to

$$\mathbb{E}[\exp[-t\xi(B)]] = \exp\left[-\rho_0(B)\int (1 - e^{-tr})\,\nu(dr)\right], \quad t \geq 0, \tag{15.6}$$

where we note that (15.5) is equivalent to $\int (1 - e^{-tr})\,\nu(dr) < \infty$ for one (and then for all) $t > 0$. In particular, ξ is *ρ_0-symmetric*; that is $\xi(B) \overset{d}{=} \xi(B')$ whenever $\rho_0(B) = \rho_0(B')$. Therefore ξ is called a *ρ_0-symmetric compound Poisson process* with *Lévy measure* ν. Since $\varepsilon\mathbf{1}\{r \geq \varepsilon\} \leq r \wedge 1$ for all $r \geq 0$ and $\varepsilon \in (0, 1)$ we obtain from (15.5) that

$$\nu([\varepsilon, \infty)) < \infty, \quad \varepsilon > 0. \tag{15.7}$$

Example 15.5 Let $\eta' = \sum_{n=1}^{\infty} \delta_{T_n}$ be a homogeneous Poisson process on $\mathbb{R}_+$ with intensity $\gamma > 0$; see Section 7.1. Let $(Z_n)_{n\geq 1}$ be a sequence of independent $\mathbb{R}_+$-valued random variables with common distribution $\mathbb{Q}$. Then $\xi := \sum_{n=1}^{\infty} Z_n \delta_{T_n}$ is a λ_+-symmetric compound Poisson process with Lévy measure $\gamma\,\mathbb{Q}$. Indeed, by the marking theorem (Theorem 5.6), $\eta := \sum_{n=1}^{\infty} \delta_{(T_n, Z_n)}$ is a Poisson process on $\mathbb{R}_+ \times (0, \infty)$ with intensity measure $\gamma\lambda_+ \otimes \mathbb{Q}$. Note that

$$\xi[0, t] = \sum_{n=1}^{\eta'[0,t]} Z_n, \quad t \in \mathbb{R}_+.$$

This piecewise constant random process is illustrated by Figure 15.1.

Example 15.6 Let ρ_0 be a σ-finite measure on $\mathbb{Y}$ and let ξ be a ρ_0-symmetric compound Poisson process with Lévy measure

$$\nu(dr) := r^{-1}e^{-br}\,dr, \tag{15.8}$$

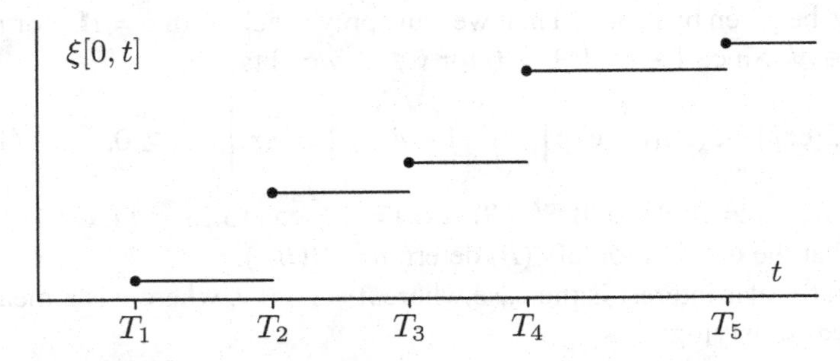

Figure 15.1 Illustration of the process $\xi[0,t]$ from Example 15.5, jumping at the times of a homogeneous Poisson process.

where $b > 0$ is a fixed parameter. Then we obtain from (15.6) and the identity

$$\int_0^\infty (1 - e^{-ur}) r^{-1} e^{-r} \, dr = \log(1 + u), \quad u \geq 0 \tag{15.9}$$

(easily checked by differentiation) that

$$\mathbb{E}[\exp(-t\xi(B))] = (1 + t/b)^{-\rho_0(B)}, \quad t \geq 0, \tag{15.10}$$

provided that $0 < \rho_0(B) < \infty$. Hence $\xi(B)$ has a Gamma distribution (see (1.28)) with shape parameter $a := \rho_0(B)$ and scale parameter b. Therefore ξ is called a *Gamma random measure* with *shape measure* ρ_0 and scale parameter b. An interesting feature of ξ comes from the fact that $\int_0^\infty r^{-1} e^{-br} \, dr = \infty$, which implies that if $\mathbb{Y}$ is a Borel space, ρ_0 is diffuse and $B \in \mathcal{Y}$ satisfies $\rho_0(B) > 0$, then almost surely $\xi\{y\} > 0$ for infinitely many $y \in B$; see Exercise 15.2.

Example 15.7 Let λ_+ denote Lebesgue measure on $[0, \infty)$ and consider a λ_+-symmetric compound Poisson process ξ with Lévy measure ν. The stochastic process $Y := (Y_t)_{t \geq 0} := (\xi[0,t])_{t \geq 0}$ is called a *subordinator* with Lévy measure ν. This process has independent increments in the sense that $Y_{t_1}, Y_{t_2} - Y_{t_1}, \dots, Y_{t_n} - Y_{t_{n-1}}$ are independent whenever $n \geq 2$ and $0 \leq t_1 < \dots < t_n$. Moreover, the increments are *homogeneous*, that is, for any $h > 0$, the distribution of $Y_{t+h} - Y_t = \xi(t, t+h]$ does not depend on $t \geq 0$. If ν is given as in Example 15.6, then Y is called a *Gamma process*. This process is almost surely strictly increasing and everywhere discontinuous.

15.2 Moments of Symmetric Compound Poisson Processes

The moments of a symmetric compound Poisson process can be expressed in terms of the moments of the associated Lévy measure. Before formulating the result, we introduce for $n \in \mathbb{N}$ and $k_1, \ldots, k_n \in \mathbb{N}_0$ the notation

$$\begin{bmatrix} n \\ k_1, \ldots, k_n \end{bmatrix} := \frac{n!}{(1!)^{k_1} k_1! (2!)^{k_2} k_2! \cdots (n!)^{k_n} k_n!}, \tag{15.11}$$

whenever $1k_1 + 2k_2 + \cdots + nk_n = n$. In all other cases this number is defined as 0. This is the number of ways to partition $\{1, \ldots, n\}$ into k_i blocks of size i for $i \in \{1, \ldots, n\}$.

Proposition 15.8 *Let ν be a measure on $(0, \infty)$ satisfying (15.5) and let ρ_0 be an σ-finite measure on $\mathbb{Y}$. Let ξ be a ρ_0-symmetric compound Poisson process with Lévy measure ν. Let $B \in \mathcal{Y}$ and $n \in \mathbb{N}$. Then*

$$\mathbb{E}[\xi(B)^n] = \sum_{k_1,\ldots,k_n \in \mathbb{N}_0} \begin{bmatrix} n \\ k_1, \ldots, k_n \end{bmatrix} \rho_0(B)^{k_1+\cdots+k_n} \prod_{i=1}^n \alpha_i^{k_i}, \tag{15.12}$$

where $\alpha_i := \int r^i \, \nu(dr)$.

Proof The proof is similar to that of Proposition 12.6 and can in fact be derived from that result. We prefer to give a direct argument. We can assume that η is proper. First we use Fubini's theorem to obtain

$$\xi(B)^n = \int_{(B\times(0,\infty))^n} r_1 \cdots r_n \, \eta^n(d((y_1, r_1), \ldots, (y_n, r_n))).$$

Consider now a partition of $[n]$. Within each of the blocks the indices of the integration variables are taken to be equal, while they are taken to be distinct in different blocks. Summing over all partitions and using the symmetry property (A.17) of factorial measures, we obtain

$$\xi(B)^n = \sum_{k_1,\ldots,k_n \in \mathbb{N}_0} \begin{bmatrix} n \\ k_1, \ldots, k_n \end{bmatrix} \int_{(B\times\mathbb{R}_+)^{k_1+\cdots+k_n}} \prod_{i=1}^n \prod_{j_i=k_1+\cdots+k_{i-1}+1}^{k_1+\cdots+k_i} r_{j_i}^i$$
$$\times \eta^{(k_1+\cdots+k_n)}(d((y_1, r_1), \ldots, (y_{k_1+\cdots+k_n}, r_{k_1+\cdots+k_n}))), \tag{15.13}$$

where $k_0 := 0$. Since the integrand in the right-hand side of (15.13) is non-negative, taking expectations, we can use the multivariate Mecke equation (Theorem 4.4) to derive (15.12). □

Corollary 15.9 *Let ρ_0, ν and ξ be as in Proposition 15.8. Suppose that $B \in \mathcal{Y}$ satisfies $0 < \rho_0(B) < \infty$ and let $n \in \mathbb{N}$. Then $\mathbb{E}[\xi(B)^n] < \infty$ if and only if $\int r^n \, \nu(dr) < \infty$.*

Proof Suppose that $\mathbb{E}[\xi(B)^n] < \infty$. Observe that the summand with $k_n = 1$ (and $k_1 = \cdots = k_{n-1} = 0$) on the right-hand side of (15.12) equals $\rho_0(B)^n \alpha_n$. Since $\rho_0(B) > 0$ we deduce that $\alpha_n < \infty$. Conversely, suppose that $\alpha_n < \infty$. Then (15.5) implies that $\alpha_i < \infty$ for each $i \in [n]$. Since $\rho_0(B) < \infty$ we hence obtain from (15.12) that $\mathbb{E}[\xi(B)^n] < \infty$. □

For $n = 1$ we obtain from (15.12) that

$$\mathbb{E}[\xi(B)] = \rho_0(B) \int r\, \nu(dr),$$

which is nothing but Campbell's formula (13.1) for this random measure. In the case $n = 2$ we have

$$\mathbb{E}[\xi(B)^2] = \rho_0(B)^2 \Big(\int r\, \nu(dr) \Big)^2 + \rho_0(B) \int r^2\, \nu(dr)$$

and therefore

$$\mathbb{V}\mathrm{ar}[\xi(B)] = \rho_0(B) \int r^2\, \nu(dr).$$

Exercises 15.4 and 15.5 give two generalisations of (15.12).

15.3 Poisson Representation of Completely Random Measures

The following generalises Definition 6.4.

Definition 15.10 A random measure ξ on $\mathbb{Y}$ is said to be *uniformly σ-finite* if there exist $B_n \in \mathcal{Y}$, $n \in \mathbb{N}$, such that $B_n \uparrow \mathbb{Y}$ as $n \to \infty$ and

$$\mathbb{P}(\xi(B_n) < \infty) = 1, \quad n \in \mathbb{N}.$$

In this case, and if $\mathbb{Y}$ is a Borel space, ξ is said to be *diffuse* if there exists $A \in \mathcal{F}$ such that $\mathbb{P}(A) = 1$ and $\xi(\omega, \{x\}) = 0$ for all $x \in \mathbb{X}$ and all $\omega \in A$.

The second part of Definition 15.10 is justified by Exercise 13.10. The following converse of Proposition 15.1 reveals the significance of Poisson processes in the study of completely independent random measures. As in the point process case, we say that two random measures ξ and ξ' on $\mathbb{Y}$ are *almost surely equal* if there is an $A \in \mathcal{F}$ with $\mathbb{P}(A) = 1$ such that $\xi(\omega) = \xi'(\omega)$ for each $\omega \in A$.

Theorem 15.11 *Suppose that $(\mathbb{Y}, \mathcal{Y})$ is a Borel space and let ξ be a uniformly σ-finite completely independent random measure on $\mathbb{Y}$. Assume that*

$$\xi\{y\} = 0, \quad \mathbb{P}\text{-}a.s., \ y \in \mathbb{Y}. \tag{15.14}$$

Then there is a Poisson process η on $\mathbb{Y} \times \mathbb{R}_+$ with diffuse σ-finite intensity measure, and a diffuse measure ν on $\mathbb{Y}$, such that almost surely

$$\xi(B) = \nu(B) + \int_{B\times(0,\infty)} r\,\eta(d(y,r)), \quad B \in \mathcal{Y}. \tag{15.15}$$

Proof By Exercise 13.10 (a version of Proposition 6.2 that applies to general finite measures) and the assumption that ξ is uniformly σ-finite, we can write

$$\xi = \chi + \sum_{n=1}^{\kappa} Z_n \delta_{Y_n}, \quad \mathbb{P}\text{-a.s.}, \tag{15.16}$$

where χ is a diffuse random measure, κ is an $\overline{\mathbb{N}}_0$-valued random variable, (Y_n) is a sequence of random elements of $\mathbb{Y}$, (Z_n) is a sequence of $(0,\infty)$-valued random variables and $\sum_{n=1}^{\kappa} \delta_{Y_n}$ is a simple point process. Since ξ is uniformly σ-finite, χ has the same property. Let $\eta := \sum_{n=1}^{\kappa} \delta_{(Y_n,Z_n)}$. This is a point process on $\mathbb{Y} \times (0,\infty)$ satisfying $\xi(B) = \chi(B) + \int_{B\times(0,\infty)} r\,\eta(d(y,r))$ for all in $B \in \mathcal{Y}$. The latter identities hold almost surely; but it is no restriction of generality to assume that they hold everywhere on Ω. For each $\varepsilon > 0$ we have

$$\varepsilon\eta(B \times [\varepsilon,\infty)) \le \sum_{n=1}^{\kappa} Z_n \mathbf{1}\{Y_n \in B\} = \xi(B) - \chi(B).$$

Therefore η is uniformly σ-finite. We need to show that χ a.s. equals its intensity measure and that η is a Poisson process. Let $C \in \mathcal{Y} \otimes \mathcal{B}((0,\infty))$ be such that $\mathbb{P}(\eta(C) < \infty) = 1$ and define the simple point process $\eta' := \eta_C(\cdot \times (0,\infty))$. Then $\mathbb{P}(\eta'(\mathbb{Y}) < \infty) = 1$ and equation (15.16) implies that

$$\eta'(B) = \int \mathbf{1}\{y \in B, (y, \xi\{y\}) \in C\}\,\xi(dy), \quad B \in \mathcal{Y},$$

where we have used that

$$\int \mathbf{1}\{(y,\xi\{y\}) \in C\}\,\chi(dy) \le \int \mathbf{1}\{\xi\{y\} > 0\}\,\chi(dy) = 0.$$

In particular, $\eta'(B)$ is a measurable function of ξ_B. Exercise 15.9 shows that $\xi_{B_1}, \dots, \xi_{B_m}$ are independent whenever $B_1, \dots, B_m \in \mathcal{Y}$ are pairwise disjoint. It follows that η' is completely independent. Moreover, since we have $\eta'\{y\} = \xi\{y\}\mathbf{1}\{(y,\xi\{y\}) \in C\}$ for each $y \in \mathbb{Y}$, (15.14) implies that η' (and also η) has a diffuse intensity measure. By Theorem 6.12, η' is a Poisson process. In particular,

$$\mathbb{P}(\eta(C) = 0) = \mathbb{P}(\eta' = 0) = \exp[-\lambda(C)],$$

where λ is the intensity measure of η and where we have used the identity $\eta(C) = \eta'(\mathbb{Y})$. Theorem 6.10 yields that η is a Poisson process. The decomposition (15.16) shows for any $B \in \mathcal{Y}$ that $\chi(B)$ depends only on the restriction ξ_B. Therefore χ is completely independent, so that Proposition 15.12 (to be proved below) shows that χ almost surely equals its intensity measure. □

The proof of Theorem 15.11 has used the following result.

Proposition 15.12 *Let ξ be a diffuse, uniformly σ-finite random measure on a Borel space $(\mathbb{Y}, \mathcal{Y})$. If ξ is completely independent, then there is a diffuse σ-finite measure ν on $\mathbb{Y}$ such that ξ and ν are almost surely equal.*

Proof Given $B \in \mathcal{Y}$, define

$$\nu(B) := -\log \mathbb{E}[\exp[-\xi(B)]],$$

using the convention $-\log 0 := \infty$. Since ξ is completely independent, the function ν is finitely additive. Moreover, if $C_n \uparrow C$ with $C_n, C \in \mathcal{Y}$, then monotone and dominated convergence show that $\nu(C_n) \uparrow \nu(C)$. Hence ν is a measure. Since ξ is diffuse, ν is diffuse. Moreover, since ξ is uniformly σ-finite, ν is σ-finite. Let η be a Cox process directed by ξ and let η' be a Poisson process with intensity measure ν. By Proposition 6.9, η' is simple. Since ξ is diffuse, we can take $A \in \mathcal{F}$ as in Definition 15.10 to obtain from (13.10) that

$$\mathbb{E}[\eta^{(2)}(D_{\mathbb{Y}})] = \mathbb{E}[\xi^2(D_{\mathbb{Y}})] = \mathbb{E}[\mathbf{1}_A \xi^2(D_{\mathbb{Y}})] = 0,$$

where $D_{\mathbb{Y}} := \{(x, y) \in \mathbb{Y}^2 : x = y\}$. Hence Proposition 6.7 shows that η is simple as well. Furthermore,

$$\mathbb{P}(\eta(B) = 0) = \mathbb{E}[\exp[-\xi(B)]] = \exp[-\nu(B)] = \mathbb{P}(\eta'(B) = 0), \quad B \in \mathcal{Y}.$$

By Theorem 6.10, $\eta \overset{d}{=} \eta'$, so that Theorem 13.7 yields $\xi \overset{d}{=} \nu$. Now let $\mathcal{H}$ be a countable π-system generating $\mathcal{Y}$. Then $A := \{\xi(B) = \nu(B) \text{ for all } B \in \mathcal{H}\}$ has full probability. As it is no restriction of generality to assume that the sets B_n from Definition 15.10 are in $\mathcal{H}$, we can apply Theorem A.5 to conclude that $\xi = \nu$ on A. □

If ξ is a random measure on a Borel space $(\mathbb{Y}, \mathcal{Y})$ and $y \in \mathbb{Y}$ is such that

$$\mathbb{P}(\xi\{y\} > 0) > 0, \tag{15.17}$$

then y is called a *fixed atom* of ξ. Theorem 15.11 does not apply to a completely independent random measure ξ with fixed atoms. Exercise 15.8

shows, however, that any such ξ is the independent superposition of a random measure with countably many fixed atoms and a completely independent random measure satisfying (15.14).

15.4 Compound Poisson Integrals

Let ξ be a compound Poisson process on a measurable space $(\mathbb{Y}, \mathcal{Y})$ as defined by (15.1). For $f \in \mathbb{R}(\mathbb{Y})$ we may then try to form the integral $\int f \, d\xi$. Expressing this as

$$\int f(z)\, \xi(dz) = \int r f(y)\, \eta(d(y,r)),$$

we can apply Proposition 12.1 (or Campbell's formula) to see that the integral converges almost surely if $\int r|f(z)|\, \lambda(d(r,z)) < \infty$. For applications it is useful to make f dependent on a parameter $x \in \mathbb{X}$, where $(\mathbb{X}, \mathcal{X})$ is another measurable space. To do so, we take a measurable function $k \in \mathbb{R}(\mathbb{X} \times \mathbb{Y})$ (known as a kernel) and define a random field $(Y(x))_{x \in \mathbb{X}}$ by

$$Y(x) := \int r k(x,y)\, \eta(d(y,r)), \quad x \in \mathbb{X}. \tag{15.18}$$

Here we can drop the assumption that the weights be positive and assume that η is a Poisson process on $\mathbb{Y} \times \mathbb{R}$ such that

$$\int |rk(x,y)|\, \lambda(d(y,r)) < \infty, \quad x \in \mathbb{X}, \tag{15.19}$$

where λ is the intensity measure of η. Then for each $x \in \mathbb{X}$ the right-hand side of (15.18) is almost surely finite and we make the convention $Y(x) := 0$ whenever it is not. Using the monotone class theorem it can be shown that the random field is measurable; see Example 13.3. By Campbell's formula (Proposition 2.7),

$$\mathbb{E}[Y(x)] = \int r k(x,y)\, \lambda(d(y,r)), \quad x \in \mathbb{X}. \tag{15.20}$$

Proposition 15.13 *Let* $(Y(x))_{x \in \mathbb{X}}$ *be the random field given by* (15.18). *Assume that* (15.19) *holds and that*

$$\int r^2 k(x,y)^2\, \lambda(d(y,r)) < \infty, \quad x \in \mathbb{X}. \tag{15.21}$$

Then

$$\mathbb{C}\mathrm{ov}[Y(x), Y(z)] = \int r^2 k(x,y) k(z,y)\, \lambda(d(y,r)), \quad x, z \in \mathbb{X}. \tag{15.22}$$

Proof The random variable $Y(x) - \mathbb{E}[Y(x)]$ takes the form of a Wiener–Itô integral; see (12.4). Hence the result follows from Lemma 12.2. □

Example 15.14 Let $d \in \mathbb{N}$. Consider the random field $(Y(x))_{x \in \mathbb{R}^d}$ given by (15.18) in the case $\mathbb{X} = \mathbb{Y} = \mathbb{R}^d$ with a kernel of the form $k(x, y) = \tilde{k}(x - y)$ for some $\tilde{k} \in L^1(\lambda_d)$. Assume also that the intensity measure λ of η is the product $\lambda_d \otimes \nu$ for some measure ν on $\mathbb{R}$ satisfying $\int |r|\, \nu(dr) < \infty$. Then

$$Y(x) = \int r\tilde{k}(x - y)\, \eta(d(y, r)),$$

and (15.20) shows that

$$\mathbb{E}[Y(x)] = \iint r\tilde{k}(x - y)\, dy\, \nu(dr).$$

The random field $(Y(x))_{x \in \mathbb{R}^d}$ is known as a Poisson driven *shot noise*, while $(Y(x) - \mathbb{E}[Y(x)])_{x \in \mathbb{R}^d}$ is known as a Poisson driven *moving average field*. A specific example is

$$\tilde{k}(x) = \mathbf{1}\{x_1 \geq 0, \ldots, x_d \geq 0\} \exp[-\langle v, x \rangle],$$

where $v \in \mathbb{R}^d$ is a fixed parameter.

Example 15.15 Let λ be a σ-finite measure on $\mathbb{Y} \times (0, \infty)$ and let η be a Poisson process with intensity measure λ. Let the random field $(Y(x))_{x \in \mathbb{X}}$ be given by (15.18), where the kernel k is assumed to be non-negative. Let ρ be a σ-finite measure on $\mathbb{X}$ such that

$$\iint \mathbf{1}\{x \in \cdot\} rk(x, y)\, \lambda(d(y, r))\, \rho(dx) \tag{15.23}$$

is a σ-finite measure on $\mathbb{X}$. Then (15.19) holds for ρ-a.e. $x \in \mathbb{X}$. A Cox process χ driven by the random measure $Y(x)\rho(dx)$ is called a *shot noise Cox process*. It has the intensity measure (15.23).

Proposition 15.16 *Let χ be a shot noise Cox process as in Example 15.15. The Laplace functional of χ is given by*

$$L_\chi(u) = \exp\left[-\int (1 - e^{-ru^*(y)})\, \lambda(d(y, r))\right], \quad u \in \mathbb{R}_+(\mathbb{X}),$$

where $u^*(y) := \int (1 - e^{-u(x)}) k(x, y)\, \rho(dx)$, $y \in \mathbb{Y}$.

Proof By (13.11) and Fubini's theorem, we have for every $u \in \mathbb{R}_+(\mathbb{X})$ that

$$L_\chi(u) = \mathbb{E}\left[\exp\left(-\iint (1 - e^{-u(x)}) rk(x, y)\, \rho(dx)\, \eta(d(y, r))\right)\right].$$

Theorem 3.9 yields the assertion. □

Example 15.17 Let χ be a shot noise Cox process as in Example 15.15 and assume that $\lambda(d(y,r)) = r^{-1}e^{-br}\rho_0(dy)dr$ as in Example 15.6. Proposition 15.16 and (15.10) yield

$$L_\chi(u) = \exp\left[-\int \log\left(1 + \frac{u^*(y)}{b}\right)\rho_0(dy)\right].$$

15.5 Exercises

Exercise 15.1 Let $n \geq 2$ and let

$$\Delta_n := \{(p_1, \dots, p_n) \in [0,1]^n : p_1 + \dots + p_n = 1\}$$

denote the *simplex* of all n-dimensional probability vectors. Lebesgue measure on Δ_n is given by the formula

$$\mu_n(C) := \int_{[0,1]^n} \mathbf{1}\{(x_1, \dots, x_{n-1}, 1 - x_1 - \dots - x_{n-1}) \in C\}\, \lambda_{n-1}(d(x_1, \dots, x_{n-1}))$$

for $C \in \mathcal{B}(\Delta_n)$. This is the Hausdorff measure $\mathcal{H}_{n-1}$ (introduced in Section A.3) restricted to the set Δ_n. The *Dirichlet distribution* with parameters $\alpha_1, \dots, \alpha_n \in (0, \infty)$ is the distribution on Δ_n with density

$$\frac{\Gamma(\alpha_1 + \dots + \alpha_n)}{\Gamma(\alpha_1) \cdots \Gamma(\alpha_n)} x_1^{\alpha_1 - 1} \cdots x_n^{\alpha_n - 1}$$

with respect to μ_n.

Let ξ be a Gamma random measure as in Example 15.6 and assume that the shape measure satisfies $0 < \rho_0(\mathbb{Y}) < \infty$. Then the random measure $\zeta := \xi(\cdot)/\xi(\mathbb{Y})$ can be considered as a random probability measure. Let $B_1, \dots, B_n$ be a measurable partition of $\mathbb{Y}$ such that $\rho_0(B_i) > 0$ for all $i \in \{1, \dots, n\}$. Show that $(\zeta(B_1), \dots, \zeta(B_n))$ has a Dirichlet distribution with parameters $\rho_0(B_1), \dots, \rho_0(B_n)$ and is independent of $\xi(\mathbb{Y})$. (Hint: Use Example 15.6 to express the expectation of a function of $(\zeta(B_1), \dots, \zeta(B_n), \xi(\mathbb{Y}))$ as a Lebesgue integral on $\mathbb{R}_+^{n+1}$ and change variables in an appropriate way.)

Exercise 15.2 Let η be a Poisson process on a Borel space $(\mathbb{X}, \mathcal{X})$ with intensity measure $\infty \cdot \nu$, where ν is a finite diffuse measure on $\mathbb{X}$. Let $B \in \mathcal{X}$ with $\nu(B) > 0$. Show that almost surely there exist infinitely many $x \in B$ with $\eta\{x\} > 0$. (Hint: Use Proposition 6.9.)

Exercise 15.3 Let ξ be a compound Poisson process as in Proposition

15.8 and let $B \in \mathcal{Y}$. Show that

$$\mathbb{E}[\xi(B)^3] = \rho_0(B)^3\Big(\int r\,\nu(dr)\Big)^3 + 3\rho_0(B)^2\Big(\int r\,\nu(dr)\Big)\Big(\int r^2\,\nu(dr)\Big) + \rho_0(B)\int r^3\,\nu(dr).$$

Exercise 15.4 Let ξ be a compound Poisson process as in Proposition 15.8. Let $n \in \mathbb{N}$ and $f \in \mathbb{R}_+(\mathbb{Y})$. Show that

$$\mathbb{E}[\xi(f)^n] = \sum_{r_1,\dots,r_n \in \mathbb{N}_0} \begin{bmatrix} n \\ r_1,\dots,r_n \end{bmatrix} \rho_0(B)^{r_1+\dots+r_n} \prod_{i=1}^{n} \Big(\int f^i\,d\rho_0\Big)^{r_i}\Big(\int r^i\,\nu(dr)\Big)^{r_i}.$$

(Hint: Generalise (15.13). You may assume that η is proper.)

Exercise 15.5 Let η be a Poisson process on $\mathbb{Y}\times\mathbb{R}$ with intensity measure $\lambda = \rho_0 \otimes \nu$, where ρ_0 is σ-finite and ν is a measure on $\mathbb{R}$ with $\nu\{0\} = 0$, and

$$\int (|r| \wedge 1)\,\nu(dr) < \infty.$$

Let $B \in \mathcal{Y}$ satisfy $\rho_0(B) < \infty$. Show that (15.1) converges almost surely. Take $n \in \mathbb{N}$ and prove that $\mathbb{E}[\xi(B)^n] < \infty$ if and only if $\int |r|^n\,\nu(dr) < \infty$ and, moreover, that (15.12) remains true in this case.

Exercise 15.6 Let $p \in (0,1)$. Show that the measure ν on $(0,\infty)$ defined by $\nu(dr) := r^{-1-p}dr$ satisfies (15.5). Let ξ be a ρ_0-symmetric compound Poisson process on $\mathbb{Y}$ with Lévy measure ν for some σ-finite measure ρ_0 on $\mathbb{Y}$. Show for all $B \in \mathcal{Y}$ that

$$\mathbb{E}[\exp[-t\xi(B)]] = \exp\big[-p^{-1}\Gamma(1-p)\rho_0(B)t^p\big], \quad t \geq 0.$$

Exercise 15.7 (Self-similar random measure) Let ξ be as in Exercise 15.6 and assume in addition that $\mathbb{Y} = \mathbb{R}^d$ for some $d \in \mathbb{N}$ and that ρ_0 is Lebesgue measure. Show that ξ is *self-similar* in the sense that $\xi_c \overset{d}{=} \xi$ for any $c > 0$, where $\xi_c(B) := c^{-d/p}\xi(cB)$, $B \in \mathcal{B}^d$. (Hint: Use Exercise 15.6.)

Exercise 15.8 Let ξ be a uniformly σ-finite random measure on a Borel space $\mathbb{Y}$. Show that the set $B \subset \mathbb{Y}$ of fixed atoms of ξ is at most countable. Assume in addition that ξ is completely independent. Show that there are independent random variables Z_x, $x \in B$, and a completely independent random measure ξ_0 without fixed atoms such that

$$\xi = \xi_0 + \sum_{x\in B} Z_x \delta_{Z_x}, \quad \mathbb{P}\text{-a.s.}$$

Exercise 15.9 Let ξ be a completely independent random measure on $(\mathbb{Y}, \mathcal{Y})$ and let $B_1, \dots, B_m \in \mathcal{Y}$ be pairwise disjoint. Show that $\xi_{B_1}, \dots, \xi_{B_m}$ are independent random measures. (Hint: Reduce first to the case $m = 2$. Then use the monotone class theorem; see the proof of Proposition 8.12.)

Exercise 15.10 Let the random field $(Y(x))_{x \in \mathbb{R}^d}$ be as in Example 15.14. Show that this field is *stationary*, that is, for any $m \in \mathbb{N}$ and $x_1, \dots, x_m \in \mathbb{R}^d$, the distribution of $(Y(x_1 + x), \dots, Y(x_m + x))$ does not depend on $x \in \mathbb{R}^d$.

Exercise 15.11 Let χ be a shot noise Cox process as in Example 15.15 and assume moreover that

$$\iint \mathbf{1}\{x \in \cdot\} r^2 k(x, y)^2 \, \lambda(d(y, r)) \, \rho(dx)$$

is a σ-finite measure on $\mathbb{X}$. Show that the second factorial moment measure α_2 of χ is given by $\alpha_2(d(x_1, x_2)) = g_2(x_1, x_2) \rho^2(d(x_1, x_2))$, where

$$\begin{aligned} g_2(x_1, x_2) := & \int r^2 k(x_1, y) k(x_2, y) \, \lambda(d(y, r)) \\ & + \Big(\int r k(x_1, y) \, \lambda(d(y, r)) \Big) \Big(\int r k(x_2, y) \, \lambda(d(y, r)) \Big). \end{aligned}$$

Compare this with the case $m = 2$ of Theorem 14.6. (Hint: Use Proposition 15.13.)

Exercise 15.12 Let χ be a shot noise Cox process as in Example 15.15 and assume that $\mathbb{Y}$ is at most countable. Show that χ is a countable superposition of independent mixed Poisson processes; see Exercise 13.4.

Exercise 15.13 A random measure ξ is said to be *infinitely divisible* if for every integer $m \in \mathbb{N}$ there are independent random measures $\xi_1, \dots, \xi_m$ with equal distribution such that $\xi \overset{d}{=} \xi_1 + \cdots + \xi_m$. Show that every compound Poisson process has this property.

Exercise 15.14 Suppose that χ is a shot noise Cox process. Show that χ is an *infinitely divisible point process*, meaning that for each $m \in \mathbb{N}$ there are independent identically distributed point processes $\chi_1, \dots, \chi_m$ such that $\chi \overset{d}{=} \chi_1 + \cdots + \chi_m$.

Exercise 15.15 Let χ be a Poisson cluster process as in Exercise 5.6. Show that χ is infinitely divisible.

16

The Boolean Model and the Gilbert Graph

The spherical Boolean model Z is a union of balls, where the centres form a stationary Poisson process η on $\mathbb{R}^d$ and the radii are obtained from an independent marking of η. The capacity functional of Z assigns to each compact set $C \subset \mathbb{R}^d$ the probability that Z intersects C. It can be expressed explicitly in terms of the intensity of η and the radius distribution, and yields formulae for contact distributions of Z. The associated Gilbert graph has vertex set η with an edge between two vertices whenever the associated balls overlap. The point process of components isomorphic to a given finite connected graph is stationary, with an intensity that can, in principle, be computed as an integral with respect to a suitable power of the radius distribution.

16.1 Capacity Functional

Let $d \in \mathbb{N}$ and let η be a stationary Poisson process on $\mathbb{R}^d$ with strictly positive intensity γ. By Corollary 6.5 there exists a sequence $X_1, X_2, \dots$ of random vectors in $\mathbb{R}^d$ such that almost surely

$$\eta = \sum_{n=1}^{\infty} \delta_{X_n}. \tag{16.1}$$

Suppose further that $(R_n)_{n\geq 1}$ is a sequence of independent and identically distributed $\mathbb{R}_+$-valued random variables, independent of η. As in (10.11), for $x \in \mathbb{R}^d$ and $r \geq 0$ set $B(x, r) := \{y \in \mathbb{R}^d : \|y - x\| \leq r\}$, where $\|\cdot\|$ is the Euclidean norm on $\mathbb{R}^d$. The union

$$Z := \bigcup_{n=1}^{\infty} B(X_n, R_n) \tag{16.2}$$

of the closed balls with centres X_n and radii $R_n \geq 0$ is a (random) subset of $\mathbb{R}^d$. The balls $B(X_n, R_n)$ are called *grains*. This is an important model of stochastic geometry:

Definition 16.1 Let $\mathbb{Q}$ be a probability measure on $\mathbb{R}_+$ and let

$$\xi = \sum_{n=1}^{\infty} \delta_{(X_n,R_n)} \tag{16.3}$$

be an independent $\mathbb{Q}$-marking of η. The random set (16.2) is called a (Poisson) *spherical Boolean model* with intensity γ and radius distribution $\mathbb{Q}$.

The Boolean model is illustrated by Figure 16.1 It is helpful to note that by the marking theorem (Theorem 5.6) the point process ξ defined by (16.3) is a Poisson process with intensity measure $\gamma\lambda_d \otimes \mathbb{Q}$.

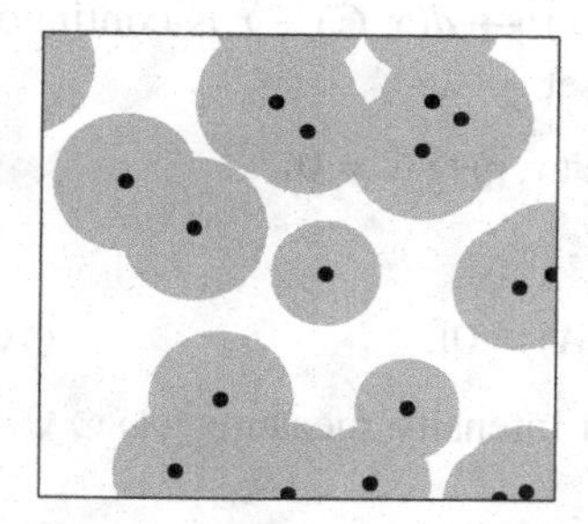
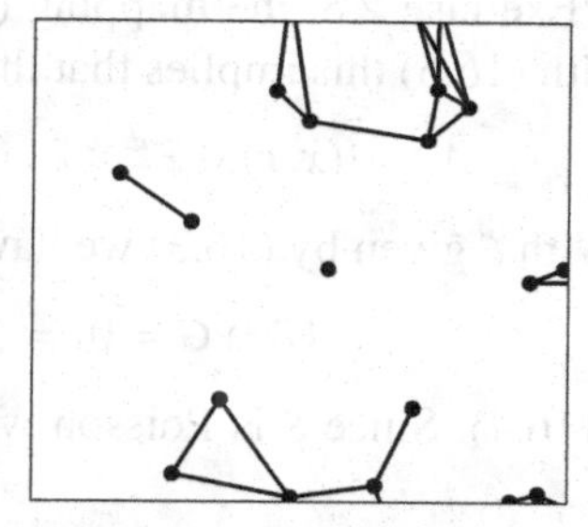

Figure 16.1 Boolean model (left) and Gilbert graph (right), based on the same system of spheres.

Formally, a Boolean model Z is the mapping $\omega \mapsto Z(\omega)$ from Ω into the space of all subsets of $\mathbb{R}^d$. We shall prove the measurability statement

$$\{Z \cap C = \emptyset\} := \{\omega \in \Omega : Z(\omega) \cap C = \emptyset\} \in \mathcal{F}, \quad C \in \mathcal{C}^d, \tag{16.4}$$

where $\mathcal{C}^d$ denotes the system of all compact subsets of $\mathbb{R}^d$. The mapping $C \mapsto \mathbb{P}(Z \cap C = \emptyset)$ is known as the *capacity functional* of Z. It determines the intensity and the radius distribution of the Boolean model; we shall prove this in a more general setting in Chapter 17.

The *Minkowski sum* $K \oplus L$ of sets $K, L \subset \mathbb{R}^d$ is given by

$$K \oplus L := \{x + y : x \in K, y \in L\}. \tag{16.5}$$

The Minkowski sum of K and the ball $B(0, r)$ centred at the origin with radius r is called the *parallel set* of K at distance r. If $K \subset \mathbb{R}^d$ is closed, then

$$K \oplus B(0, r) = \{x \in \mathbb{R}^d : d(x, K) \leq r\} = \{x \in \mathbb{R}^d : B(x, r) \cap K \neq \emptyset\}, \tag{16.6}$$

where

$$d(x,K) := \inf\{\|y-x\| : y \in K\} \tag{16.7}$$

is the Euclidean distance of $x \in \mathbb{R}^d$ from a set $K \subset \mathbb{R}^d$ and $\inf \emptyset := \infty$. We now give a formula for the capacity functional of Z.

Theorem 16.2 *Let Z be a Boolean model with intensity γ and radius distribution $\mathbb{Q}$. Then* (16.4) *holds and moreover*

$$\mathbb{P}(Z \cap C = \emptyset) = \exp\left[-\gamma \int \lambda_d(C \oplus B(0,r))\,\mathbb{Q}(dr)\right], \quad C \in \mathcal{C}^d. \tag{16.8}$$

Proof Let $C \in \mathcal{C}^d$. We may assume that $C \neq \emptyset$. In view of the Lipschitz property in Exercise 2.8, the mapping $(x,r) \mapsto d(x,C) - r$ is continuous. Together with (16.6) this implies that the set

$$A := \{(x,r) \in \mathbb{R}^d \times \mathbb{R}_+ : B(x,r) \cap C \neq \emptyset\}$$

is closed. With ξ given by (16.3) we have

$$\{Z \cap C = \emptyset\} = \{\xi(A) = 0\}, \tag{16.9}$$

and hence (16.4). Since ξ is Poisson with intensity measure $\gamma\lambda_d \otimes \mathbb{Q}$, we have that

$$\mathbb{P}(Z \cap C = \emptyset) = \exp[-\gamma(\lambda_d \otimes \mathbb{Q})(A)]. \tag{16.10}$$

Using (16.6) we obtain

$$\begin{aligned}(\lambda_d \otimes \mathbb{Q})(A) &= \iint \mathbf{1}\{B(x,r) \cap C \neq \emptyset\}\,dx\,\mathbb{Q}(dr) \\ &= \int \lambda_d(C \oplus B(0,r))\,\mathbb{Q}(dr),\end{aligned} \tag{16.11}$$

and hence (16.8). □

16.2 Volume Fraction and Covering Property

Let Z be a Boolean model with fixed intensity γ and radius distribution $\mathbb{Q}$. Let R_0 be a random variable with distribution $\mathbb{Q}$. By (16.8),

$$\mathbb{P}(Z \cap C \neq \emptyset) = 1 - \exp(-\gamma\,\mathbb{E}[\lambda_d(C \oplus B(0,R_0))]), \quad C \in \mathcal{C}^d. \tag{16.12}$$

Taking $C = \{x\}$ we obtain

$$\mathbb{P}(x \in Z) = 1 - \exp\left(-\gamma\kappa_d \mathbb{E}[R_0^d]\right), \quad x \in \mathbb{R}^d, \tag{16.13}$$

where $\kappa_d := \lambda_d(B(0,1))$ is the volume of the unit ball in $\mathbb{R}^d$. Because of the next result, the number $p := \mathbb{P}(0 \in Z)$ is called the *volume fraction* of Z.

Proposition 16.3 *The mapping $(\omega, x) \mapsto \mathbf{1}_{Z(\omega)}(x)$ is measurable and*

$$\mathbb{E}[\lambda_d(Z \cap B)] = p\lambda_d(B), \quad B \in \mathcal{B}(\mathbb{R}^d). \tag{16.14}$$

Proof The asserted measurability follows from the identity

$$1 - \mathbf{1}_{Z(\omega)}(x) = \prod_{n=1}^{\infty} \mathbf{1}\{\|x - X_n(\omega)\| > R_n(\omega)\}, \quad (\omega, x) \in \Omega \times \mathbb{R}^d.$$

Take $B \in \mathcal{B}(\mathbb{R}^d)$. By (16.13) and Fubini's theorem,

$$\mathbb{E}[\lambda_d(Z \cap B)] = \mathbb{E}\Big[\int \mathbf{1}_Z(x)\mathbf{1}_B(x)\,dx\Big] = \int_B \mathbb{E}[\mathbf{1}\{x \in Z\}]\,dx = p\lambda_d(B),$$

as asserted. □

The next result gives a necessary and sufficient condition for Z to cover all of $\mathbb{R}^d$. For $A' \subset \Omega$ we write $\mathbb{P}(A') = 1$ if there exists $A \in \mathcal{F}$ with $A \subset A'$ and $\mathbb{P}(A) = 1$. This is in accordance with our terminology on the almost sure equality of point processes and random measures.

Theorem 16.4 *We have $\mathbb{P}(Z = \mathbb{R}^d) = 1$ if and only if $\mathbb{E}[R_0^d] = \infty$.*

Proof Assume that $A \in \mathcal{F}$ satisfies $A \subset \{Z = \mathbb{R}^d\}$ and $\mathbb{P}(A) = 1$. Then $\mathbb{P}(0 \in Z) = 1$ so that (16.13) implies $\mathbb{E}[R_0^d] = \infty$.

Assume, conversely, that $\mathbb{E}[R_0^d] = \infty$. As a preliminary result we first show for any $n \in \mathbb{N}$ that

$$\lambda_d \otimes \mathbb{Q}(\{(x, r) \in \mathbb{R}^d \times \mathbb{R}_+ : B(0, n) \subset B(x, r)\}) = \infty. \tag{16.15}$$

Since $B(0, n) \subset B(x, r)$ if and only if $r \geq \|x\| + n$, the left-hand side of (16.15) equals

$$\iint \mathbf{1}\{r \geq \|x\| + n\}\,dx\,\mathbb{Q}(dr) = \kappa_d \int_{[n,\infty)} (r - n)^d\,\mathbb{Q}(dr).$$

This is bounded below by

$$\kappa_d \int_{[2n,\infty)} \Big(\frac{r}{2}\Big)^d \mathbb{Q}(dr) = \kappa_d 2^{-d}\,\mathbb{E}[\mathbf{1}\{R_0 \geq 2n\}R_0^d],$$

proving (16.15). Since ξ is a Poisson process with intensity $\gamma\lambda_d \otimes \mathbb{Q}$, the ball $B(0, n)$ is almost surely covered even by infinitely many of the balls $B(x, r)$, $(x, r) \in \xi$. Since n is arbitrary, it follows that $\mathbb{P}(Z = \mathbb{R}^d) = 1$. □

16.3 Contact Distribution Functions

This section is concerned with the random variables given by the distance from the origin 0 to the Boolean model Z, and the distance from 0 to Z in a specified direction. We assume throughout that

$$\mathbb{E}[R_0^d] < \infty.$$

By Exercise 16.1 it is no restriction of generality to assume that $Z(\omega)$ is closed for each $\omega \in \Omega$. Define

$$X_\circ := \inf\{r \geq 0 : Z \cap B(0,r) \neq \emptyset\} = d(0,Z).$$

Since Z is closed and closed balls are compact we have

$$\{X_\circ \leq t\} = \{Z \cap B(0,t) \neq \emptyset\}, \quad t \geq 0, \tag{16.16}$$

and by (16.4), $X_\circ$ is a random variable. Since $\mathbb{P}(X_\circ = 0) = \mathbb{P}(0 \in Z) = p$, the distribution of $X_\circ$ has an atom at 0. Therefore it is convenient to consider the conditional distribution of $X_\circ$ given that the origin is not covered by Z. The function

$$H_\circ(t) := \mathbb{P}(X_\circ \leq t \mid 0 \notin Z), \quad t \geq 0, \tag{16.17}$$

is called the *spherical contact distribution function* of Z.

Proposition 16.5 *The spherical contact distribution function of Z is given by*

$$H_\circ(t) = 1 - \exp\left(-\gamma\kappa_d \sum_{j=1}^{d} \binom{d}{j} t^j\, \mathbb{E}[R_0^{d-j}]\right), \quad t \geq 0. \tag{16.18}$$

Proof Take $t \geq 0$. Then, by (16.16) and (16.12),

$$\begin{aligned}\mathbb{P}(X_\circ > t) &= \mathbb{P}(Z \cap B(0,t) = \emptyset) = \exp\left(-\gamma\, \mathbb{E}[\lambda_d(B(0, R_0 + t))]\right)\\ &= \exp\left(-\gamma\kappa_d\, \mathbb{E}[(R_0 + t)^d]\right).\end{aligned}$$

Since

$$1 - H_\circ(t) = \mathbb{P}(X_\circ > 0)^{-1}\mathbb{P}(X_\circ > t),$$

we can use the binomial formula to derive the result. □

For $x, y \in \mathbb{R}^d$ let $[x,y] := \{x + s(y - x) : 0 \leq s \leq 1\}$ denote the line segment between x and y. Let $u \in \mathbb{R}^d$ with $\|u\| = 1$ and let

$$X_{[u]} := \inf\{r \geq 0 : Z \cap [0, ru] \neq \emptyset\}$$

denote the linear distance of the origin from Z in direction u. The function

$$H_{[u]}(t) := \mathbb{P}(X_{[u]} \le t \mid 0 \notin Z), \quad t \ge 0, \tag{16.19}$$

is called the *linear contact distribution function* of Z in direction u. The next result shows that if $\mathbb{P}(R_0 = 0) < 1$ then $H_{[u]}$ is the distribution function of an exponential distribution with a mean that does not depend on the direction u. We set $\kappa_0 := 1$.

Proposition 16.6 *Let $u \in \mathbb{R}^d$ with $\|u\| = 1$. The linear contact distribution function of Z in direction u is given by*

$$H_{[u]}(t) = 1 - \exp\big(-\gamma t \kappa_{d-1}\, \mathbb{E}[R_0^{d-1}]\big), \quad t \ge 0. \tag{16.20}$$

Proof Let $t \ge 0$. As at (16.16) it follows that

$$\{X_{[u]} \le t\} = \{Z \cap [0, tu] \ne \emptyset\}.$$

Hence (16.12) implies

$$\mathbb{P}(X_{[u]} > t) = \exp\big(-\gamma\, \mathbb{E}[\lambda_d(B(0, R_0) \oplus [0, tu])]\big).$$

It is a well-known geometric fact (elementary in the cases $d = 1, 2, 3$) that

$$\lambda_d(B(0, R_0) \oplus [0, tu]) = \lambda_d(B(0, R_0)) + t\kappa_{d-1} R_0^{d-1}.$$

Since

$$1 - H_{[u]}(t) = (1 - p)^{-1} \mathbb{P}(X_{[u]} > t),$$

the result follows from (16.13). □

16.4 The Gilbert Graph

We need to introduce some graph terminology. An (undirected) *graph* is a pair $G = (V, E)$, where V is a set of *vertices* and $E \subset \{\{x, y\} : x, y \in V, x \ne y\}$ is the set of *edges*. The number card V is known as the *order* of G. An edge $\{x, y\} \in E$ is thought of as connecting its endpoints x and y. Two distinct points $x, y \in V$ are said to be *connected* if there exist $m \in \mathbb{N}$ and $x_0, \dots, x_m \in V$ such that $x_0 = x$, $x_m = y$ and $\{x_{i-1}, x_i\} \in E$ for all $i \in [m]$. The graph G itself is said to be *connected* if any two of its vertices are connected. Two graphs $G = (V, E)$ and $G' = (V', E')$ are said to be *isomorphic* if there is a bijection $T: V \to V'$ such that $\{x, y\} \in E$ if and only if $\{T(x), T(y)\} \in E'$ for all $x, y \in V$ with $x \ne y$. In this case we write $G \simeq G'$.

The remainder of this chapter is concerned with the *Gilbert graph*, a close relative of the spherical Boolean model. We continue to work with the Poisson processes η from (16.1) and the point process ξ given by (16.3), that is

$$\eta = \sum_{n=1}^{\infty} \delta_{X_n}, \qquad \xi = \sum_{n=1}^{\infty} \delta_{(X_n,R_n)}, \tag{16.21}$$

where $(R_n)_{n\geq 1}$ is a sequence of independent $\mathbb{R}_+$-valued random variables with common distribution $\mathbb{Q}$. Suppose that two points $X_m, X_n \in \eta$, $m \neq n$, are connected by an edge whenever the associated balls overlap, that is, $B(X_m, R_m) \cap B(X_n, R_n) \neq \emptyset$. This yields an undirected (random) graph with vertex set η; see Figure 16.1

For a formal definition of the Gilbert graph we introduce the space $\mathbb{R}^{[2d]}$ of all sets $e \subset \mathbb{R}^d$ containing exactly two elements. Any $e \in \mathbb{R}^{[2d]}$ is a potential edge of the graph. When equipped with a suitable metric, $\mathbb{R}^{[2d]}$ becomes a separable metric space; see Exercise 17.5.

Definition 16.7 Let ξ be an independent marking of a stationary Poisson process on $\mathbb{R}^d$ as in (16.21). Define the point process χ on $\mathbb{R}^{[2d]}$ by

$$\chi := \int \mathbf{1}\{\{x,y\} \in \cdot, x < y\}\mathbf{1}\{B(x,r) \cap B(y,s) \neq \emptyset\}\, \xi^2(d((x,r),(y,s))),$$

where $x < y$ means that x is lexicographically strictly smaller than y. Then we call the pair (η,χ) the *Gilbert graph* (based on η) with radius distribution $\mathbb{Q}$. In the special case where $\mathbb{Q}$ is concentrated on a single positive value (all balls have a fixed radius), it is also known as the *random geometric graph.*

Given distinct points $x_1, \dots, x_k \in \eta$ we let $G(x_1, \dots, x_k, \chi)$ denote the graph with vertex set $\{x_1, \dots, x_k\}$ and edges induced by χ, that is such that $\{x_i, x_j\}$ is an edge if and only if $\{x_i, x_j\} \in \chi$. This graph is called a *component* (of the Gilbert graph) if it is connected and none of the x_i is connected to a point in $\eta - \delta_{x_1} - \cdots - \delta_{x_k}$. Let G be a connected graph with $k \geq 2$ vertices. The point process η_G of all components isomorphic to G is then defined by

$$\begin{aligned}\eta_G := \int & \mathbf{1}\{x_1 \in \cdot, x_1 < \cdots < x_k\} \\ & \times \mathbf{1}\{G(x_1, \dots, x_k, \chi) \text{ is a component isomorphic to } G\}\, \eta^k(d(x_1, \dots, x_k)).\end{aligned} \tag{16.22}$$

Hence a component isomorphic to G contributes to $\eta_G(C)$ if its lexicographic minimum lies in C. The indicator $\mathbf{1}\{x_1 < \cdots < x_k\}$ ensures that each component is counted only once. Given distinct $x_1, \dots, x_k \in \mathbb{R}^d$ and given $r_1, \dots, r_k \in \mathbb{R}_+$ we define a graph $\Gamma_k(x_1, r_1, \dots, x_k, r_k)$ with vertex set $\{x_1, \dots, x_k\}$ by taking $\{x_i, x_j\}$ as an edge whenever $B(x_i, r_i) \cap B(x_j, r_j) \neq \emptyset$.

The following theorem shows that η_G is stationary and yields a formula for its intensity. To ease notation, for each $k \in \mathbb{N}$ we define a function $h_k \in \mathbb{R}_+((\mathbb{R}^d \times \mathbb{R}_+)^k)$ by

$$h_k(x_1, r_1, \dots, x_k, r_k) := \mathbb{E}\Big[\lambda_d\Big(\bigcup_{j=1}^k B(x_j, R_0 + r_j)\Big)\Big],$$

where R_0 has distribution $\mathbb{Q}$.

Theorem 16.8 *Let $k \in \mathbb{N}$ with $k \geq 2$ and suppose that G is a connected graph with k vertices. Then the point process η_G is stationary with intensity*

$$\gamma_G := \gamma^k \iint \mathbf{1}\{0 < y_2 < \cdots < y_k\}\mathbf{1}\{\Gamma_k(y_1, r_1, \dots, y_k, r_k) \simeq G\} \tag{16.23}$$
$$\times \exp[-\gamma h_k(y_1, r_1, \dots, y_k, r_k)]\, d(y_2, \dots, y_k)\, \mathbb{Q}^k(d(r_1, \dots, r_k)),$$

where $y_1 := 0$.

Proof Let $\mathbf{N}^*$ denote the measurable set of all $\mu \in \mathbf{N}(\mathbb{R}^d \times \mathbb{R}_+)$ such that $\mu(\cdot \times \mathbb{Y}) \in \mathbf{N}_{ls}$ (the space of all locally finite simple counting measures on $\mathbb{R}^d$). It is no restriction of generality to assume that ξ is a random element of $\mathbf{N}^*$ and that η is a random element of $\mathbf{N}_{ls}$. We construct a measurable mapping $T_G \colon \mathbf{N}^* \to \mathbf{N}_{ls}$ as follows. Given $\mu \in \mathbf{N}^*$ and $(x_1, r_1) \in \mathbb{R}^d \times \mathbb{R}_+$, define $f_G(x_1, r_1, \mu) := 1$ if and only if $(x_1, r_1) \in \mu$, there are $(x_2, r_2), \dots, (x_k, r_k) \in \mu$ such that $\Gamma_k(x_1, r_1, \dots, x_k, r_k) \simeq G$, $x_1 < \cdots < x_k$ and $B(x_i, r_i) \cap B(x, r) = \emptyset$ for all $i \in \{1, \dots, k\}$ and all $(x, r) \in \mu - \delta_{(x_1, r_1)} - \cdots - \delta_{(x_k, r_k)}$; otherwise set $f_G(x_1, r_1, \mu) := 0$. Then set

$$T_G(\mu) := \int \mathbf{1}\{x_1 \in \cdot\} f_G(x_1, r_1, \mu)\, \mu(d(x_1, r_1))$$

and note that $\eta_G = T_G(\xi)$. The mapping T_G has the (covariance) property

$$T_G(\theta_x^* \mu) = \theta_x T_G(\mu), \quad (x, \mu) \in \mathbb{R}^d \times \mathbf{N}^*, \tag{16.24}$$

where θ_x is given by (8.1) (see also (8.2)) and $\theta_x^* \mu \in \mathbf{N}^*$ is defined by

$$\theta_x^* \mu := \int \mathbf{1}\{(y - x, r) \in \cdot\}\, \mu(d(x, r)).$$

By Exercise 16.5 we have

$$\theta_x^* \xi \overset{d}{=} \xi, \quad x \in \mathbb{R}^d. \tag{16.25}$$

Combining this fact with (16.24) shows for all $x \in \mathbb{R}^d$ that

$$\theta_x \eta_G = \theta_x T_G(\xi) = T_G(\theta_x^* \xi) \overset{d}{=} T_G(\xi) = \eta_G. \tag{16.26}$$

Thus, η_G is stationary.

Let $(X_{i_1}, R_{i_1}), \dots, (X_{i_k}, R_{i_k})$ be distinct points of ξ. The graph

$$\Gamma' := G(X_{i_1}, \dots, X_{i_k}, \chi) = \Gamma_k(X_{i_1}, R_{i_1}, \dots, X_{i_k}, R_{i_k})$$

is a component isomorphic to G if and only if $\Gamma' \simeq G$ and none of the X_{i_j} is connected to any point in $\eta - \delta_{X_{i_1}} - \cdots - \delta_{X_{i_k}}$. Let $C \in \mathcal{B}^d$ with $\lambda_d(C) = 1$. By the multivariate Mecke equation for ξ (Theorem 4.5),

$$\mathbb{E}[\eta_G(C)] = \gamma^k \iint \mathbf{1}\{x_1 \in C, x_1 < \cdots < x_k\} \mathbf{1}\{\Gamma_k(x_1, r_1, \dots, x_k, r_k) \simeq G\}$$
$$\times \mathbb{P}(\xi(B_{x_1, r_1, \dots, x_k, r_k}) = 0)\, d(x_1, \dots, x_k)\, \mathbb{Q}^k(d(r_1, \dots, r_k)),$$

where

$$B_{x_1, r_1, \dots, x_k, r_k} := \Big\{(y, r) \in \mathbb{R}^d \times [0, \infty) : B(y, r) \cap \bigcup_{j=1}^k B(x_j, r_j) \neq \emptyset\Big\}.$$

It follows that

$$\mathbb{E}[\eta_G(C)] = \gamma^k \iint \mathbf{1}\{x_1 \in C, x_1 < \cdots < x_k\} \mathbf{1}\{\Gamma_k(x_1, r_1, \dots, x_k, r_k) \simeq G\}$$
$$\times \exp\Big[-\gamma \int h'_k(y, x_1, r_1, \dots, x_k, r_k)\, dy\Big]\, d(x_1, \dots, x_k)\, \mathbb{Q}^k(d(r_1, \dots, r_k)),$$

where

$$h'_k(y, x_1, r_1, \dots, x_k, r_k) := \mathbb{P}\Big(B(y, R_0) \cap \bigcup_{j=1}^k B(x_j, r_j) \neq \emptyset\Big).$$

Since $B(y, R_0) \cap \bigcup_{j=1}^k B(x_j, r_j) \neq \emptyset$ if and only if $y \in \bigcup_{j=1}^k B(x_j, R_0 + r_j)$, we obtain from Fubini's theorem that

$$\int h'_k(y, x_1, r_1, \dots, x_k, r_k)\, dy = h_k(x_1, r_1, \dots, x_k, r_k).$$

Note that

$$\Gamma_k(x_1, r_1, \dots, x_k, r_k) \simeq \Gamma_k(0, r_1, x_2 - x_1, r_2, \dots, x_k - x_1, r_k)$$

and that $x_1 < \cdots < x_k$ if and only if $0 < x_2 - x_1 < \cdots < x_k - x_1$. Moreover,

$$h_k(x_1, r_1, \dots, x_k, r_k) = h_k(0, r_1, x_2 - x_1, r_2, \dots, x_k - x_1, r_k).$$

Performing the change of variables $y_i := x_i - x_1$ for $i \in \{2, \dots, k\}$ and using the fact that $\lambda_d(C) = 1$ gives the asserted formula (16.23). □

If the graph G has only one vertex, then $\eta_1 := \eta_G$ is the point process of *isolated points* of the Gilbert graph (η, χ). In this case (a simplified version of) the proof of Theorem 16.8 yields that η_1 is a stationary point process with intensity

$$\gamma_1 := \gamma \int \exp(-\gamma\, \mathbb{E}[\lambda_d(B(0, R_0 + r))])\, \mathbb{Q}(dr)$$
$$= \gamma \int \exp\left(-\gamma \kappa_d\, \mathbb{E}[(R_0 + r)^d]\right) \mathbb{Q}(dr), \tag{16.27}$$

where $\kappa_d := \lambda_d(B(0,1))$ is the volume of the unit ball.

For $k \in \mathbb{N}$ let $\mathbf{G}_k$ denote a set of connected graphs with k vertices containing exactly one member of each isomorphism equivalence class. Thus for any connected graph G with k vertices there is exactly one $G' \in \mathbf{G}_k$ such that $G \simeq G'$. Then $\sum_{G \in \mathbf{G}_k} \eta_G$ is a stationary point process counting the *k-components* of (η, χ), that is the components with k vertices.

Example 16.9 The set $\mathbf{G}_2$ contains one graph, namely one with two vertices and one edge. By Theorem 16.8, the intensity of 2-components is given by

$$\frac{\gamma^2}{2} \iint \mathbf{1}\{\|z\| \le r_1 + r_2\}$$
$$\times \exp\left(-\gamma\, \mathbb{E}[\lambda_d(B(0, R_0 + r_1) \cup B(z, R_0 + r_2))]\right) dz\, \mathbb{Q}^2(d(r_1, r_2)).$$

For $x \in \eta$ denote by $C(x)$ the set of vertices in the component containing x. This set consists of x and all vertices $y \in \eta$ connected to x in the Gilbert graph. For $k \in \mathbb{N}$ we let

$$\eta_k := \int \mathbf{1}\{x \in \cdot, \operatorname{card} C(x) = k\}\, \eta(dx) \tag{16.28}$$

denote the point process of all points of η that belong to a component of order k. Note that η_1 is the point process of isolated points introduced previously.

Theorem 16.10 *Let $k \ge 2$. Then η_k is a stationary point process with intensity*

$$\gamma_k := \frac{\gamma^k}{(k-1)!} \iint \mathbf{1}\{\Gamma(y_1, r_1, \dots, y_k, r_k) \text{ is connected}\}$$
$$\times \exp[-\gamma h_k(y_1, r_1, \dots, y_k, r_k)]\, d(y_2, \dots, y_k)\, \mathbb{Q}^k(d(r_1, \dots, r_k)), \tag{16.29}$$

where $y_1 := 0$.

Proof Stationarity of η_k follows as at (16.26).

Let $G \in \mathbf{G}_k$ and $j \in \{1, \dots, k\}$. In the definition of η_G we used the lexicographically smallest point to label a component. Using instead the j-smallest point yields a stationary point process $\eta_G^{(j)}$. Exactly as in the proof of Theorem 16.8 it follows that $\eta_G^{(j)}$ has intensity

$$\gamma_G^{(j)} := \gamma^k \int_{B_j} \mathbf{1}\{\Gamma_k(y_1, r_1, \dots, y_k, r_k) \simeq G\}$$
$$\times \exp[-\gamma h_k(y_1, r_1, \dots, y_k, r_k)]\, d(y_2, \dots, y_k)\, \mathbb{Q}^k(d(r_1, \dots, r_k)),$$

where $y_1 := 0$ and B_j denotes the set of all $(y_2, \dots, y_k) \in (\mathbb{R}^d)^{k-1}$ such that $y_2 < \cdots < y_k$ and $0 < y_2$ for $j = 1$, $y_{j-1} < 0 < y_j$ for $j \in \{2, \dots, k-1\}$ and $y_k < 0$ for $j = k$. Since clearly $\gamma_k = \sum_{G \in \mathbf{G}_k} \sum_{j=1}^k \gamma_G^{(j)}$ we obtain

$$\gamma_k = \gamma^k \int \mathbf{1}\{y_2 < \cdots < y_k\} \mathbf{1}\{\Gamma_k(y_1, r_1, \dots, y_k, r_k) \text{ is connected}\}$$
$$\times \exp[-\gamma h_k(y_1, r_1, \dots, y_k, r_k)]\, d(y_2, \dots, y_k)\, \mathbb{Q}^k(d(r_1, \dots, r_k)),$$

so that the symmetry of the integrand (without the first indicator) implies the asserted identity (16.29). □

The quantity γ_k/γ is the fraction of Poisson points that belong to a component of order k. Hence it can be interpreted as probability that a typical point of η belongs to a component of order k. This interpretation can be deepened by introducing the point process $\eta^0 := \eta + \delta_0$ and the Gilbert graph (η^0, χ^0), where $\eta^0 := \eta + \delta_0$ and χ^0 is a point process on $\mathbb{R}^{[2d]}$ that is defined (in terms of an independent marking of η^0) as before. Then

$$\mathbb{P}(\operatorname{card} C^0(0) = k) = \gamma_k/\gamma, \quad k \in \mathbb{N}, \tag{16.30}$$

where $C^0(x)$ is the component of $x \in \eta^0$ in the Gilbert graph (η^0, χ^0); see Exercise 16.6.

16.5 The Point Process of Isolated Nodes

In this section we compute the pair correlation function of the point process η_1 of isolated nodes.

Proposition 16.11 *The pair correlation function ρ_2 of η_1 is given, for λ_d-a.e. $x \in \mathbb{R}^d$, by*

$$\rho_2(x) = \frac{\gamma^2}{\gamma_1^2} \int \mathbf{1}\{\|x\| > r + s\} \tag{16.31}$$
$$\times \exp(-\gamma\, \mathbb{E}[\lambda_d(B(0, R_0 + r) \cup B(x, R_0 + s))])\, \mathbb{Q}^2(d(r, s)).$$

Proof Let $m, n \in \mathbb{N}$ with $m \neq n$. Then X_m and X_n are isolated if and only if $\|X_m - X_n\| > R_m + R_n$ and

$$(B(X_m, R_m) \cup B(X_n, R_n)) \cap \bigcup_{k \neq m,n} B(X_k, R_k) = \emptyset.$$

Hence we obtain from the bivariate Mecke equation (Theorem 4.5) that the reduced second factorial moment measure of η_1 (see Definition 8.6) is given by

$$\alpha_2^!(B) = \gamma^2 \iint \mathbf{1}\{x_1 \in [0,1]^d, x_2 - x_1 \in B\}\mathbf{1}\{\|x_2 - x_1\| > r + s\}$$
$$\times \mathbb{P}((B(x_1, r) \cup B(x_2, s)) \cap Z = \emptyset)\, d(x_1, x_2)\, \mathbb{Q}^2(d(r, s)), \quad B \in \mathcal{B}^d.$$

From (16.12), a change of variable and translation invariance of Lebesgue measure it follows that

$$\alpha_2^!(B) = \gamma^2 \iint \mathbf{1}\{x \in B\}\mathbf{1}\{\|x\| > r + s\}$$
$$\times \exp(-\gamma\, \mathbb{E}[\lambda_d(B(0, R_0 + r) \cup B(x, R_0 + s))])\, \mathbb{Q}^2(d(r, s))\, dx.$$

Hence the assertion follows from Definition 8.9. □

In the case of deterministic radii we have the following result.

Corollary 16.12 *Assume that $\mathbb{Q} = \delta_{s/2}$ for some $s \geq 0$. Then the pair correlation function ρ_2 of η_1 is given, for λ_d-a.e. $x \in \mathbb{R}^d$, by*

$$\rho_2(x) = \mathbf{1}\{\|x\| > s\} \exp[\gamma\, \lambda_d(B(0, s) \cap B(x, s))].$$

Proof By the additivity and invariance property of Lebesgue measure the right-hand side of (16.31) equals

$$\frac{\gamma^2}{\gamma_1^2}\mathbf{1}\{\|x\| > s\} \exp[-2\gamma\lambda_d(B(0, s))] \exp[\gamma\lambda_d(B(0, s) \cap B(x, s))].$$

Inserting here the formula (16.27) for γ_1 yields the result. □

16.6 Exercises

Exercise 16.1 Let Z be a Boolean model as in Definition 16.1 and assume that $\mathbb{E}[R_1^d] < \infty$. For $n \in \mathbb{N}$ let $Y_n := \xi(\{(x, r) : B(x, r) \cap B(0, n) \neq \emptyset\})$. Show that $\mathbb{E}[Y_n] < \infty$. Let $A := \bigcap_{n \geq 1}\{Y_n < \infty\}$. Show that $\mathbb{P}(A) = 1$ and that $Z(\omega)$ is closed for each $\omega \in A$.

Exercise 16.2 Given a Boolean model based on the Poisson process η, let us say that a point $x \in \eta$ is *visible* if x is contained in exactly one of the balls $B(X_n, R_n)$, $n \in \mathbb{N}$. Show that the point process η_v of visible points is stationary with intensity

$$\gamma_v := \gamma\, \mathbb{E}[\exp(-\gamma \kappa_d\, \mathbb{E}[R_1^d])].$$

Now let $c > 0$ and consider the class of all Boolean models with the same intensity γ and radius distributions $\mathbb{Q}$ satisfying $\int r^d\, \mathbb{Q}(dr) = c$. Show that the intensity of visible points is then minimised by the distribution $\mathbb{Q}$ concentrated on a single point. (Hint: Use Jensen's inequality to prove the second part.)

Exercise 16.3 Let Z be a Boolean model with intensity γ and assume $\mathbb{E}[R_0^d] < \infty$. Let L be a one-dimensional line embedded in $\mathbb{R}^d$ (e.g., the first coordinate axis). Show that $Z \cap L$ is a one-dimensional Boolean model with intensity $\gamma \kappa_{d-1} \mathbb{E}[R_0^{d-1}]$. Use this to obtain an alternative proof of Proposition 16.6. What is the radius distribution of the Boolean model $Z \cap L$?

Exercise 16.4 Consider the Gilbert graph under the assumption that R_0^d has an infinite mean. Show that $\gamma_G = 0$ for any connected graph G (with a finite number of vertices).

Exercise 16.5 Prove the stationarity relation (16.25).

Exercise 16.6 Prove (16.30).

Exercise 16.7 Consider the Gilbert graph under the assumption that R_0^d has a finite mean. Show that the pair correlation function ρ_2 of the point process η_1 of isolated nodes satisfies $\lim_{\|x\| \to \infty} \rho_2(x) = 1$.

Exercise 16.8 Consider the Gilbert graph under the assumption that R_0^d has a finite mean. Given an edge e of the Gilbert graph, let the *left endpoint* of e be the first of its endpoints in the lexicographic ordering. Define a point process ξ on $\mathbb{R}^d$ by setting $\xi(B)$ to be the number of edges of the Gilbert graph having left endpoint in B, for each Borel $B \subset R^d$. Show that ξ is a stationary point process with intensity $(\gamma^2/2) \int \mathbb{P}(R_0 + R_1 \geq \|x\|)\, dx$, where R_1 is independent of R_0 and has the same distribution. Is ξ a simple point process?

17

The Boolean Model with General Grains

The spherical Boolean model Z is generalised so as to allow for arbitrary random compact grains. The capacity functional of Z can again be written in an exponential form involving the intensity and the grain distribution. This implies an explicit formula for the covariance of Z. Moreover, in the case of convex grains, the Steiner formula of convex geometry leads to a formula for the spherical contact distribution function involving the mean intrinsic volumes of a typical grain. In the general case the capacity functional determines the intensity and the grain distribution up to a centring.

17.1 Capacity Functional

Let $\mathcal{C}^{(d)}$ denote the space of non-empty compact subsets of $\mathbb{R}^d$ and define the *Hausdorff distance* between sets $K, L \in \mathcal{C}^{(d)}$ by

$$\delta(K, L) := \inf\{\varepsilon \geq 0 : K \subset L \oplus B(0, \varepsilon), L \subset K \oplus B(0, \varepsilon)\}. \tag{17.1}$$

It is easy to check that $\delta(\cdot, \cdot)$ is a metric. By Theorem A.26, $\mathcal{C}^{(d)}$ is a CSMS. We equip $\mathcal{C}^{(d)}$ with the associated Borel σ-field $\mathcal{B}(\mathcal{C}^{(d)})$. For $B \subset \mathbb{R}^d$ and $x \in \mathbb{R}^d$ we recall the notation $B + x := \{y + x : y \in B\}$.

Definition 17.1 Let η be a stationary Poisson process on $\mathbb{R}^d$ with intensity $\gamma > 0$, given as in (16.1). Let $\mathbb{Q}$ be a probability measure on $\mathcal{C}^{(d)}$ and let

$$\xi = \sum_{n=1}^{\infty} \delta_{(X_n, Z_n)} \tag{17.2}$$

be an independent $\mathbb{Q}$-marking of η. Then

$$Z := \bigcup_{n=1}^{\infty} (Z_n + X_n) \tag{17.3}$$

is called the *Boolean model* with intensity γ and *grain distribution* $\mathbb{Q}$ (or the *Boolean model induced by* ξ for short).

As in Chapter 16, Z is a short-hand notation for the mapping $\omega \mapsto Z(\omega)$. We wish to generalise Theorem 16.2 for the spherical Boolean model and give a formula for the *capacity functional* $C \mapsto \mathbb{P}(Z \cap C \neq \emptyset)$ of Z. As preparation we need to identify useful generators of the Borel σ-field $\mathcal{B}(\mathcal{C}^{(d)})$. For $B \subset \mathbb{R}^d$ define

$$\mathcal{C}_B := \{K \in \mathcal{C}^{(d)} : K \cap B \neq \emptyset\}; \quad \mathcal{C}^B := \{K \in \mathcal{C}^{(d)} : K \cap B = \emptyset\}. \tag{17.4}$$

Lemma 17.2 *The σ-field $\mathcal{B}(\mathcal{C}^{(d)})$ is generated by $\{\mathcal{C}_B : B \in \mathcal{C}^{(d)}\}$.*

Proof It is a quick consequence of the definition of the Hausdorff distance that $\mathcal{C}^B$ is open whenever $B \in \mathcal{C}^{(d)}$. Hence the σ-field $\mathcal{H}$ generated by $\{\mathcal{C}_B : B \in \mathcal{C}^{(d)}\}$ is contained in $\mathcal{B}(\mathcal{C}^{(d)})$.

To prove $\mathcal{B}(\mathcal{C}^{(d)}) \subset \mathcal{H}$ we first note that $\mathcal{C}^{(d)}$, equipped with the Hausdorff distance, is a *separable* metric space, that is has a countable dense set; see Exercise 17.3. It follows from elementary properties of separable metric spaces that any open set in $\mathcal{C}^{(d)}$ is either empty or a countable union of closed balls. Hence it is sufficient to show that for any $K \in \mathcal{C}^{(d)}$ and $\varepsilon > 0$ the closed ball

$$B(K, \varepsilon) = \{L \in \mathcal{C}^{(d)} : L \subset K \oplus B(0, \varepsilon), K \subset L \oplus B(0, \varepsilon)\}$$

is in $\mathcal{H}$. Since $L \subset K \oplus B(0, \varepsilon)$ is equivalent to $L \cap (\mathbb{R}^d \setminus (K \oplus B(0, \varepsilon))) = \emptyset$ we have

$$\{L \in \mathcal{C}^{(d)} : L \subset K \oplus B(0, \varepsilon)\} = \bigcap_{n \in \mathbb{N}} \{L \in \mathcal{C}^{(d)} : L \cap (B_n \setminus (K \oplus B(0, \varepsilon))) = \emptyset\}, \tag{17.5}$$

where B_n is the interior of the ball $B(0, n)$. Since $A_n := B_n \setminus K \oplus B(0, \varepsilon)$ is open, it is easy to prove that $\mathcal{C}^{A_n} \in \mathcal{H}$, so that the right-hand side of (17.5) is in $\mathcal{H}$ as well. It remains to show that $\mathcal{C}_{K,\varepsilon} := \{L \in \mathcal{C}^{(d)} : K \subset L \oplus B(0, \varepsilon)\}$ is in $\mathcal{H}$. To this end we take a countable dense set $D \subset K$ (see Lemma A.22) and note that $K \not\subset L \oplus B(0, \varepsilon)$ if and only if there exists $x \in D$ such that $B(x, \varepsilon) \cap L = \emptyset$. (Use (16.6) and a continuity argument.) Therefore $\mathcal{C}^{(d)} \setminus \mathcal{C}_{K,\varepsilon}$ is a countable union of sets of the form $\mathcal{C}^B$, where B is a closed ball. Hence $\mathcal{C}_{K,\varepsilon} \in \mathcal{H}$ and the proof is complete. □

For $C \subset \mathbb{R}^d$ let $C^* := \{-x : x \in C\}$ denote the *reflection* of C in the origin.

Theorem 17.3 *Let Z be a Boolean model with intensity γ and grain distribution $\mathbb{Q}$. Then* (16.4) *holds and, moreover,*

$$\mathbb{P}(Z \cap C = \emptyset) = \exp\left[-\gamma \int \lambda_d(K \oplus C^*)\, \mathbb{Q}(dK)\right], \quad C \in \mathcal{C}^{(d)}. \tag{17.6}$$

Proof We can follow the proof of Theorem 16.2. By Exercise 17.6, the mapping $(x, K) \mapsto K + x$ from $\mathbb{R}^d \times C^{(d)}$ to $C^{(d)}$ is continuous and hence measurable. Take $C \in C^{(d)}$. By Lemma 17.2,

$$A := \{(x, K) \in \mathbb{R}^d \times C^{(d)} : (K + x) \cap C \neq \emptyset\}$$

is a measurable set. Since (16.9) holds, (16.4) follows. Moreover, since ξ is a Poisson process with intensity measure $\gamma\lambda_d \otimes \mathbb{Q}$ we again obtain (16.10). Using the fact that

$$\{x \in \mathbb{R}^d : (K + x) \cap C \neq \emptyset\} = C \oplus K^*, \tag{17.7}$$

together with the reflection invariance of Lebesgue measure, we obtain the assertion (17.6). □

Taking $B = \{x\}$ in (17.6) yields, as in (16.13), that

$$\mathbb{P}(x \in Z) = 1 - \exp\left(-\gamma \int \lambda_d(K)\, \mathbb{Q}(dK)\right), \quad x \in \mathbb{R}^d. \tag{17.8}$$

The quantity $p := \mathbb{P}(0 \in Z)$ is the *volume fraction* of Z.

Proposition 17.4 *The mapping* $(\omega, x) \mapsto \mathbf{1}_{Z(\omega)}(x)$ *is measurable and*

$$\mathbb{E}[\lambda_d(Z \cap B)] = p\lambda_d(B), \quad B \in \mathcal{B}(\mathbb{R}^d). \tag{17.9}$$

Proof By (17.2),

$$1 - \mathbf{1}_Z(x) = \prod_{n=1}^{\infty} \mathbf{1}\{x \notin Z_n + X_n\}, \quad x \in \mathbb{R}^d.$$

Therefore the asserted measurability follows from the fact that the mappings $(x, K) \mapsto \mathbf{1}\{x \notin K\}$ and $(x, K) \mapsto K + x$ are measurable on $\mathbb{R}^d \times C^{(d)}$; see Exercises 17.6 and 17.7. Equation (17.9) can then be proved in the same manner as (16.14). □

In what follows we shall always assume that

$$\int \lambda_d(K \oplus B(0, r))\, \mathbb{Q}(dK) < \infty, \quad r \geq 0. \tag{17.10}$$

By Exercise 17.1 we can assume, as in Section 16.3, that $Z(\omega)$ is a closed set for each $\omega \in \Omega$. By the next result, Z is a random element of the space $\mathcal{F}^d$ of closed subsets of $\mathbb{R}^d$, equipped with the σ-field $\mathcal{B}(\mathcal{F}^d)$ generated by the Fell topology; see Section A.3.

Proposition 17.5 *Assume that* (17.10) *holds. Then* Z *is a random element of* $\mathcal{F}^d$.

Proof The assertion follows from Theorem 17.3 and Lemma A.28. □

By Lemma A.27 the mapping $(F, x) \mapsto F + x$ from $\mathcal{F}^d \times \mathbb{R}^d$ to $\mathcal{F}^d$ is continuous and hence measurable. In particular, $Z + x$ is, for each $x \in \mathbb{R}^d$, again a random element of $\mathcal{F}^d$. The next result says that Z is *stationary.*

Proposition 17.6 *Assume that* (17.10) *holds. Let* $x \in \mathbb{R}^d$. *Then* $Z + x \overset{d}{=} Z$.

Proof We need to prove that $\mathbb{P}(Z + x \in A) = \mathbb{P}(Z \in A)$ holds for each $A \in \mathcal{B}(\mathcal{F}^d)$. By Lemma A.28 and Theorem A.5 it is sufficient to show for each $C \in \mathcal{C}^{(d)}$ that $\mathbb{P}((Z+x) \cap C = \emptyset) = \mathbb{P}(Z \cap C = \emptyset)$. Since $(Z+x) \cap C = \emptyset$ if and only if $Z \cap (C - x) = \emptyset$, the assertion follows from Theorem 17.3 and translation invariance of Lebesgue measure. □

17.2 Spherical Contact Distribution Function and Covariance

In this section we first give a more general version of Proposition 16.5 under an additional assumption on $\mathbb{Q}$. Let $\mathcal{K}^{(d)}$ denote the system of all convex $K \in \mathcal{C}^{(d)}$. By Theorem A.26, $\mathcal{K}^{(d)}$ is a closed and hence measurable subset of $\mathcal{C}^{(d)}$. Recall from Section A.3 the definition of the intrinsic volumes $V_0, \dots, V_d$ as non-negative continuous functions on $\mathcal{K}^{(d)}$. If the grain distribution $\mathbb{Q}$ is concentrated on $\mathcal{K}^{(d)}$ (i.e. $\mathbb{Q}(\mathcal{K}^{(d)}) = 1$), then we can define

$$\phi_i := \int V_i(K)\,\mathbb{Q}(dK), \quad i = 0, \dots, d. \tag{17.11}$$

Note that $\phi_0 = 1$. The Steiner formula (A.22) implies that $\phi_i < \infty$ for all $i \in \{0, \dots, d\}$ if and only if (17.10) holds. The *spherical contact distribution function* $H_\circ$ of Z is defined by (16.17).

Proposition 17.7 *Suppose that* (17.10) *holds and that* $\mathbb{Q}$ *is concentrated on* $\mathcal{K}^{(d)}$. *Then the spherical contact distribution function of the Boolean model* Z *is given by*

$$H_\circ(t) = 1 - \exp\Big[-\sum_{j=0}^{d-1} t^{d-j}\kappa_{d-j}\gamma\phi_j\Big], \quad t \geq 0, \tag{17.12}$$

where $\phi_0, \dots, \phi_d$ *are defined by* (17.11).

Proof Let $t \geq 0$. Similarly to the proof of Proposition 16.5 we obtain from (17.6) that

$$1 - H_\circ(t) = \exp\Big[-\gamma \int (\lambda_d(K \oplus B(0,t)) - \lambda_d(K))\,\mathbb{Q}(dK)\Big].$$

By assumption on $\mathbb{Q}$ we can use the Steiner formula (A.22) (recall that $V_d = \lambda_d$) to simplify the exponent and to conclude the proof of (17.12). □

Next we deal with second order properties of the Boolean model. The function $(x, y) \mapsto \mathbb{P}(x \in Z, y \in Z)$ is called the *covariance* (or *two point correlation function*) of Z. It can be expressed in terms of the function

$$\beta_d(x) := \int \lambda_d(K \cap (K + x))\, \mathbb{Q}(dK), \quad x \in \mathbb{R}^d, \tag{17.13}$$

as follows.

Theorem 17.8 *Suppose that* (17.10) *holds. The covariance of Z is given by*

$$\mathbb{P}(x \in Z, y \in Z) = p^2 + (1-p)^2(e^{\gamma\beta_d(x-y)} - 1), \quad x, y \in \mathbb{R}^d.$$

Proof Let Z_0 have distribution $\mathbb{Q}$ and let $x, y \in \mathbb{R}^d$. By (17.6),

$$\begin{aligned}\mathbb{P}(Z \cap \{x, y\} = \emptyset) &= \exp(-\gamma\, \mathbb{E}[\lambda_d((Z_0 - x) \cup (Z_0 - y))]) \\ &= \exp(-\gamma\, \mathbb{E}[\lambda_d(Z_0 \cup (Z_0 + x - y))]).\end{aligned}$$

By additivity of λ_d and linearity of expectation we obtain

$$\begin{aligned}\mathbb{P}(Z \cap \{x, y\} = \emptyset) &= \exp(-2\gamma\, \mathbb{E}[\lambda_d(Z_0)]) \exp(\gamma\, \mathbb{E}[\lambda_d(Z_0 \cap (Z_0 + x - y))]) \\ &= (1-p)^2 \exp[\gamma\beta_d(x-y)],\end{aligned}$$

where we have used (17.8). By the additivity of probability,

$$\mathbb{P}(Z \cap \{x, y\} = \emptyset) = \mathbb{P}(x \notin Z, y \notin Z) = \mathbb{P}(x \in Z, y \in Z) + 1 - 2p$$

and the result follows. □

17.3 Identifiability of Intensity and Grain Distribution

In this section we shall prove that the capacity functional of a Boolean model Z determines the intensity and the centred grain distribution of the underlying marked Poisson process. To this end we need the following lemma, which is of some independent interest.

Lemma 17.9 *Let ν be a measure on $\mathcal{C}^{(d)}$ satisfying*

$$\nu(\mathcal{C}_B) < \infty, \quad B \in \mathcal{C}^{(d)}. \tag{17.14}$$

Then ν is determined by its values on $\{\mathcal{C}_B : B \in \mathcal{C}^{(d)}\}$.

Proof For $m \in \mathbb{N}$ and $B_0, \dots, B_m \in \mathcal{C}^d = \mathcal{C}^{(d)} \cup \{\emptyset\}$ let

$$C^{B_0}_{B_1,\dots,B_m} := C^{B_0} \cap C_{B_1} \cap \cdots \cap C_{B_m}. \tag{17.15}$$

Since $C^{B_0} \cap C^{B'_0} = C^{B_0 \cup B'_0}$ for all $B_0, B'_0 \in \mathcal{C}^d$, the sets of the form (17.15) form a π-system. By Lemma 17.2 this is a generator of $\mathcal{B}(\mathcal{C}^{(d)})$. Since (17.14) easily implies that ν is σ-finite, the assertion follows from Theorem A.5 once we have shown that

$$\nu(C^{B_0}_{B_1,\dots,B_m}) = \sum_{j=0}^{m} (-1)^{j+1} \sum_{1 \le i_1 < \dots < i_j \le m} \nu(C_{B_0 \cup B_{i_1} \cup \dots \cup B_{i_j}}). \tag{17.16}$$

In fact, we only need the case with $B_0 = \emptyset$, but it is simpler to prove the more general case. Moreover, the identity (17.16) is of some independent interest. For $m = 1$ the identity means that

$$\nu(C^{B_0}_{B_1}) = \nu(C_{B_0 \cup B_1}) - \nu(C_{B_0}),$$

a direct consequence of the equality $C^{B_0}_{B_1} = C_{B_0 \cup B_1} \setminus C_{B_0}$. In the general case we can use the equality

$$C^{B_0}_{B_1,\dots,B_m} = C^{B_0}_{B_1,\dots,B_{m-1}} \setminus C^{B_0 \cup B_m}_{B_1,\dots,B_{m-1}}$$

and induction. □

We need to fix a *centre function* $c\colon \mathcal{C}^{(d)} \to \mathbb{R}^d$. This is a measurable function satisfying

$$c(K + x) = c(K) + x, \quad (x, K) \in \mathbb{R}^d \times \mathcal{C}^{(d)}. \tag{17.17}$$

An example is the centre of the (uniquely determined) *circumball* of K, that is the smallest ball containing K.

Theorem 17.10 *Let Z and Z' be Boolean models with respective intensities γ and γ' and grain distributions $\mathbb{Q}$ and $\mathbb{Q}'$. Assume that $\mathbb{Q}$ satisfies (17.10). If*

$$\mathbb{P}(Z \cap B = \emptyset) = \mathbb{P}(Z' \cap B = \emptyset), \quad B \in \mathcal{C}^{(d)}, \tag{17.18}$$

then $\gamma = \gamma'$ and $\mathbb{Q}(\{K : K - c(K) \in \cdot\}) = \mathbb{Q}'(\{K : K - c(K) \in \cdot\})$.

Proof Define a measure ν on $(\mathcal{C}^{(d)}, \mathcal{B}(\mathcal{C}^{(d)}))$ by

$$\nu(\cdot) := \gamma \iint \mathbf{1}\{K + x \in \cdot\}\, \mathbb{Q}(dK)\, dx. \tag{17.19}$$

Similarly define a measure ν' by replacing $(\gamma, \mathbb{Q})$ with $(\gamma', \mathbb{Q}')$. By (17.10) for $B \in \mathcal{C}^{(d)}$ we have $K \oplus B^* = \{x \in \mathbb{R}^d : K + x \in C_B\}$. Theorem 17.3 then shows that

$$\nu(C_B) = -\log \mathbb{P}(Z \cap B = \emptyset), \quad B \in \mathcal{C}^{(d)}.$$

Assuming that (17.18) holds, we hence obtain from assumption (17.10) that

$$\nu(C_B) = \nu'(C_B) < \infty, \quad B \in \mathcal{C}^{(d)}. \tag{17.20}$$

In particular, both measures satisfy (17.14). By Lemma 17.9 we conclude that $\nu = \nu'$.

Take a measurable set $A \subset \mathcal{C}^{(d)}$ and $B \in \mathcal{B}^d$ with $\lambda_d(B) = 1$. Then, using property (17.17) of a centre function,

$$\begin{aligned}
&\int \mathbf{1}\{K - c(K) \in A, c(K) \in B\}\, \nu(dK) \\
&= \gamma \iint \mathbf{1}\{K + x - c(K + x) \in A, c(K + x) \in B\}\, \mathbb{Q}(dK)\, dx \\
&= \gamma \iint \mathbf{1}\{K - c(K) \in A, c(K) + x \in B\}\, dx\, \mathbb{Q}(dK) \\
&= \gamma\, \mathbb{Q}(\{K : K - c(K) \in A\}).
\end{aligned}$$

Since ν' satisfies a similar equation and $\nu = \nu'$, the assertion follows. □

17.4 Exercises

Exercise 17.1 Let Z be a Boolean model whose grain distribution $\mathbb{Q}$ satisfies (17.10). Prove that almost surely any compact set is intersected by only a finite number of the grains $Z_n + X_n$, $n \in \mathbb{N}$. Show then that Z is almost surely a closed set.

Exercise 17.2 Let $m \in \mathbb{N}$ and assume that $\int \lambda_d(K \oplus B(0, \varepsilon))^m\, \mathbb{Q}(dK) < \infty$ for some (fixed) $\varepsilon > 0$. Prove that $\int \lambda_d(K \oplus B(0, r))^m\, \mathbb{Q}(dK) < \infty$ for each $r > 0$.

Exercise 17.3 Show that the space $\mathcal{C}^{(d)}$ equipped with the Hausdorff distance (17.1) is a separable metric space. (Hint: Use a dense countable subset of $\mathbb{R}^d$.)

Exercise 17.4 Let $m \in \mathbb{N}$ and let $\mathcal{C}_m \subset \mathcal{C}^{(d)}$ be the space of all compact non-empty subsets of $\mathbb{R}^d$ with at most m points. Show that $\mathcal{C}_m$ is closed (with respect to the Hausdorff distance).

Exercise 17.5 For $i \in \{1,2\}$, let $e_i := \{x_i, y_i\}$, where $x_1, x_2, y_1, y_2 \in \mathbb{R}^d$ satisfy $x_1 \neq x_2$ and $y_1 \neq y_2$. Show that the Hausdorff distance between e_1 and e_2 is given by

$$\delta(e_1, e_2) = (\|x_1 - y_1\| \vee \|x_2 - y_2\|) \wedge (\|x_1 - y_2\| \vee \|x_2 - y_1\|).$$

Show that $\mathbb{R}^{[2d]}$ is not closed in $C^{(d)}$. Use Exercise 17.4 to show that $\mathbb{R}^{[2d]}$ is a measurable subset of $C^{(d)}$. Show finally that a set $C \subset \mathbb{R}^{[2d]}$ is bounded if and only if $U(C)$ is bounded, where $U(C)$ is the union of all $e \in C$.

Exercise 17.6 Prove that the mapping $(x, K) \mapsto K + x$ from $\mathbb{R}^d \times C^{(d)}$ to $C^{(d)}$ is continuous. Prove also that the mapping $(K, L) \mapsto K \oplus L$ is continuous on $C^{(d)} \times C^{(d)}$. Why is this a more general statement?

Exercise 17.7 Prove that the mapping $(x, K) \mapsto \mathbf{1}_K(x)$ from $\mathbb{R}^d \times C^{(d)}$ to $\mathbb{R}$ is measurable. (Hint: Show that $\{(x, K) \in \mathbb{R}^d \times C^{(d)} : x \in K\}$ is closed.)

Exercise 17.8 Let Z be a Boolean model whose grain distribution $\mathbb{Q}$ satisfies (17.10) as well as the equation $\mathbb{Q}(\{K \in C^{(d)} : \lambda_d(\partial K) = 0\}) = 1$, where ∂K is the boundary of a set K. Show that

$$\lim_{x \to 0} \mathbb{P}(0 \in Z, x \in Z) = \mathbb{P}(0 \in Z).$$

Exercise 17.9 Let ν be a measure on $C^{(d)}$ satisfying (17.14). Show that ν is locally finite. Show also that the measure

$$\nu := \int_0^\infty \mathbf{1}\{B(0, r) \in \cdot\}\, dr$$

is locally finite but does not satisfy (17.14).

Exercise 17.10 Consider a Boolean model whose grain distribution $\mathbb{Q}$ satisfies

$$\int \lambda_d(K)^2\, \mathbb{Q}(dK) < \infty. \tag{17.21}$$

Prove that $\int (e^{\gamma \beta_d(x)} - 1)\, dx < \infty$. (Hint: Use that $e^t - 1 \leq te^t$, $t \geq 0$.)

Exercise 17.11 Let $W \subset \mathbb{R}^d$ be a Borel set with $0 < \lambda_d(W) < \infty$ such that the boundary of W has Lebesgue measure 0. Show that

$$\lim_{r \to \infty} \lambda_d(rW)^{-1} \lambda_d(rW \cap (rW + x)) = 1$$

for all $x \in \mathbb{R}^d$, where $rW := \{rx : x \in W\}$. (Hint: Decompose W into its interior and $W \cap \partial W$; see Section A.2.)

18

Fock Space and Chaos Expansion

The difference operator is the increment of a measurable function of a counting measure, upon adding an extra point. It can be iterated to yield difference operators of higher orders. Each square integrable function $f(\eta)$ of a Poisson process η determines an infinite sequence of expected difference operators. This sequence is an element of a direct sum of Hilbert spaces, called the Fock space associated with the intensity measure of η. The second moment of $f(\eta)$ coincides with the squared norm of this Fock space representation. A consequence is the Poincaré inequality for the variance of $f(\eta)$. A deeper result is the orthogonal decomposition of $f(\eta)$ into a series of Wiener–Itô integrals, known as the chaos expansion.

18.1 Difference Operators

Throughout this chapter we consider a Poisson process η on an arbitrary measurable space $(\mathbb{X}, \mathcal{X})$ with σ-finite intensity measure λ. Let $\mathbb{P}_\eta$ denote the distribution of η, a probability measure on $\mathbf{N} := \mathbf{N}(\mathbb{X})$. Let $f \in \mathbb{R}(\mathbf{N})$. For $x \in \mathbb{X}$ define the function $D_x f \in \mathbb{R}(\mathbf{N})$ by

$$D_x f(\mu) := f(\mu + \delta_x) - f(\mu), \quad \mu \in \mathbf{N}. \tag{18.1}$$

Iterating this definition, we define $D^n_{x_1,\dots,x_n} f \in \mathbb{R}(\mathbf{N})$ for each $n \geq 2$ and $(x_1, \dots, x_n) \in \mathbb{X}^n$ inductively by

$$D^n_{x_1,\dots,x_n} f := D^1_{x_1} D^{n-1}_{x_2,\dots,x_n} f, \tag{18.2}$$

where $D^1 := D$ and $D^0 f = f$. Observe that

$$D^n_{x_1,\dots,x_n} f(\mu) = \sum_{J \subset \{1,2,\dots,n\}} (-1)^{n-|J|} f\Big(\mu + \sum_{j \in J} \delta_{x_j}\Big), \tag{18.3}$$

where $|J|$ denotes the number of elements of J. This shows that $D^n_{x_1,\dots,x_n} f$ is symmetric in $x_1, \dots, x_n$ and that $(\mu, x_1, \dots, x_n) \mapsto D^n_{x_1,\dots,x_n} f(\mu)$ is measurable. As a function of f these operators are linear.

Example 18.1 Assume that $\mathbb{X} = C^{(d)}$ is the space of all non-empty compact subsets of $\mathbb{R}^d$, as in Definition 17.1. For $\mu \in \mathbf{N}$ let

$$Z(\mu) := \bigcup_{K \in \mu} K \tag{18.4}$$

whenever μ is locally finite (with respect to the Hausdorff distance); otherwise let $Z(\mu) := \emptyset$. Let ν be a finite measure on $\mathbb{R}^d$ and define $f\colon \mathbf{N} \to \mathbb{R}_+$ by

$$f(\mu) := \nu(Z(\mu)). \tag{18.5}$$

Thanks to Theorem A.26 we can apply Proposition 6.3. Therefore we obtain for all $x \in \mathbb{R}^d$ and all locally finite $\mu \in \mathbf{N}$ that

$$1 - \mathbf{1}_{Z(\mu)}(x) = \prod_{n=1}^{\mu(C^{(d)})} \mathbf{1}\{x \notin \pi_n(\mu)\}.$$

Hence Exercise 17.7 shows that $(x, \mu) \mapsto \mathbf{1}_{Z(\mu)}(x)$ is measurable on $\mathbb{R}^d \times \mathbf{N}$. In particular, f is a measurable mapping. For each locally finite $\mu \in \mathbf{N}$ and each $K \in C^{(d)}$, we have

$$f(\mu + \delta_K) = \nu(Z(\mu) \cup K) = \nu(Z(\mu)) + \nu(K) - \nu(Z(\mu) \cap K),$$

that is

$$D_K f(\mu) = \nu(K) - \nu(Z(\mu) \cap K) = \nu(K \cap Z(\mu)^c).$$

It follows by induction that

$$D^n_{K_1,\dots,K_n} f(\mu) = (-1)^{n+1}\nu(K_1 \cap \cdots \cap K_n \cap Z(\mu)^c), \quad \mu \in \mathbf{N},$$

for all $n \in \mathbb{N}$ and $K_1, \dots, K_n \in C^{(d)}$.

The next lemma yields further insight into the difference operators. For $h \in \mathbb{R}(\mathbb{X})$ we set $h^{\otimes 0} := 1$ and recall from Chapter 12 that for $n \in \mathbb{N}$ the function $h^{\otimes n} \in \mathbb{R}(\mathbb{X}^n)$ is defined by

$$h^{\otimes n}(x_1, \dots, x_n) := \prod_{i=1}^{n} h(x_i), \quad x_1, \dots, x_n \in \mathbb{X}. \tag{18.6}$$

Lemma 18.2 *Let $\upsilon \in \mathbb{R}_+(\mathbb{X})$ and define $f \in \mathbb{R}_+(\mathbf{N})$ by $f(\mu) = \exp[-\mu(\upsilon)]$, $\mu \in \mathbf{N}$. Let $n \in \mathbb{N}$. Then*

$$D^n_{x_1,\dots,x_n} f(\mu) = \exp[-\mu(\upsilon)](e^{-\upsilon} - 1)^{\otimes n}(x_1, \dots, x_n), \quad x_1, \dots, x_n \in \mathbb{X}. \tag{18.7}$$

Proof For each $\mu \in \mathbf{N}$ and $x \in \mathbb{X}$ we have

$$f(\mu + \delta_x) = \exp\left[-\int v(y)\,(\mu + \delta_x)(dy)\right] = \exp[-\mu(v)]\exp[-v(x)],$$

so that

$$D_x f(\mu) = \exp[-\mu(v)](\exp[-v(x)] - 1).$$

Iterating this identity yields for all $n \in \mathbb{N}$ and all $x_1, \dots, x_n \in \mathbb{X}$ that

$$D^n_{x_1,\dots,x_n} f(\mu) = \exp[-\mu(v)] \prod_{i=1}^{n} (\exp[-v(x_i)] - 1)$$

and hence the assertion. □

18.2 Fock Space Representation

Theorem 18.6 below is the main result of this chapter. To formulate it, we need to introduce some notation. For $n \in \mathbb{N}$ and $f \in \mathbb{R}(\mathbf{N})$ we define the symmetric measurable function $T_n f \colon \mathbb{X}^n \to \mathbb{R}$ by

$$T_n f(x_1, \dots, x_n) := \mathbb{E}[D^n_{x_1,\dots,x_n} f(\eta)], \tag{18.8}$$

and set $T_0 f := \mathbb{E}[f(\eta)]$.

The inner product of $u, v \in L^2(\lambda^n)$ for $n \in \mathbb{N}$ is denoted by

$$\langle u, v \rangle_n := \int uv \, d\lambda^n.$$

Denote by $\|\cdot\|_n := \langle \cdot, \cdot \rangle_n^{1/2}$ the associated norm. For $n \in \mathbb{N}$ let $\mathbf{H}_n$ be the space of symmetric functions in $L^2(\lambda^n)$, and let $\mathbf{H}_0 := \mathbb{R}$. The *Fock space* $\mathbf{H}$ is the set of all sequences $(u_n)_{n\ge 0} \in \times_{n=0}^{\infty} \mathbf{H}_n$ such that $\langle (u_n)_{n\ge0}, (u_n)_{n\ge0} \rangle_{\mathbf{H}} < \infty$, where, for $(v_n)_{n\ge0} \in \times_{n=0}^{\infty} \mathbf{H}_n$, we set

$$\langle (u_n)_{n\ge0}, (v_n)_{n\ge0} \rangle_{\mathbf{H}} := \sum_{n=0}^{\infty} \frac{1}{n!} \langle u_n, v_n \rangle_n$$

and $\langle a, b \rangle_0 := ab$ for $a, b \in \mathbb{R}$. The space $\mathbf{H}$ is a vector space under componentwise addition and scalar multiplication and $\langle \cdot, \cdot \rangle_{\mathbf{H}}$ is bilinear. It is a well-known analytic fact (the reader is invited to prove this as an exercise) that $\mathbf{H}$ is a Hilbert space, that is $\mathbf{H}$ is a complete metric space with respect to the metric $((u_n)_{n\ge0}, (v_n)_{n\ge0}) \mapsto (\langle (u_n - v_n)_{n\ge0}, (u_n - v_n)_{n\ge0} \rangle_{\mathbf{H}})^{1/2}$. In this section we prove that the mapping $f \mapsto (T_n(f))_{n\ge0}$ is an *isometry* from $L^2(\mathbb{P}_\eta)$ to $\mathbf{H}$.

Let $\mathcal{X}_0$ be the system of all measurable $B \in \mathcal{X}$ having $\lambda(B) < \infty$. Let

$\mathbb{R}_0(\mathbb{X})$ be the space of all functions $v \in \mathbb{R}_+(\mathbb{X})$ such that v is bounded and $\{x \in \mathbb{X} : v(x) > 0\} \in \mathcal{X}_0$. Let $\mathbf{G}$ denote the space of all functions $g \colon \mathbf{N} \to \mathbb{R}$ of the form

$$g(\mu) = a_1 e^{-\mu(v_1)} + \cdots + a_n e^{-\mu(v_n)}, \quad \mu \in \mathbf{N}, \tag{18.9}$$

where $n \in \mathbb{N}$, $a_1, \dots, a_n \in \mathbb{R}$ and $v_1, \dots, v_n \in \mathbb{R}_0(\mathbb{X})$. All such functions are bounded and measurable.

For each $f \in L^2(\mathbb{P}_\eta)$ define $Tf := (T_n f)_{n \geq 0}$, where $T_n f$ is given at (18.8). By the next result, $Tf \in \mathbf{H}$ for $f \in \mathbf{G}$ and

$$\mathbb{E}[f(\eta)g(\eta)] = \langle Tf, Tg \rangle_{\mathbf{H}}, \quad f, g \in \mathbf{G}. \tag{18.10}$$

Later we shall see that these assertions remain true for all $f, g \in L^2(\mathbb{P}_\eta)$.

Lemma 18.3 *The mapping T is linear on $\mathbf{G}$ and $T(f) \in \mathbf{H}$ for all $f \in \mathbf{G}$. Furthermore, equation* (18.10) *holds for all $f, g \in \mathbf{G}$.*

Proof Let $v \in \mathbb{R}_0(\mathbb{X})$ and define $f \in \mathbf{G}$ by $f(\mu) = \exp[-\mu(v)]$, $\mu \in \mathbf{N}$. From (18.7) and Theorem 3.9 we obtain

$$T_n f = \exp[-\lambda(1 - e^{-v})](e^{-v} - 1)^{\otimes n}. \tag{18.11}$$

Since $v \in \mathbb{R}_0(\mathbb{X})$ it follows that $T_n f \in \mathbf{H}_n$, $n \geq 0$. Since the difference operators are linear, this remains true for every $f \in \mathbf{G}$. Moreover, the mapping T is linear on $\mathbf{G}$.

By linearity, it is now sufficient to prove (18.10) in the case

$$f(\mu) = \exp[-\mu(v)], \quad g(\mu) = \exp[-\mu(w)], \quad \mu \in \mathbf{N},$$

for $v, w \in \mathbb{R}_0(\mathbb{X})$. Using Theorem 3.9 again, we obtain

$$\mathbb{E}[f(\eta)g(\eta)] = \exp[-\lambda(1 - e^{-(v+w)})]. \tag{18.12}$$

On the other hand, we have from (18.11) that

$$\begin{aligned}
&\sum_{n=0}^{\infty} \frac{1}{n!} \langle T_n f, T_n g \rangle_n \\
&\quad = \exp[-\lambda(1 - e^{-v})] \exp[-\lambda(1 - e^{-w})] \sum_{n=0}^{\infty} \frac{1}{n!} \lambda^n (((e^{-v} - 1)(e^{-w} - 1))^{\otimes n}) \\
&\quad = \exp[-\lambda(2 - e^{-v} - e^{-w})] \exp[\lambda((e^{-v} - 1)(e^{-w} - 1))].
\end{aligned}$$

This equals the right side of (18.12). Choosing $f = g$ yields $T(f) \in \mathbf{H}$. □

To extend (18.10) to general $f, g \in L^2(\mathbb{P}_\eta)$ we need two lemmas.

Lemma 18.4 *The set $\mathbf{G}$ is dense in $L^2(\mathbb{P}_\eta)$.*

Proof Let $\mathbf{W}$ be the space of all bounded measurable $g: \mathbf{N} \to \mathbb{R}$ that can be approximated in $L^2(\mathbb{P}_\eta)$ by functions in $\mathbf{G}$. This space is closed under monotone uniformly bounded convergence and under uniform convergence. Also it contains the constant functions. The space $\mathbf{G}$ is closed under multiplication. Let $\mathcal{N}' := \sigma(\mathbf{G})$ denote the σ-field generated by the elements of $\mathbf{G}$. A functional version of the monotone class theorem (Theorem A.4) shows that $\mathbf{W}$ contains every bounded $\mathcal{N}'$-measurable g. On the other hand, we have for each $C \in \mathcal{X}_0$ and each $t \geq 0$ that $\mu \mapsto e^{-t\mu(C)}$ is in $\mathbf{G}$ so is $\mathcal{N}'$-measurable, and therefore, since

$$\mu(C) = \lim_{t \to 0+} t^{-1}(1 - e^{-t\mu(C)}), \quad \mu \in \mathbf{N},$$

also $\mu \mapsto \mu(C)$ is $\mathcal{N}'$-measurable. Since λ is a σ-finite measure, for any $C \in \mathcal{X}$ there is a monotone sequence $C_k \in \mathcal{X}_0$, $k \in \mathbb{N}$, with union C, so that $\mu \mapsto \mu(C)$ is $\mathcal{N}'$-measurable. Hence $\mathcal{N} \subset \mathcal{N}'$ and it follows that $\mathbf{W}$ contains all bounded measurable functions. Hence $\mathbf{W}$ is dense in $L^2(\mathbb{P}_\eta)$ and the proof is complete. □

Lemma 18.5 *Suppose that $f, f^1, f^2, \ldots \in L^2(\mathbb{P}_\eta)$ satisfy $f^k \to f$ in $L^2(\mathbb{P}_\eta)$ as $k \to \infty$. Let $n \in \mathbb{N}$ and $C \in \mathcal{X}_0$. Then*

$$\lim_{k \to \infty} \int_{C^n} \mathbb{E}[|D^n_{x_1,\ldots,x_n} f(\eta) - D^n_{x_1,\ldots,x_n} f^k(\eta)|]\, \lambda^n(d(x_1, \ldots, x_n)) = 0. \quad (18.13)$$

Proof By (18.3) it suffices to prove that

$$\lim_{n \to \infty} \int_{C^n} \mathbb{E}\Big[\Big|f\Big(\eta + \sum_{i=1}^m \delta_{x_i}\Big) - f^k\Big(\eta + \sum_{i=1}^m \delta_{x_i}\Big)\Big|\Big]\, \lambda^n(d(x_1, \ldots, x_n)) = 0 \quad (18.14)$$

for all $m \in \{0, \ldots, n\}$. For $m = 0$ this follows from Jensen's inequality. Suppose $m \in \{1, \ldots, n\}$. By the multivariate Mecke equation (see (4.11)), the integral in (18.14) equals

$$\begin{aligned}
&\lambda(C)^{n-m}\mathbb{E}\Big[\int_{C^m} \Big|f\Big(\eta + \sum_{i=1}^m \delta_{x_i}\Big) - f^k\Big(\eta + \sum_{i=1}^m \delta_{x_i}\Big)\Big|\, \lambda^m(d(x_1, \ldots, x_m))\Big] \\
&= \lambda(C)^{n-m}\mathbb{E}\Big[\int_{C^m} |f(\eta) - f^k(\eta)|\, \eta^{(m)}(d(x_1, \ldots, x_m))\Big] \\
&= \lambda(C)^{n-m}\mathbb{E}[|f(\eta) - f^k(\eta)|\, \eta^{(m)}(C^m)].
\end{aligned}$$

By the Cauchy–Schwarz inequality the last is bounded above by

$$\lambda(C)^{n-m}(\mathbb{E}[(f(\eta) - f^k(\eta))^2])^{1/2}(\mathbb{E}[(\eta^{(m)}(C^m))^2])^{1/2}.$$

Since the Poisson distribution has moments of all orders, we obtain (18.14) and hence the lemma. □

Theorem 18.6 (Fock space representation) *The mapping $f \mapsto (T_n(f))_{n\geq 0}$ is linear on $L^2(\mathbb{P}_\eta)$ and takes values in $\mathbf{H}$. Furthermore, we have*

$$\mathbb{E}[f(\eta)g(\eta)] = (\mathbb{E}[f(\eta)])(\mathbb{E}[g(\eta)]) + \sum_{n=1}^{\infty} \frac{1}{n!}\langle T_n f, T_n g\rangle_n \tag{18.15}$$

for all $f, g \in L^2(\mathbb{P}_\eta)$. In particular,

$$\mathbb{E}[f(\eta)^2] = (\mathbb{E}[f(\eta)])^2 + \sum_{n=1}^{\infty} \frac{1}{n!}\|T_n f\|_n^2. \tag{18.16}$$

Proof We first prove (18.16) for $f \in L^2(\mathbb{P}_\eta)$. By Lemma 18.4 there exist $f^k \in \mathbf{G}$, defined for $k \in \mathbb{N}$, satisfying $f^k \to f$ in $L^2(\mathbb{P}_\eta)$ as $k \to \infty$. By Lemma 18.3, for $k, l \in \mathbb{N}$ we have

$$\langle Tf^k - Tf^l, Tf^k - Tf^l\rangle_{\mathbf{H}} = \mathbb{E}[(f^k(\eta) - f^l(\eta))^2]$$

so that Tf^k, $k \in \mathbb{N}$, is a Cauchy sequence in $\mathbf{H}$. Let $\tilde{f} = (\tilde{f}_n)_{n\geq 0} \in \mathbf{H}$ be the limit, meaning that

$$\lim_{k\to\infty} \sum_{n=0}^{\infty} \frac{1}{n!}\|T_n f^k - \tilde{f}_n\|_n^2 = 0. \tag{18.17}$$

Taking the limit in the identity $\mathbb{E}[f^k(\eta)^2] = \langle Tf^k, Tf^k\rangle_{\mathbf{H}}$ yields $\mathbb{E}[f(\eta)^2] = \langle \tilde{f}, \tilde{f}\rangle_{\mathbf{H}}$. Since $f^k \to f \in L^2(\mathbb{P}_\eta)$, we have $\mathbb{E}[f^k(\eta)] \to \mathbb{E}[f(\eta)]$ as $k \to \infty$. Hence equation (18.17) implies that $\tilde{f}_0 = \mathbb{E}[f(\eta)] = T_0 f$. It remains to show for all $n \geq 1$ that

$$\tilde{f}_n = T_n f, \quad \lambda^n\text{-a.e.} \tag{18.18}$$

Let $C \in \mathcal{X}_0$ and, as in Section 4.3, let λ_C^n denote the restriction of the measure λ^n to C^n. By (18.17), then $T_n f^k$ converges in $L^2(\lambda_C^n)$ (and hence in $L^1(\lambda_C^n)$) to $\tilde{f}_n$, while, by the definition (18.8) of T_n and (18.13), $T_n f^k$ converges in $L^1(\lambda_C^n)$ to $T_n f$. Hence these L^1-limits must be the same almost everywhere, so that $\tilde{f}_n = T_n f$ λ^n-a.e. on C^n. Since λ is assumed to be σ-finite, this implies (18.18) and hence (18.16); in particular, $Tf \in \mathbf{H}$.

To see that T is linear, we take $f, g \in L^2(\mathbb{P}_\eta)$ and $a, b \in \mathbb{R}$. As above we can approximate f (resp. g) by a sequence $f^k \in \mathbf{G}$ (resp. $g^k \in \mathbf{G}$), $k \in \mathbb{N}$. Then $af^k + bg^k \to af + bg$ in $L^2(\mathbb{P}_\eta)$ as $k \to \infty$. Since T is linear on $\mathbf{G}$ (Lemma 18.3), we have $T(af^k + bg^k) = aT(f^k) + bT(g^k)$ for all $k \in \mathbb{N}$. In the first part of the proof we have shown that the left-hand side of this equation tends to $T(af + bg)$, while the right-hand side tends to $aT(f) + bT(g)$.

To prove (18.15) we can now use linearity, the *polarisation* identity

$$4\langle u, v\rangle_{\mathbf{H}} = \langle u + v, u + v\rangle_{\mathbf{H}} - \langle u - v, u - v\rangle_{\mathbf{H}}, \quad u, v \in \mathbf{H},$$

and its counterpart in $L^2(\mathbb{P}_\eta)$. □

18.3 The Poincaré Inequality

As a first consequence of the Fock space representation we derive an upper bound for the variance of functions of η in terms of the expected squared difference operator, known as the *Poincaré inequality*.

Theorem 18.7 *Suppose $f \in L^2(\mathbb{P}_\eta)$. Then*

$$\mathbb{V}\mathrm{ar}[f(\eta)] \le \int \mathbb{E}[(D_x f(\eta))^2]\,\lambda(dx). \tag{18.19}$$

Proof We can assume that the right-hand side of (18.19) is finite. In particular, $D_x f \in L^2(\mathbb{P}_\eta)$ for λ-a.e. x. By (18.16),

$$\begin{aligned}
\mathbb{V}\mathrm{ar}[f(\eta)] &= \int (\mathbb{E}[D_x f(\eta)])^2\,\lambda(dx)\\
&\quad + \sum_{n=2}^{\infty} \frac{1}{n!} \iint (\mathbb{E}[D^{n-1}_{x_1,\dots,x_{n-1}} D_x f(\eta)])^2\,\lambda^{n-1}(d(x_1,\dots,x_{n-1}))\,\lambda(dx)\\
&\le \int (\mathbb{E}[D_x f(\eta)])^2\,\lambda(dx)\\
&\quad + \sum_{m=1}^{\infty} \frac{1}{m!} \iint (\mathbb{E}[D^{m}_{x_1,\dots,x_m} D_x f(\eta)])^2\,\lambda^{m}(d(x_1,\dots,x_m))\,\lambda(dx).
\end{aligned}$$

Applying (18.16) to $D_x f$ shows that the preceding upper bound for the variance of $f(\eta)$ is equal to $\int \mathbb{E}[(D_x f(\eta))^2]\,\lambda(dx)$, as required. □

The Poincaré inequality is sharp. Indeed, let $f(\mu) := \mu(B)$ for $B \in \mathcal{X}$ with $\lambda(B) < \infty$. Then $D_x f(\mu) = \mathbf{1}\{x \in B\}$ for all $(x,\mu) \in \mathbb{X} \times \mathbf{N}$ and the right-hand side of (18.19) equals $\lambda(B)$, the variance of F.

Later we shall need the following L^1 version of the Poincaré inequality.

Corollary 18.8 Let $f \in L^1(\mathbb{P}_\eta)$. Then

$$\mathbb{E}[f(\eta)^2] \le (\mathbb{E}[f(\eta)])^2 + \int \mathbb{E}[(D_x f(\eta))^2]\,\lambda(dx).$$

Proof Let $r > 0$. Applying Theorem 18.7 with $f_r := (f \wedge r) \vee (-r)$ gives

$$\mathbb{E}[f_r(\eta)^2] \le (\mathbb{E}[f_r(\eta)])^2 + \int \mathbb{E}[(D_x f(\eta)^2)]\,\lambda(dx),$$

where we have used Exercise 18.4. Monotone (resp. dominated) convergence applied to $\mathbb{E}[f_r(\eta)^2]$ (resp. to $\mathbb{E}[f_r(\eta)]$) yields the result. □

18.4 Chaos Expansion

Let $f \in L^2(\mathbb{P}_\eta)$. In this section we prove that

$$f(\eta) = \sum_{n=0}^{\infty} \frac{1}{n!} I_n(T_n f), \quad \text{in } L^2(\mathbb{P}), \tag{18.20}$$

where the Wiener–Itô integrals $I_n(\cdot)$, $n \in \mathbb{N}$, are defined in Definition 12.10 and $I_0(c) := c$ for each $c \in \mathbb{R}$. This is known as the *chaos expansion* of $f(\eta)$. The following special case is the key for the proof.

Lemma 18.9 *Let $f(\mu) := e^{-\mu(v)}$, $\mu \in \mathbf{N}$, where $v \in \mathbb{R}_+(\mathbb{X})$ and v vanishes outside a set $B \in \mathcal{X}_0$. Then* (18.20) *holds.*

Proof By Theorem 3.9 and (18.11) the right-hand side of (18.20) equals the formal sum

$$I := \exp[-\lambda(1 - e^{-v})] + \exp[-\lambda(1 - e^{-v})] \sum_{n=1}^{\infty} \frac{1}{n!} I_n((e^{-v} - 1)^{\otimes n}). \tag{18.21}$$

Using the pathwise identity (12.12) we obtain that almost surely

$$\begin{aligned} I &= \exp[-\lambda(1 - e^{-v})] \sum_{n=0}^{\infty} \frac{1}{n!} \sum_{k=0}^{n} \binom{n}{k} \eta^{(k)}((e^{-v} - 1)^{\otimes k})(\lambda(1 - e^{-v}))^{n-k} \\ &= \exp[-\lambda(1 - e^{-v})] \sum_{k=0}^{\infty} \frac{1}{k!} \eta^{(k)}((e^{-v} - 1)^{\otimes k}) \sum_{n=k}^{\infty} \frac{1}{(n-k)!} (\lambda(1 - e^{-v}))^{n-k} \\ &= \sum_{k=0}^{N} \frac{1}{k!} \eta^{(k)}((e^{-v} - 1)^{\otimes k}), \end{aligned} \tag{18.22}$$

where $N := \eta(B)$. Assume now that η is proper and write $\delta_{X_1} + \cdots + \delta_{X_N}$ for the restriction of η to B. Then we have almost surely that

$$I = \sum_{J \subset \{1,\ldots,N\}} \prod_{i \in J} (e^{-v(X_i)} - 1) = \prod_{i=1}^{N} e^{-v(X_i)} = e^{-\eta(v)},$$

and hence (18.20) holds with almost sure convergence of the series. To demonstrate that convergence also holds in $L^2(\mathbb{P})$, let $I(m)$ be the partial sum given by the right-hand side of (18.21) with the series terminated at $n = m$. Then since $\lambda(1 - e^{-v})$ is non-negative and $|1 - e^{-v(y)}| \leq 1$ for all y, a similar argument to (18.22) yields

$$|I(m)| \leq \sum_{k=0}^{\min(N,m)} \frac{1}{k!} |\eta^{(k)}((e^{-v} - 1)^{\otimes k})| \leq \sum_{k=0}^{N} \frac{(N)_k}{k!} = 2^N.$$

Since 2^N has finite moments of all orders, by dominated convergence the series (18.21) (and hence (18.20)) converges in $L^2(\mathbb{P})$.

Since (18.20) concerns only the distribution of η, by Proposition 3.2 it has been no restriction of generality to assume that η is proper. □

Theorem 18.10 (Chaos expansion) *Let $f \in L^2(\mathbb{P}_\eta)$. Then* (18.20) *holds.*

Proof By (12.19) and Theorem 18.6,

$$\sum_{n=0}^{\infty} \mathbb{E}\Big[\Big(\frac{1}{n!} I_n(T_n f)\Big)^2\Big] = \sum_{n=0}^{\infty} \frac{1}{n!} \|T_n f\|_n^2 = \mathbb{E}[f(\eta)^2] < \infty.$$

Hence the infinite series of orthogonal terms $S_f := \sum_{n=0}^{\infty} \frac{1}{n!} I_n(T_n f)$ converges in $L^2(\mathbb{P})$. Let $h \in \mathbf{G}$, where $\mathbf{G}$ was defined at (18.9). By Lemma 18.9 and linearity of $I_n(\cdot)$ the sum $\sum_{n=0}^{\infty} \frac{1}{n!} I_n(T_n h)$ converges in $L^2(\mathbb{P})$ to $h(\eta)$. Using (12.19) followed by Theorem 18.6 yields

$$\begin{aligned}\mathbb{E}[(h(\eta) - S_f)^2] &= \sum_{n=0}^{\infty} \frac{1}{(n!)^2} \mathbb{E}[(I_n(T_n h - T_n f))^2] \\ &= \sum_{n=0}^{\infty} \frac{1}{n!} \|T_n h - T_n f\|_n = \mathbb{E}[(f(\eta) - h(\eta))]^2.\end{aligned}$$

Hence if $\mathbb{E}[(f(\eta) - h(\eta))^2]$ is small, then so is $\mathbb{E}[(f(\eta) - S_f)^2]$. Since $\mathbf{G}$ is dense in $L^2(\mathbb{P}_\eta)$ by Lemma 18.4, it follows from the Minkowski inequality that $f(\eta) = S_f$ almost surely. □

18.5 Exercises

Exercise 18.1 Let $v \in \mathbb{R}(\mathbb{X})$ and define $f \in \mathbb{R}_+(\mathbf{N})$ by $f(\mu) := \int v \, d\mu$ if $\int |v| \, d\mu < \infty$ and by $f(\mu) := 0$ otherwise. Show for all $x \in \mathbb{X}$ and all $\mu \in \mathbf{N}$ with $\mu(|v|) < \infty$ that $D_x f(\mu) = v(x)$.

Exercise 18.2 Let $f, g \in \mathbb{R}(\mathbf{N})$ and $x \in \mathbb{X}$. Show that

$$D_x(fg) = (D_x f)g + f(D_x g) + (D_x f)(D_x g).$$

Exercise 18.3 Let $f, \tilde{f} : \mathbf{N} \to \mathbb{R}$ be measurable functions such that $f(\eta) = \tilde{f}(\eta)$ $\mathbb{P}$-a.s. Show for all $n \in \mathbb{N}$ that

$$D^n_{x_1,\dots,x_n} f(\eta) = D^n_{x_1,\dots,x_n} \tilde{f}(\eta), \quad \lambda^n\text{-a.e. } (x_1, \dots, x_n), \ \mathbb{P}\text{-a.s.}$$

(Hint: Use the multivariate Mecke equation (4.11).)

Exercise 18.4 Let $f \in \mathbb{R}(\mathbf{N})$ and $r \geq 0$. Define $f_r \in \mathbb{R}(\mathbf{N})$ by $f_r := (f \wedge r) \vee (-r)$. Show that $|D_x f_r(\mu)| \leq |D_x f(\mu)|$ for each $x \in \mathbb{X}$ and $\mu \in \mathbf{N}$.

Exercise 18.5 Let $\mathbb{X} = C^{(d)}$ as in Example 18.1 and define the function f by (18.5). Let the measure λ be given by the right-hand side of (17.19) (with $\gamma = 1$), where $\mathbb{Q}$ is assumed to satisfy (17.10) and

$$\int (\nu(K+z))^2\, dz\, \mathbb{Q}(dK) < \infty. \tag{18.23}$$

Show that $f \in L^2(\mathbb{P}_\eta)$ and, moreover, that

$$T_n f(K_1, \dots, K_n) = (-1)^{n+1}(1-p)\nu(K_1 \cap \dots \cap K_n),$$

where $p = \mathbb{P}(0 \in Z(\eta))$ is the volume fraction of the Boolean model $Z(\eta)$. Also show that (18.23) is implied by (17.21) whenever $\nu(dx) = \mathbf{1}_W(x)dx$ for some $W \in \mathcal{B}^d$ with $\lambda_d(W) < \infty$.

Exercise 18.6 Let $\mathbb{X} = C^{(d)}$ and let λ be as in Exercise 18.5. Let ν_1, ν_2 be two finite measures on $\mathbb{R}^d$ satisfying (18.23) and define, for $i \in \{1, 2\}$, $f_i(\mu) := \nu_i(Z(\mu))$, $\mu \in \mathbf{N}$; see (18.5). Use Fubini's theorem and Theorem 17.8 to prove that

$$\mathbb{C}\mathrm{ov}(f_1(\eta), f_2(\eta)) = (1-p)^2 \iint (e^{\beta_d(x_1-x_2)} - 1)\, \nu_1(dx_1)\, \nu_2(dx_2),$$

where β_d is given by (17.13). Confirm this result using Theorem 18.6 and Exercise 18.5.

Exercise 18.7 Let $v \in L^1(\lambda) \cap L^2(\lambda)$ and define the function $f \in \mathbb{R}(\mathbf{N})$ as in Exercise 18.1. Show that (18.19) is an equality in this case.

Exercise 18.8 Let $f \in L^2(\mathbb{P}_\eta)$ and $n \in \mathbb{N}$. Show that

$$\mathbb{V}\mathrm{ar}[f(\eta)] \geq \int (\mathbb{E}[D_{x_1,\dots,x_n} f(\eta)])^2\, \lambda^n(d(x_1, \dots, x_n)).$$

Exercise 18.9 Suppose that $f \in L^2(\mathbb{P}_\eta)$ and $g_n \in L^2_s(\lambda^n)$, $n \in \mathbb{N}_0$, such that $f(\eta) = \sum_{n=0}^\infty \frac{1}{n!} I_n(g_n)$ in $L^2(\mathbb{P})$. Show that $g_0 = \mathbb{E}[f(\eta)]$ and $g_n = T_n f$, λ^n-a.e. for all $n \in \mathbb{N}$. (Hint: Let $n \in \mathbb{N}$ and $h \in L^2_s(\lambda^n)$ and use Theorem 18.10 to show that $\mathbb{E}[f(\eta) I_n(h)] = n! \langle T_n f, h \rangle_n$.)

Exercise 18.10 Let $g \in L^2(\lambda)$ and $h \in L^1(\lambda^2) \cap L^2(\lambda^2)$. Define $F := \int g(x) I(h_x)\, \lambda(dx)$, where $h_x := h(x, \cdot)$, $x \in \mathbb{X}$. Prove that $\mathbb{E}[F] = 0$ and

$$\mathbb{E}[F^2] = \int g(x_1) g(x_2) h(x_1, z) h(x_2, z)\, \lambda^3(d(x_1, x_2, z)).$$

(Hint: Prove that $\mathbb{E}[\int |g(x)||I(h_x)|\, \lambda(dx)] < \infty$ using the Cauchy–Schwarz inequality. Then use $f(\mu) = \int g(x)(\mu(h_x) - \lambda(h_x))\, \lambda(dx)$, $\mu \in \mathbf{N}$, as a representative of F and apply (18.16).)

19

Perturbation Analysis

The expectation of a function of a Poisson process can be viewed as a function of the intensity measure. Under first moment assumptions the (suitably defined) directional derivatives of this function can be expressed in terms of the difference operator. This can be applied to geometric functionals of a Boolean model Z with convex grains, governed by a Poisson process with intensity $t \geq 0$. The expectation of an additive functional of the restriction of Z to a convex observation window (viewed as a function of t) satisfies a linear differential equation. As examples we derive explicit formulae for the expected surface content of a general Boolean model with convex grains and the expected Euler characteristic of a planar Boolean model with an isotropic grain distribution concentrated on convex sets.

19.1 A Perturbation Formula

In this chapter we consider an arbitrary measurable space $(\mathbb{X}, \mathcal{X})$ and a Poisson process η_λ on $\mathbb{X}$ with s-finite intensity measure λ. We study the effect of a *perturbation* of the intensity measure λ on the expectation of a fixed function of η.

To explain the idea we take a finite measure ν on $\mathbb{X}$, along with a bounded measurable function $f\colon \mathbf{N}(\mathbb{X}) \to \mathbb{R}$, and study the behaviour of $\mathbb{E}[f(\eta_{\lambda+t\nu})]$ as $t \downarrow 0$. Here and later, given any s-finite measure ρ on $\mathbb{X}$, we let η_ρ denote a Poisson process with this intensity measure. By the superposition theorem (Theorem 3.3) we can write $\mathbb{E}[f(\eta_{\lambda+t\nu})] = \mathbb{E}[f(\eta_\lambda + \eta'_{t\nu})]$, where $\eta'_{t\nu}$ is a Poisson process with intensity measure $t\nu$, independent of η_λ. Then

$$\begin{aligned}\mathbb{E}&[f(\eta_{\lambda+t\nu})] \\ &= \mathbb{E}[f(\eta_\lambda)]\,\mathbb{P}(\eta'_{t\nu}(\mathbb{X}) = 0) + \mathbb{E}[f(\eta_\lambda + \eta'_{t\nu}) \mid \eta'_{t\nu}(\mathbb{X}) = 1]\,\mathbb{P}(\eta'_{t\nu}(\mathbb{X}) = 1) \\ &\quad + \mathbb{E}[f(\eta_\lambda + \eta'_{t\nu}) \mid \eta'_{t\nu}(\mathbb{X}) \geq 2]\,\mathbb{P}(\eta'_{t\nu}(\mathbb{X}) \geq 2).\end{aligned}$$

The measure ν can be written as $\nu = \gamma\,\mathbb{Q}$, where $\gamma \in \mathbb{R}_+$ and $\mathbb{Q}$ is a probability measure on $\mathbb{X}$. Using Proposition 3.5 (and the independence of η_λ and $\eta'_{t\nu}$) to rewrite the second term in the expression on the right-hand side of the preceding equation, we obtain

$$\mathbb{E}[f(\eta_{\lambda+t\nu})] = e^{-t\gamma}\,\mathbb{E}[f(\eta_\lambda)] + \gamma t e^{-t\gamma} \int \mathbb{E}[f(\eta_\lambda + \delta_x)]\,\mathbb{Q}(dx) + R_t, \quad (19.1)$$

where

$$R_t := (1 - e^{-t\gamma} - \gamma t e^{-t\gamma})\mathbb{E}[f(\eta_\lambda + \eta'_{t\nu}) \mid \eta'_{t\nu}(\mathbb{X}) \geq 2].$$

Since f is bounded, $|R_t| \leq ct^2$ for some $c > 0$ and it follows that

$$\lim_{t\downarrow 0} t^{-1}(\mathbb{E}[f(\eta_{\lambda+t\nu})] - \mathbb{E}[f(\eta_\lambda)]) = -\gamma\,\mathbb{E}[f(\eta_\lambda)] + \gamma \int \mathbb{E}[f(\eta_\lambda + \delta_x)]\,\mathbb{Q}(dx).$$

Therefore, the right derivative of $\mathbb{E}[f(\eta_{\lambda+t\nu})]$ at $t = 0$ is given by

$$\frac{d^+}{dt}\mathbb{E}[f(\eta_{\lambda+t\nu})]\bigg|_{t=0} = \int \mathbb{E}[D_x f(\eta_\lambda)]\,\nu(dx). \quad (19.2)$$

The following results elaborate on (19.2). Recall from Theorem A.9 the Hahn–Jordan decomposition $\nu = \nu_+ - \nu_-$ of a finite signed measure ν on $\mathbb{X}$. We also recall from Section A.1 that the integral $\int f\,d\nu$ is defined as $\int f\,d\nu_+ - \int f\,d\nu_-$, whenever this makes sense. Given a finite signed measure ν, we denote by $I(\lambda, \nu)$ the set of all $t \in \mathbb{R}$ such that $\lambda + t\nu$ is a measure. Then $0 \in I(\lambda, \nu)$ and it is easy to see that $I(\lambda, \nu)$ is a (possibly infinite) closed interval. We abbreviate $\mathbf{N} := \mathbf{N}(\mathbb{X})$.

Theorem 19.1 (Perturbation formula) *Let ν be a finite signed measure on $\mathbb{X}$ such that $I(\lambda, \nu) \neq \{0\}$ and suppose that $f \in \mathbb{R}(\mathbf{N})$ is bounded. Then $t \mapsto \mathbb{E}[f(\eta_{\lambda+t\nu})]$ is infinitely differentiable on $I(\lambda, \nu)$ and, for $n \in \mathbb{N}$,*

$$\frac{d^n}{dt^n}\mathbb{E}[f(\eta_{\lambda+t\nu})] = \int \mathbb{E}[D^n_{x_1,\dots,x_n} f(\eta_{\lambda+t\nu})]\,\nu^n(d(x_1, \dots, x_n)), \quad t \in I(\lambda, \nu). \quad (19.3)$$

(For t in the boundary of $I(\lambda, \nu)$ these are one-sided derivatives.)

Proof We start by proving the case $n = 1$, that is

$$\frac{d}{dt}\mathbb{E}[f(\eta_{\lambda+t\nu})] = \int \mathbb{E}[D_x f(\eta_{\lambda+t\nu})]\,\nu(dx), \quad t \in I(\lambda, \nu). \quad (19.4)$$

We first assume that ν is a measure. It is enough to prove (19.4) for $t = 0$ since then for general $t \in I(\lambda, \nu)$ we can apply this formula with λ replaced by $\lambda + t\nu$. Assume that $-s \in I(\lambda, \nu)$ for all sufficiently small $s > 0$. For such s we let $\eta'_{s\nu}$ be a Poisson process with intensity measure $s\nu$, independent of $\eta_{\lambda-s\nu}$. By the superposition theorem (Theorem 3.3) we can then assume without loss of generality that $\eta_\lambda = \eta_{\lambda-s\nu} + \eta'_{s\nu}$. Then it follows exactly as at (19.1) that

$$\mathbb{E}[f(\eta_\lambda)] = e^{-s\gamma}\mathbb{E}[f(\eta_{\lambda-s\nu})] + \gamma s e^{-s\gamma} \int \mathbb{E}[f(\eta_{\lambda-s\nu} + \delta_x)]\,\mathbb{Q}(dx) + R_s,$$

where $|R_s| \le cs^2$ for some $c > 0$. Therefore

$$\begin{aligned}-s^{-1}(\mathbb{E}[f(\eta_{\lambda-s\nu})] - \mathbb{E}[f(\eta_\lambda)]) &= s^{-1}(e^{-s\gamma} - 1)\mathbb{E}[f(\eta_{\lambda-s\nu})] \\ &\quad + \gamma e^{-s\gamma} \int \mathbb{E}[f(\eta_{\lambda-s\nu} + \delta_x)]\,\mathbb{Q}(dx) + s^{-1}R_s. \end{aligned} \tag{19.5}$$

Since ν is a finite measure

$$\mathbb{P}(\eta_\lambda \ne \eta_{\lambda-s\nu}) = \mathbb{P}(\eta'_{s\nu} \ne 0) \to 0$$

as $s \downarrow 0$. Since f is bounded it follows that $\mathbb{E}[f(\eta_{\lambda-s\nu})] \to \mathbb{E}[f(\eta_\lambda)]$ as $s \downarrow 0$. Similarly $\mathbb{E}[f(\eta_{\lambda-s\nu}+\delta_x)]$ tends to $\mathbb{E}[f(\eta_\lambda+\delta_x)]$ for all $x \in \mathbb{X}$. By dominated convergence, even the integrals with respect to $\mathbb{Q}$ converge. By (19.5), the left derivative of $\mathbb{E}[f(\eta_{\lambda+t\nu})]$ at $t = 0$ coincides with the right-hand side of (19.2). Hence (19.4) follows.

By dominated convergence the right-hand side of (19.4) is a continuous function of $t \in I(\lambda, \nu)$. Therefore we obtain from the fundamental theorem of calculus for each $t \in I(\lambda, \nu)$ that

$$\mathbb{E}[f(\eta_{\lambda+t\nu})] = \mathbb{E}[f(\eta_\lambda)] + \int_0^t \int_{\mathbb{X}} \mathbb{E}[D_x f(\eta_{\lambda+s\nu})]\,\nu(dx)\,ds, \tag{19.6}$$

where we use the convention $\int_0^t := -\int_t^0$ for $t < 0$.

We now consider the case where $\nu = \nu_+ - \nu_-$ is a general finite signed measure. Suppose first that $a \in I(\lambda, \nu)$ for some $a > 0$. Then, by (19.6), for $0 \le t \le a$ we have

$$\mathbb{E}[f(\eta_\lambda)] - \mathbb{E}[f(\eta_{\lambda-t\nu_-})] = \int_0^t \int_{\mathbb{X}} \mathbb{E}[D_x f(\eta_{\lambda+(u-t)\nu_-})]\,\nu_-(dx)\,du \tag{19.7}$$

and

$$\mathbb{E}[f(\eta_{\lambda-t\nu_-+t\nu_+})] - \mathbb{E}[f(\eta_{\lambda-t\nu_-})] = \int_0^t \int_{\mathbb{X}} \mathbb{E}[D_x f(\eta_{\lambda-t\nu_-+u\nu_+})]\,\nu_+(dx)\,du. \tag{19.8}$$

For $s \ge 0$, let η_s^- be a Poisson process with intensity measure $s\nu_-$ independent of $\eta_{\lambda-s\nu_-}$. By the superposition theorem we can assume for all $s \ge 0$

that $\eta_\lambda = \eta_{\lambda-s\nu} + \eta_s^-$. Then it follows as before that

$$\mathbb{P}(\eta_\lambda \neq \eta_{\lambda-s\nu_-}) = \mathbb{P}(\eta_s^- \neq 0) \to 0$$

as $s \downarrow 0$, since ν_- is a finite measure. Since also f is bounded we have $\mathbb{E}[D_x f(\eta_{\lambda-s\nu_-})] \to \mathbb{E}[D_x f(\eta_\lambda)]$ as $s \downarrow 0$, so the right-hand side of (19.7) is asymptotic to $t \int \mathbb{E}[D_x f(\eta_\lambda)]\, \nu_-(dx)$ as $t \downarrow 0$. Similarly we have that $\mathbb{E}[f(\eta_{\lambda-t\nu_-+u\nu_+})] - \mathbb{E}[f(\eta_{\lambda-t\nu_-})] \to 0$ as $t, u \downarrow 0$, so the right-hand side of (19.8) is asymptotic to $t \int \mathbb{E} D_x f(\eta_\lambda)\, \nu_+(dx)$ as $t \downarrow 0$. Then we can deduce (19.2) from (19.7) and (19.8).

If $\lambda - a\nu$ is a measure for some $a > 0$, then applying the same argument with $-\nu$ instead of ν gives the differentiability at $t = 0$ of $\mathbb{E}[f(\eta_{\lambda+t\nu})]$. For an arbitrary $t \in I(\lambda, \nu)$ we can apply this result to the measure $\lambda + t\nu$ (instead of λ) to obtain (19.4).

We can now prove (19.3) by induction. Assume that $t \mapsto \mathbb{E}[f(\eta_{\lambda+t\nu})]$ is n times differentiable on $I(\lambda, \nu)$ for each bounded $f \in \mathbb{R}(\mathbf{N})$. For a given f we apply (19.4) to the bounded function $g \in \mathbb{R}(\mathbf{N})$ defined by

$$g(\mu) := \int D^n_{x_1,\dots,x_n} f(\mu)\, \nu^n(d(x_1, \dots, x_n)), \quad \mu \in \mathbf{N}.$$

By linearity of integration we have for each $x \in \mathbb{X}$ that

$$D_x g(\mu) = \int D^{n+1}_{x_1,\dots,x_n,x} f(\mu)\, \nu^n(d(x_1, \dots, x_n)),$$

so that we can conclude the proof from Fubini's theorem. □

19.2 Power Series Representation

Given an interval $I \subset \mathbb{R}$ containing the origin, we say that a function $f\colon I \to \mathbb{R}$ has a *power series representation* (on I) if there is a sequence $a_n \in \mathbb{R}$, $n \in \mathbb{N}$, such that $f(t) = \sum_{n=0}^{\infty} a_n t^n$ for each $t \in I$. In this section we show that $t \mapsto \mathbb{E}[f(\eta_{\lambda+t\nu})]$ has this property under certain assumptions on the function f, the measure λ and the finite signed measure ν.

Theorem 19.2 *Suppose that the assumptions of Theorem 19.1 hold. Then*

$$\mathbb{E}[f(\eta_{\lambda+t\nu})] = \sum_{n=0}^{\infty} \frac{t^n}{n!} \int \mathbb{E}[D^n_{x_1,\dots,x_n} f(\eta_\lambda)]\, \nu^n(d(x_1, \dots, x_n)) \tag{19.9}$$

for all $t \in I(\lambda, \nu)$, where for $n = 0$ the summand is interpreted as $\mathbb{E}[f(\eta_\lambda)]$.

Proof By Theorem 19.1 and Taylor's theorem we have for each $t \in I(\lambda, \nu)$ and each $m \in \mathbb{N}$ that

$$\mathbb{E}[f(\eta_{\lambda+t\nu})] = \sum_{n=0}^{m} \frac{t^n}{n!} \int \mathbb{E}[D^n_{x_1,\dots,x_n} f(\eta_\lambda)]\, \nu^n(d(x_1, \dots, x_n)) + R_m(t),$$

where $|R_m(t)| \le (\nu_+(\mathbb{X}) + \nu_-(\mathbb{X}))c2^{m+1}|t|^{m+1}/(m+1)!$, with c being an upper bound of $|f|$. The result follows. □

The preceding result required the function f to be bounded. For some applications in stochastic geometry this assumption is too strong. The following results apply to more general functions. For a finite signed measure ν with Hahn–Jordan decomposition $\nu = \nu_+ - \nu_-$ we denote by $|\nu| = \nu_+ + \nu_-$ the *total variation measure* of ν.

Theorem 19.3 *Let ν be a finite signed measure on $\mathbb{X}$. Suppose that $f \in \mathbb{R}(\mathbf{N})$ and $t \in I(\lambda, \nu)$ satisfy $\mathbb{E}[|f(\eta_{\lambda+|t||\nu|})|] < \infty$. Then*

$$\sum_{n=0}^{\infty} \frac{|t|^n}{n!} \int \mathbb{E}[|D^n_{x_1,\dots,x_n} f(\eta_\lambda)|]\, |\nu|^n(d(x_1, \dots, x_n)) < \infty \tag{19.10}$$

and (19.9) *holds.*

Proof It suffices to treat the case $t = 1$, since then we could replace ν with $t\nu$. For all $k \in \mathbb{N}$ we define a bounded function $f_k := (f \wedge k) \vee (-k) \in \mathbb{R}(\mathbf{N})$, as in Exercise 18.4. By Theorem 19.2,

$$\mathbb{E}[f_k(\eta_{\lambda+\nu})] = \sum_{n=0}^{\infty} \frac{1}{n!} \int \mathbb{E}[D^n f_k(\eta_\lambda)]\, (h_+ - h_-)^{\otimes n}\, d|\nu|^n, \tag{19.11}$$

where h_- (resp. h_+) is a Radon–Nikodým derivative of ν_- (resp. ν_+) with respect to $|\nu| = \nu_- + \nu_+$ (see Theorem A.10) and where we recall the definition (18.6) of $(h_+ - h_-)^{\otimes n}$. Since $\nu_- \le |\nu|$ and $\nu_+ \le |\nu|$ we have that $h_-(x) \le 1$ and $h_+(x) \le 1$ for $|\nu|$-a.e. x. Since ν_- and ν_+ are mutually singular we also have $h_-(x)h_+(x) = 0$ for $|\nu|$-a.e. x. Therefore,

$$|(h_+(x) - h_-(x))| = h_+(x) + h_-(x) = 1, \quad |\nu|\text{-a.e. } x \in \mathbb{X}.$$

Now let $k \to \infty$ in (19.11). By Exercise 3.8 we have $\mathbb{E}[|f(\eta_\lambda)|] < \infty$. Dominated convergence shows that the left-hand side of (19.11) tends to $\mathbb{E}[f(\eta_{\lambda+\nu})]$. Also $D^n_{x_1,\dots,x_n} f_k(\eta_\lambda)$ tends to $D^n_{x_1,\dots,x_n} f(\eta_\lambda)$ for all $n \in \mathbb{N}_0$, all $x_1, \dots, x_n \in \mathbb{X}$ and everywhere on Ω. Furthermore, by (18.3),

$$|D^n_{x_1,\dots,x_n} f_k(\eta_\lambda)| \le \sum_{J \subset \{1,\dots,n\}} \Big| f\Big(\eta_\lambda + \sum_{j \in J} \delta_{x_j}\Big)\Big|.$$

We shall show that

$$I := \sum_{n=0}^{\infty} \frac{1}{n!} \int \mathbb{E}\Big[\sum_{J\subset\{1,\dots,n\}} \Big| f\Big(\eta_\lambda + \sum_{j\in J} \delta_{x_j}\Big)\Big|\Big] |\nu|^n(d(x_1,\dots,x_n)) < \infty,$$

so that (19.10) follows. Moreover, we can then deduce (19.9) from (19.11) and dominated convergence.

By symmetry, I equals

$$\sum_{n=0}^{\infty} \frac{1}{n!} \sum_{m=0}^{n} \binom{n}{m} |\nu|(\mathbb{X})^{n-m} \int \mathbb{E}[|f(\eta_\lambda + \delta_{x_1} + \cdots + \delta_{x_m})|]\, |\nu|^m(d(x_1,\dots,x_m)).$$

Swapping the order of summation yields that I equals

$$\exp[|\nu|(\mathbb{X})] \sum_{m=0}^{\infty} \frac{1}{m!} \int \mathbb{E}[|f(\eta_\lambda + \delta_{x_1} + \cdots + \delta_{x_m})|]\, |\nu|^m(d(x_1,\dots,x_m))$$
$$= \exp[2|\nu|(\mathbb{X})]\, \mathbb{E}[|f(\eta_\lambda + \eta'_{|\nu|})|],$$

where $\eta'_{|\nu|}$ is a Poisson process with intensity measure $|\nu|$, independent of η_λ, and where we have used Exercise 3.7 (or Proposition 3.5) to achieve the equality. By the superposition theorem (Theorem 3.3) we obtain

$$I = \exp[2|\nu|(\mathbb{X})]\, \mathbb{E}[|f(\eta_{\lambda+|\nu|})|],$$

which is finite by assumption. □

For $f \in \mathbb{R}(\mathbf{N})$ and ν a finite signed measure on $\mathbb{X}$ we let $I_f(\lambda,\nu)$ denote the set of all $t \in I(\lambda,\nu)$ such that $\mathbb{E}[|f(\eta_{\lambda+|t||\nu|})|] < \infty$. If $I_f(\lambda,\nu) \neq \emptyset$ then Exercise 3.8 shows that $I(\lambda,\nu)$ is an interval containing 0. Using Theorem 19.3 we can generalise (19.3) to potentially unbounded functions f.

Theorem 19.4 *Let ν be a finite signed measure on $\mathbb{X}$ and let $f \in \mathbb{R}(\mathbf{N})$. Then the function $t \mapsto \mathbb{E}[f(\eta_{\lambda+t\nu})]$ is infinitely differentiable on the interior I^0 of $I_f(\lambda,\nu)$ and, for all $n \in \mathbb{N}$ and $t \in I^0$,*

$$\frac{d^n}{dt^n}\mathbb{E}[f(\eta_{\lambda+t\nu})] = \int \mathbb{E}[D^n_{x_1,\dots,x_n} f(\eta_{\lambda+t\nu})]\, \nu^n(d(x_1,\dots,x_n)). \tag{19.12}$$

Proof The asserted differentiability is a consequence of (19.9) and well-known properties of power series. The same is true for (19.3) in the case $t = 0$. For general $t \in I^0$ we can apply this formula with λ replaced by $\lambda + t\nu$. □

19.3 Additive Functions of the Boolean Model

As an application of the perturbation formulae, consider the Boolean model in $\mathbb{R}^d$, as in Definition 17.1. Let η_t be a stationary Poisson process on $\mathbb{R}^d$ with intensity $t \geq 0$ and let $\mathbb{Q}$ be a grain distribution satisfying (17.10). Recall that $C^{(d)} = C^d \setminus \{\emptyset\}$ is the system of all non-empty compact subsets of $\mathbb{R}^d$ equipped with the Hausdorff distance (17.1). We assume that $\mathbb{Q}$ is concentrated on the system $\mathcal{K}^{(d)}$ of all convex $K \in C^{(d)}$. Let ξ_t be an independent $\mathbb{Q}$-marking of η_t and define

$$Z_t := \bigcup_{(K,x)\in\xi_t} (K + x). \tag{19.13}$$

Given compact $W \subset \mathbb{R}^d$, Proposition 17.5 and Lemma A.30 show that $Z_t \cap W$ is a random element of C^d. In fact, by Exercise 17.1 we can assume without loss of generality that for all $\omega \in \Omega$ and all compact convex $W \subset \mathbb{R}^d$ the set $Z_t(\omega) \cap W$ is a finite (possibly empty) union of convex sets. We define the *convex ring* $\mathcal{R}^d$ to be the system of all such unions. By Theorem A.26, $\mathcal{R}^d$ is a Borel subset of C^d. Therefore, $Z_t \cap W$ is a random element of $\mathcal{R}^d$, whenever W is a compact convex set.

A measurable function $\varphi\colon \mathcal{R}^d \to \mathbb{R}$ is said to be *locally finite* if

$$\sup\{|\varphi(K)| : K \in \mathcal{K}^d, K \subset W\} < \infty, \quad W \in \mathcal{K}^d. \tag{19.14}$$

Recall that a function $\varphi\colon \mathcal{R}^d \to \mathbb{R}$ is said to be *additive* if $\varphi(\emptyset) = 0$ and $\varphi(K \cup L) = \varphi(K) + \varphi(L) - \varphi(K \cap L)$ for all $K, L \in \mathcal{R}^d$.

Proposition 19.5 *Let $\varphi\colon \mathcal{R}^d \to \mathbb{R}$ be measurable, additive and locally finite. For $W \in \mathcal{K}^d$ let $S_{\varphi,W}(t) := \mathbb{E}[\varphi(Z_t \cap W)]$. Then $S_{\varphi,W}(\cdot)$ has a power series representation on $\mathbb{R}_+$ and the derivative is given by*

$$\begin{aligned} S'_{\varphi,W}(t) = &\iint \varphi(W \cap (K + x))\, dx\, \mathbb{Q}(dK) \\ &- \iint \mathbb{E}[\varphi(Z_t \cap W \cap (K + x))]\, dx\, \mathbb{Q}(dK). \end{aligned} \tag{19.15}$$

Proof We aim to apply Theorem 19.4 with

$$\mathbb{X} := \{(x, K) \in \mathbb{R}^d \times \mathcal{K}^{(d)} : (K + x) \cap W \neq \emptyset\},$$

$\lambda = 0$ and ν the restriction of $\lambda_d \otimes \mathbb{Q}$ to $\mathbb{X}$. By assumption (17.10) and (16.11), this measure is finite. Define the function $f\colon \mathbf{N}(\mathbb{X}) \to \mathbb{R}_+$ by

$$f(\mu) := \varphi(Z(\mu) \cap W),$$

where $Z(\mu) := \bigcup_{(x,K)\in\mu}(K + x)$ if $\mu(\mathbb{X}) < \infty$ and $Z(\mu) := \emptyset$ otherwise.

By Theorem A.26 the space $C^{(d)}$ is a CSMS and, by Lemma A.24, so is $\mathbb{R}^d \times C^{(d)}$. Hence Proposition 6.2 applies and it follows as in Proposition 17.5 that $\mu \mapsto Z(\mu)$ is a measurable mapping taking values in the convex ring $\mathcal{R}^d$. Since φ is a measurable function, so is f. To show that $\mathbb{E}[f(\eta_{tv})] < \infty$, we write η_{tv} in the form

$$\eta_{tv} = \sum_{n=1}^{\kappa} \delta_{(X_n, Z_n')},$$

where κ is Poisson distributed with parameter $\lambda(\mathbb{X})$, (X_n) is a sequence of random vectors in $\mathbb{R}^d$ and (Z_n') is a sequence of random elements of $\mathcal{K}^{(d)}$. Let $Y_n := Z_n' + X_n$. By the inclusion–exclusion principle (A.30),

$$\begin{aligned} f(\eta_{tv}) &= f\Big(\bigcup_{n=1}^{\kappa} Y_n \cap W\Big) \\ &= \sum_{n=1}^{\kappa} (-1)^{n-1} \sum_{1 \le i_1 < \cdots < i_n \le \kappa} \varphi(W \cap Y_{i_1} \cdots \cap Y_{i_n}). \end{aligned} \tag{19.16}$$

Using (19.14) we get

$$|f(\eta_{tv})| \le \sum_{n=1}^{\kappa} \binom{\kappa}{n} c_W \le 2^{\kappa} c_W,$$

where c_W is the supremum in (19.14). It follows that $\mathbb{E}[|f(\eta_{tv})|] < \infty$.

By Theorem 19.3 the function $S_{\varphi,W}(\cdot)$ has a power series representation on $\mathbb{R}_+$. By Theorem 19.4 the derivative is given by

$$\begin{aligned} S'_{\varphi,W}(t) &= \iint \mathbb{E}[D_{(x,K)} f(\eta_{tv})]\, dx\, \mathbb{Q}(dK) \\ &= \iint \mathbb{E}[\varphi((Z_t \cup (K + x)) \cap W) - \varphi(Z_t \cap W)]\, dx\, \mathbb{Q}(dK). \end{aligned}$$

From the additivity property (A.29) and linearity of integrals we obtain (19.15), provided that

$$\iint (|\varphi(W \cap (K + x))| + \mathbb{E}[|\varphi(Z_t \cap W \cap (K + x))|])\, dx\, \mathbb{Q}(dK) < \infty.$$

Since $\varphi(\emptyset) = 0$, we have

$$\iint |\varphi(W \cap (K + x))|\, dx\, \mathbb{Q}(dK) \le c_W \iint \mathbf{1}\{W \cap (K + x) \neq \emptyset\}\, dx\, \mathbb{Q}(dK),$$

which is finite by (17.7) and assumption (17.10). Using (19.16) with W

replaced by $W \cap (K + x)$ gives $\mathbb{E}[|\varphi(Z_t \cap W \cap (K+x))|] \le c_W \mathbb{E}[2^\kappa]$, uniformly in $(K, x) \in \mathcal{K}^d \times \mathbb{R}^d$. Hence we obtain as before that

$$\iint \mathbb{E}[|\varphi(Z_t \cap W \cap (K + x))|]\, dx\, \mathbb{Q}(dK)$$
$$\le c_W \mathbb{E}[2^\kappa] \iint \mathbf{1}\{W \cap (K + x) \ne \emptyset\}\, dx\, \mathbb{Q}(dK) < \infty,$$

and the proposition is proved. □

Sometimes the right-hand side of (19.15) satisfies the assumptions of the following theorem.

Theorem 19.6 *Let $m \in \mathbb{N}$. For $j \in \{1, \dots, m\}$ let $\psi_j \colon \mathcal{R}^d \to \mathbb{R}$ be a measurable, additive and locally finite function. For $W \in \mathcal{K}^d$ let $S_{j,W}(t) := \mathbb{E}[\psi_j(Z_t \cap W)]$. Suppose that*

$$\iint \psi_1(A \cap (K + x))\, dx\, \mathbb{Q}(dK) = \sum_{j=1}^{m} c_j \psi_j(A), \quad A \in \mathcal{R}^d, \tag{19.17}$$

for certain constants $c_j \in \mathbb{R}$ depending on $\mathbb{Q}$ but not on A. Then $S_{1,W}$ is a differentiable function satisfying

$$S'_{1,W}(t) = \sum_{j=1}^{m} c_j \psi_j(W) - \sum_{j=1}^{m} c_j S_{j,W}(t). \tag{19.18}$$

Proof Applying (19.15) with $\varphi = \psi_1$ and using (19.17) twice (the second time with $Z_t \cap W$ in place of W), we obtain

$$S'_{1,W}(t) = \sum_{j=1}^{m} c_j \psi_j(W) - \mathbb{E}\Big[\sum_{j=1}^{m} c_j \psi_j(Z_t \cap W)\Big],$$

where we have also used Fubini's theorem. This yields the assertion. □

Now we consider the intrinsic volumes $V_i \colon \mathcal{R}^d \to \mathbb{R}$, $i \in \{0, \dots, d\}$, as defined in Section A.3. As mentioned after (A.26) these functionals are increasing on $\mathcal{K}^d$ with respect to set inclusion; therefore they are locally finite. The distribution $\mathbb{Q}$ is said to be *isotropic* if $\mathbb{Q}(\{\rho K : K \in A\}) = \mathbb{Q}(A)$ for all measurable $A \subset \mathcal{K}^{(d)}$ and (proper) rotations $\rho \colon \mathbb{R}^d \to \mathbb{R}^d$, where $\rho K := \{\rho(x) : x \in K\}$. If $\mathbb{Q}$ is isotropic then there are coefficients $c_{i,j} \in \mathbb{R}$ (for $i, j \in \{0, \dots, d\}$) such that for each $i \in \{0, \dots, d\}$

$$\iint V_i(A \cap (K + x))\, dx\, \mathbb{Q}(dK) = \sum_{j=0}^{d} c_{i,j} V_j(A), \quad A \in \mathcal{R}^d. \tag{19.19}$$

This can be established with Hadwiger's characterisation theorem (Theorem A.25), just as in the special case $i = 0$ in the forthcoming proof of Theorem 19.8. In fact, for $i \in \{d-1, d\}$ isotropy is not needed; see (22.11) for the case $i = d$ and (A.24) for the case $i = d-1$. Moreover, it is possible to show that $c_{i,j} = 0$ for $i > j$. Now, Theorem 19.6 shows for $W \in \mathcal{K}^d$ and $i \in \{0, \dots, d\}$ that

$$\frac{d}{dt}\mathbb{E}[V_i(Z_t \cap W)] = \sum_{j=0}^{d} c_{i,j} V_j(W) - \sum_{j=0}^{d} c_{i,j}\mathbb{E}[V_j(Z_t \cap W)], \quad t \geq 0,$$

which is a system of linear differential equations. Using methods from *integral geometry* it is possible to determine the coefficients c_{ij}. We shall not pursue this general case any further. In the following two sections we shall instead discuss the (simple) case $i = d-1$ (without isotropy assumption) and the case $i = 0$ for isotropic planar Boolean models.

19.4 Surface Density of the Boolean Model

In this section we shall give a formula for the *surface density*

$$S_W(t) := \mathbb{E}[V_{d-1}(Z_t \cap W)], \quad t \geq 0,$$

where $W \in \mathcal{K}^d$ and where we refer to (A.23) for a geometric interpretation. It turns out that $S_W(t)$ can be easily expressed in terms of ϕ_d and ϕ_{d-1}, where as in (17.11) we define $\phi_i := \int V_i(K)\,\mathbb{Q}(dK)$ for $i \in \{0, \dots, d\}$.

Theorem 19.7 *Let Z_t be as in* (19.13) *and let $W \in \mathcal{K}^d$. Then*

$$S_W(t) = \phi_{d-1} t e^{-t\phi_d} V_d(W) + (1 - e^{-t\phi_d}) V_{d-1}(W), \quad t \geq 0. \tag{19.20}$$

Proof By Proposition 19.5 and (A.24),

$$\begin{aligned} S'_W(t) = &\int V_d(W)V_{d-1}(K)\,\mathbb{Q}(dK) + \int V_{d-1}(W)V_d(K)\,\mathbb{Q}(dK) \\ &- \int \mathbb{E}[V_d(Z_t \cap W)]V_{d-1}(K)\,\mathbb{Q}(dK) \\ &- \int \mathbb{E}[V_{d-1}(Z_t \cap W)]V_d(K)\,\mathbb{Q}(dK). \end{aligned}$$

By Proposition 17.4 we have for all $t \geq 0$ that

$$\mathbb{E}[V_d(Z_t \cap W)] = (1 - e^{-t\phi_d})V_d(W), \tag{19.21}$$

and therefore

$$S'_W(t) = V_d(W)\phi_{d-1} + V_{d-1}(W)\phi_d - (1 - e^{-t\phi_d})V_d(W)\phi_{d-1} - S_W(t)\phi_d.$$

Note that $S_W(0) = 0$. It is easily checked that this linear differential equation is (uniquely) solved by the right-hand side of the asserted formula (19.20). □

The right-hand side of (19.20) admits a clear geometric interpretation. The first term is the mean surface content in W of all grains that are not covered by other grains. Indeed, $\phi_{d-1}tV_d(W)$ is the mean surface content of all grains ignoring overlapping, while $e^{-t\phi_d}$ can be interpreted as the probability that a point on the boundary of a contributing grain is not covered by other grains; see (19.21). The second term is the contribution of that part of the boundary of W which is covered by the Boolean model.

19.5 Mean Euler Characteristic of a Planar Boolean Model

Finally in this chapter we deal with the *Euler characteristic* V_0 in the case $d = 2$; see Section A.3 for the definition and a geometric interpretation.

Theorem 19.8 *Let Z_t be a Boolean model in $\mathbb{R}^2$ with intensity $t \geq 0$ and with an isotropic grain distribution $\mathbb{Q}$ concentrated on $\mathcal{K}^{(2)}$ and satisfying (17.10). Then, for all $W \in \mathcal{K}^2$,*

$$\mathbb{E}[V_0(Z_t \cap W)] = (1 - e^{-t\phi_2})V_0(W) + \frac{2}{\pi}te^{-t\phi_2}\phi_1 V_1(W) + te^{-t\phi_2}V_2(W) - \frac{1}{\pi}t^2e^{-t\phi_2}\phi_1^2V_2(W). \quad (19.22)$$

Proof In the first part of the proof we work in general dimensions. We plan to apply Theorem 19.6 to the intrinsic volumes $V_0, \ldots, V_d$. To do so, we need to establish (19.17), that is

$$\iint V_0(A \cap (K + x))\, dx\, \mathbb{Q}(dK) = \sum_{j=0}^{d} c_j V_j(A), \quad A \in \mathcal{R}^d, \quad (19.23)$$

for certain constants $c_0, \ldots, c_d \in \mathbb{R}$. Since both sides of this equation are additive in A, we can by (A.30) assume that $A \in \mathcal{K}^d$. Then the left-hand side of (19.23) simplifies to

$$\varphi(A) := \iint \mathbf{1}\{A \cap (K + x) \neq \emptyset\}\, dx\, \mathbb{Q}(dK) = \int V_d(K \oplus A^*)\, \mathbb{Q}(dK),$$

where we recall that $A^* := \{-x : x \in A\}$. By Exercise 19.7 the function φ is invariant under translations and rotations. We now prove that φ is continuous on $\mathcal{K}^{(d)}$ (with respect to Hausdorff distance). If $A_n \in \mathcal{K}^{(d)}$ converge to some $A \in \mathcal{K}^{(d)}$, then $(A_n)^*$ converges to A^*. Hence, for any $K \in \mathcal{K}^{(d)}$, by Exercise 17.6 the sets $K \oplus (A_n)^*$ converge to $K \oplus A^*$, so that the continuity of V_d on $\mathcal{K}^{(d)}$ (mentioned in Section A.3) shows that $V_d(K \oplus (A_n)^*)$ tends to $V_d(K \oplus A^*)$. Moreover, the definition of Hausdorff distance implies that the A_n are all contained in some (sufficiently large) ball. By assumption (17.10) we can apply dominated convergence to conclude that $\varphi(A_n) \to \varphi(A)$ as $n \to \infty$.

Hadwiger's characterisation (Theorem A.25) shows that (19.23) holds. To determine the coefficients $c_0, \dots, c_d$ we take $A = B(0, r)$ for $r \geq 0$. The Steiner formula (A.22) and the definition (17.11) of ϕ_i show that

$$\varphi(B(0,r)) = \sum_{j=0}^{d} r^j \kappa_j \int V_{d-j}(K)\, \mathbb{Q}(dK) = \sum_{j=0}^{d} r^j \kappa_j \phi_{d-j}. \tag{19.24}$$

On the other hand, by (A.25) and (A.26), for $A = B(0, r)$ the right-hand side of (19.23) equals

$$\sum_{j=0}^{d} c_j r^j \binom{d}{j} \frac{\kappa_d}{\kappa_{d-j}},$$

where we recall that $\kappa_0 = 1$. It follows that

$$c_j = \phi_{d-j} \frac{j!\kappa_j (d-j)!\kappa_{d-j}}{d!\kappa_d}, \qquad j = 0, \dots, d.$$

In the remainder of the proof we assume that $d = 2$. Then (19.23) reads

$$\iint V_0(A \cap (K + x))\, dx\, \mathbb{Q}(dK_2) = \phi_2 V_0(A) + \frac{2\phi_1}{\pi} V_1(A) + V_2(A), \tag{19.25}$$

for each $A \in \mathcal{R}^2$. Inserting (19.20) and (19.21) into (19.18) yields

$$\begin{aligned}\frac{d}{dt}\mathbb{E}[V_0(Z_t \cap W)] = \sum_{j=0}^{2} c_j V_j(W) - c_0\, \mathbb{E}[V_0(Z_t \cap W)] - c_1 \phi_1 t e^{-t\phi_2} V_2(W) \\ - c_1(1 - e^{-t\phi_2}) V_1(W) - c_2(1 - e^{-t\phi_2}) V_2(W).\end{aligned}$$

A simple calculation shows that this differential equation is indeed solved by the right-hand side of (19.22). □

19.6 Exercises

Exercise 19.1 Let λ be an s-finite measure on $\mathbb{X}$ and ν a finite measure on $\mathbb{X}$. Suppose that $f \in \mathbb{R}(\mathbf{N})$ satisfies $\mathbb{E}[|f(\eta_{\lambda+t|\nu|})|] < \infty$ for some $t > 0$. Prove by a direct calculation that (19.9) holds. (Hint: Use the calculation in the proof of Theorem 19.3.)

Exercise 19.2 Let ν be a finite measure on $\mathbb{X}$ and let $f \in \mathbb{R}(\mathbf{N})$ satisfy $\mathbb{E}[|f(\eta_{a\nu})|] < \infty$ for some $a > 0$. Show that

$$\frac{d}{dt}\mathbb{E}[f(\eta_{t\nu})] = t^{-1}\mathbb{E}\int (f(\eta_{t\nu}) - f(\eta_{t\nu} \setminus \delta_x))\,\eta_{t\nu}(dx), \quad t \in [0, a].$$

Exercise 19.3 Let ν be a finite measure on $\mathbb{X}$ and let $A \in \mathcal{N}$ be *increasing*, that is $\mu \in A$ implies $\mu + \delta_x \in A$ for all $x \in \mathbb{X}$. Let

$$N_A(t) := \int \mathbf{1}\{\eta_{t\nu} \in A, \eta_{t\nu} \setminus \delta_x \notin A\}\,\eta_{t\nu}(dx)$$

denote the number of points of $\eta_{t\nu}$ that are *pivotal* for A. Show that

$$\frac{d}{dt}\mathbb{P}(\eta_{t\nu} \in A) = t^{-1}\mathbb{E}[N_A(t)], \quad t > 0.$$

Exercise 19.4 Let ν be a finite signed measure on $\mathbb{X}$ and let $t > 0$ be such that $\lambda + t\nu$ is a measure. Let $f \in \mathbb{R}(\mathbf{N})$ satisfy $\mathbb{E}[|f(\eta_{\lambda+t\nu})|] < \infty$, $\mathbb{E}[|f(\eta_\lambda)|] < \infty$ and

$$\int_0^t \int_{\mathbb{X}} \mathbb{E}[|D_x f(\eta_{\lambda+s\nu})|]\,|\nu|(dx)\,ds < \infty. \tag{19.26}$$

Prove that then (19.6) holds. (Hint: Apply Theorem 19.1 to a suitably truncated function f and apply dominated convergence.)

Exercise 19.5 Let ν be a finite signed measure and $t > 0$ such that $\lambda + t\nu$ is a measure. Suppose that $f \in \mathbb{R}_+(\mathbf{N})$ satisfies (19.26). Show that $\mathbb{E}[f(\eta_{\lambda+t\nu})] < \infty$ if and only if $\mathbb{E}[f(\eta_\lambda)] < \infty$. (Hint: Use Exercise 19.4.)

Exercise 19.6 Let $W \in \mathcal{K}^d$ such that $\lambda_d(W) > 0$ and let $S_W(t)$ be as in Theorem 19.7. Show that

$$\lim_{r\to\infty} \frac{S_{rW}(t)}{\lambda_d(rW)} = \phi_{d-1} t e^{-t\phi_d}.$$

Formulate and prove an analogous result in the setting of Theorem 19.8.

Exercise 19.7 Let $\mathbb{Q}$ be a distribution on $\mathcal{K}^{(d)}$ satisfying (17.10) and let $i \in \{0, \dots, d\}$. Show that the function $A \mapsto \iint V_i(A \cap (K + x))\,dx\,\mathbb{Q}(dK)$ from $\mathcal{K}^d$ to $\mathbb{R}$ is invariant under translations. Assume in addition that $\mathbb{Q}$ is invariant under rotations; show that then φ has the same property.

Exercise 19.8 Let η be a proper Poisson process on $\mathbb{X}$ with finite intensity measure λ. For $p \in [0, 1]$ let $\mathbb{Q}_p$ be the probability measure on $\{0, 1\}$ given by $\mathbb{Q}_p := p\delta_1 + (1-p)\delta_0$. Let ξ_p be an independent $\mathbb{Q}_p$-marking of η. Suppose that $f \in \mathbb{R}(\mathbf{N}(\mathbb{X} \times \{0,1\}))$ is bounded and show that

$$\frac{d}{dp}\mathbb{E}[f(\xi_p)] = \int \mathbb{E}[f(\xi_p + \delta_{(x,1)}) - f(\xi_p + \delta_{(x,0)})]\,\lambda(dx).$$

(Hint: Use Theorem 19.1 with λ replaced by $\lambda \otimes \delta_0$ and with ν replaced by $\lambda \otimes \delta_1 - \lambda \otimes \delta_0$.)

Exercise 19.9 Suppose that $f \in \mathbb{R}(\mathbf{N}(\mathbb{X} \times \{0,1\}))$. For each $x \in \mathbb{X}$ define $\Delta_x f \in \mathbb{R}(\mathbf{N}(\mathbb{X} \times \{0,1\}))$ by $\Delta_x f(\mu) := f(\mu + \delta_{(x,1)}) - f(\mu + \delta_{(x,0)})$. For $n \in \mathbb{N}$ and $x_1, \dots, x_n \in \mathbb{X}$ define $\Delta^n_{x_1,\dots,x_n} f \in \mathbb{R}(\mathbf{N}(\mathbb{X} \times \{0,1\}))$ recursively by $\Delta^1_{x_1} f := \Delta_{x_1} f$ and

$$\Delta^n_{x_1,\dots,x_n} f := \Delta_{x_n} \Delta^{n-1}_{x_1,\dots,x_{n-1}} f, \quad n \geq 2.$$

Show for all $\mu \in \mathbf{N}(\mathbb{X} \times \{0,1\})$ that

$$\Delta^n_{x_1,\dots,x_n} f(\mu) = \sum_{(i_1,\dots,i_n)\in\{0,1\}^n} (-1)^{n-i_1-\cdots-i_n} f(\mu + \delta_{(x_1,i_1)} + \cdots + \delta_{(x_n,i_n)}).$$

Exercise 19.10 Suppose that η and ξ_p, $p \in [0,1]$, are as in Exercise 19.8 and let $f \in \mathbb{R}(\mathbf{N}(\mathbb{X} \times \{0,1\}))$ be bounded. Let $n \in \mathbb{N}$ and show that

$$\frac{d^n}{dp^n}\mathbb{E}[f(\xi_p)] = \int \mathbb{E}[\Delta^n_{x_1,\dots,x_n} f(\xi_p)]\,\lambda^n(d(x_1,\dots,x_n)), \quad p \in [0,1].$$

Deduce from Taylor's theorem for all $p, q \in [0,1]$ that

$$\mathbb{E}[f(\xi_q)] = \mathbb{E}[f(\xi_p)] + \sum_{n=1}^{\infty} \frac{(q-p)^n}{n!} \int \mathbb{E}[\Delta^n_{x_1,\dots,x_n} f(\xi_p)]\,\lambda^n(d(x_1,\dots,x_n)).$$

(Hint: Use Exercise 19.8 and induction as in the final part of the proof of Theorem 19.1. The second assertion can be proved as Theorem 19.2.)

Exercise 19.11 Let η and ξ_p, $p \in [0,1]$, be as in Exercise 19.8. Assume that $\mathbb{X}$ is a Borel space and that λ is diffuse. Let $\mu \in \mathbf{N}(\mathbb{X} \times \{0,1\})$ and $x_1, \dots, x_n \in \mathbb{X}$. Define $\mu^!_{x_1,\dots,x_n} := \mu - \delta_{(x_1,j_1)} - \cdots - \delta_{(x_n,j_n)}$ whenever $\mu(\cdot \times \{0,1\})$ is simple, $x_1, \dots, x_n$ are pairwise distinct and $(x_1, j_1), \dots, (x_n, j_n) \in \mu$ for some $j_1, \dots, j_n \in \{0,1\}$. In all other cases define $\mu^!_{x_1,\dots,x_n} := \mu$. Let $f \in \mathbb{R}(\mathbf{N}(\mathbb{X} \times \{0,1\}))$ be bounded. Show for all $p, q \in [0,1]$ that $\mathbb{P}$-a.s.

$$\mathbb{E}[f(\xi_q) \mid \eta] = \mathbb{E}[f(\xi_p) \mid \eta] + \sum_{n=1}^{\infty} \frac{(q-p)^n}{n!} \int \mathbb{E}\Big[\Delta^n_{x_1,\dots,x_n} f((\xi_p)^!_{x_1,\dots,x_n}) \;\Big|\; \eta\Big]\,\eta^{(n)}(d(x_1,\dots,x_n)).$$

(Hint: Use Exercise 19.10 and the multivariate Mecke equation and note that η is simple.)

20

Covariance Identities

A measurable function of a Poisson process is called Poisson functional. Given a square integrable Poisson functional F, and given $t \in [0, 1]$, the Poisson functional $P_t F$ is defined by a combination of t-thinning and independent superposition. The family $P_t F$ interpolates between the expectation of F and F. The Fock space series representation of the covariance between two Poisson functionals can be rewritten as an integral equation involving only the first order difference operator and the operator P_t. This identity will play a key role in Chapter 21 on normal approximation. A corollary is the Harris–FKG correlation inequality.

20.1 Mehler's Formula

In this chapter we consider a proper Poisson process η on a measurable space $(\mathbb{X}, \mathcal{X})$ with σ-finite intensity measure λ and distribution $\mathbb{P}_\eta$. Let L^0_η be the space of all $\overline{\mathbb{R}}$-valued random variables (Poisson functionals) F such that $F = f(\eta)$ $\mathbb{P}$-a.s. for some measurable $f\colon \mathbf{N} \to \mathbb{R}$. This f is called a *representative* of F. If f is a (fixed) representative of F we define

$$D^n_{x_1,\dots,x_n} F := D^n_{x_1,\dots,x_n} f(\eta), \quad n \in \mathbb{N}, x_1, \dots, x_n \in \mathbb{X}.$$

Exercise 18.3 shows that the choice of the representative f does not affect $D^n_{x_1,\dots,x_n} F(\omega)$ up to sets of $\mathbb{P} \otimes \lambda^n$-measure 0. We also write $D_x F := D_x f(\eta)$, $x \in \mathbb{X}$. For ease of exposition we shall also allow for representatives f with values in $[-\infty, \infty]$. In this case we apply the previous definitions to the representative $\mathbf{1}\{|f| < \infty\} f$.

By assumption we can represent η as in (2.4), that is $\eta = \sum_{n=1}^{\kappa} \delta_{X_n}$. Let $U_1, U_2, \dots$ be independent random variables, uniformly distributed on $[0, 1]$ and independent of $(\kappa, (X_n)_{n\ge1})$. Define

$$\eta_t := \sum_{n=1}^{\kappa} \mathbf{1}\{U_n \le t\} \delta_{X_n}, \quad t \in [0, 1]. \tag{20.1}$$

Then η_t is a t-thinning of η. Note that $\eta_0 = 0$ and $\eta_1 = \eta$. For $q > 0$ let L^q_η denote the space of all $F \in L^0_\eta$ such that $\mathbb{E}[|F|^q] < \infty$. For $F \in L^1_\eta$ with representative f we define

$$P_t F := \mathbb{E}\Big[\int f(\eta_t + \mu)\,\Pi_{(1-t)\lambda}(d\mu)\,\Big|\,\eta\Big], \quad t \in [0,1], \tag{20.2}$$

where we recall that $\Pi_{\lambda'}$ denotes the distribution of a Poisson process with intensity measure λ'. By the superposition and thinning theorems (Theorem 3.3 and Corollary 5.9),

$$\Pi_\lambda = \mathbb{E}\Big[\int \mathbf{1}\{\eta_t + \mu \in \cdot\}\,\Pi_{(1-t)\lambda}(d\mu)\Big]. \tag{20.3}$$

Hence the definition of $P_t F$ does not depend on the representative of F up to almost sure equality. Moreover, Lemma B.16 shows that

$$P_t F = \int \mathbb{E}[f(\eta_t + \mu) \mid \eta]\,\Pi_{(1-t)\lambda}(d\mu), \quad \mathbb{P}\text{-a.s.}, \quad t \in [0,1]. \tag{20.4}$$

We also note that

$$P_t F = \mathbb{E}[f(\eta_t + \eta'_{1-t}) \mid \eta], \tag{20.5}$$

where η'_{1-t} is a Poisson process with intensity measure $(1-t)\lambda$, independent of the pair (η, η_t). Exercise 20.1 yields further insight into the properties of the operator P_t.

By (20.5),

$$\mathbb{E}[P_t F] = \mathbb{E}[F], \quad F \in L^1_\eta, \tag{20.6}$$

while the (conditional) Jensen inequality (Proposition B.1) shows for all $p \geq 1$ the contractivity property

$$\mathbb{E}[|P_t F|^p] \leq \mathbb{E}[|F|^p], \quad t \in [0,1],\ F \in L^p_\eta. \tag{20.7}$$

The proof of the covariance identity in Theorem 20.2 below is based on the following result, which is of independent interest.

Lemma 20.1 (Mehler's formula) *Let $F \in L^2_\eta$, $n \in \mathbb{N}$ and $t \in [0,1]$. Then*

$$D^n_{x_1,\dots,x_n}(P_t F) = t^n P_t D^n_{x_1,\dots,x_n} F, \quad \lambda^n\text{-a.e. } (x_1,\dots,x_n) \in \mathbb{X}^n,\ \mathbb{P}\text{-a.s.} \tag{20.8}$$

In particular,

$$\mathbb{E}[D^n_{x_1,\dots,x_n}(P_t F)] = t^n \mathbb{E}[D^n_{x_1,\dots,x_n} F], \quad \lambda^n\text{-a.e. } (x_1,\dots,x_n) \in \mathbb{X}^n. \tag{20.9}$$

Proof Let f be a representative of F. We first assume that $f(\mu) = e^{-\mu(v)}$ for some $v \in \mathbb{R}_0(\mathbb{X})$, where $\mathbb{R}_0(\mathbb{X})$ is as in the text preceding (18.9). It follows from the definition (20.1) (see also Exercise 5.4) that

$$\mathbb{E}[e^{-\eta_t(v)} \mid \eta] = \exp\Big[\int \log(1 - t + te^{-v(y)})\,\eta(dy)\Big], \quad \mathbb{P}\text{-a.s.} \tag{20.10}$$

Hence, by (20.5) and Theorem 3.9, the following function f_t is a representative of P_tF:

$$f_t(\mu) := \exp\Big[-(1-t)\int (1 - e^{-v})\,d\lambda\Big]\exp\Big[\int \log((1-t) + te^{-v(y)})\,\mu(dy)\Big]. \tag{20.11}$$

Let $x \in \mathbb{X}$. Since

$$\exp\Big[\int \log(1 - t + te^{-v(y)})\,(\mu + \delta_x)(dy)\Big]$$
$$= \exp\Big[\int \log(1 - t + te^{-v(y)})\,\mu(dy)\Big](1 - t + te^{-v(x)}),$$

we obtain $\mathbb{P}$-a.s. and for λ-a.e. $x \in \mathbb{X}$ that

$$D_xP_tF = f_t(\eta + \delta_x) - f_t(\eta) = t(e^{-v(x)} - 1)f_t(\eta) = t(e^{-v(x)} - 1)P_tF.$$

This identity can be iterated to yield for all $n \in \mathbb{N}$ and λ^n-a.e. $(x_1, \dots, x_n)$ and $\mathbb{P}$-a.s. that

$$D^n_{x_1,\dots,x_n}P_tF = t^n \prod_{i=1}^n (e^{-v(x_i)} - 1)P_tF.$$

On the other hand, we have from (18.7) that $\mathbb{P}$-a.s.

$$P_tD^n_{x_1,\dots,x_n}F = P_t\prod_{i=1}^n (e^{-v(x_i)} - 1)F = \prod_{i=1}^n (e^{-v(x_i)} - 1)P_tF,$$

so that (20.8) holds for Poisson functionals of the given form.

By linearity, (20.8) extends to all F with a representative in the set $\mathbf{G}$ defined at (18.9). By Lemma 18.4 there exist functions $f^k \in \mathbf{G}$, $k \in \mathbb{N}$, satisfying $F^k := f^k(\eta) \to F = f(\eta)$ in $L^2(\mathbb{P})$ as $k \to \infty$. Therefore we obtain from the contractivity property (20.7) that

$$\mathbb{E}[(P_tF^k - P_tF)^2] = \mathbb{E}[(P_t(F^k - F))^2] \le \mathbb{E}[(F^k - F)^2] \to 0$$

as $k \to \infty$. Taking $B \in \mathcal{X}$ with $\lambda(B) < \infty$, it therefore follows from Lemma 18.5 that

$$\mathbb{E}\Big[\int_{B^n} |D^n_{x_1,\dots,x_n}P_tF_k - D^n_{x_1,\dots,x_n}P_tF|\,\lambda^n(d(x_1,\dots,x_n))\Big] \to 0$$

as $k \to \infty$. On the other hand, we obtain from the Fock space representation (18.16) that $\mathbb{E}[|D^n_{x_1,\dots,x_n}F|] < \infty$ for λ^n-a.e. $(x_1,\dots,x_n) \in \mathbb{X}^n$, so that the linearity of P_t and (20.7) together imply

$$\mathbb{E}\Big[\int_{B^n} |P_t D^n_{x_1,\dots,x_n}F_k - P_t D^n_{x_1,\dots,x_n}F|\,\lambda^n(d(x_1,\dots,x_n))\Big]$$
$$\leq \int_{B^n} \mathbb{E}[|D^n_{x_1,\dots,x_n}(F_k - F)|]\,\lambda^n(d(x_1,\dots,x_n)).$$

Again by Lemma 18.5, this latter integral tends to 0 as $k \to \infty$. Since (20.8) holds for each F_k we obtain from the triangle inequality that

$$\mathbb{E}\Big[\int_{B^n} |D^n_{x_1,\dots,x_n}P_t F - t^n P_t D^n_{x_1,\dots,x_n}F|\,\lambda^n(d(x_1,\dots,x_n))\Big] = 0.$$

Therefore (20.8) holds $\mathbb{P} \otimes (\lambda_B)^n$-a.e., and hence, since λ is σ-finite, also $\mathbb{P} \otimes \lambda^n$-a.e.

Taking the expectation in (20.8) and using (20.6) proves (20.9). □

20.2 Two Covariance Identities

For $F \in L^2_\eta$ we denote by DF the mapping $(\omega, x) \mapsto (D_x F)(\omega)$. The next theorem requires the additional assumption $DF \in L^2(\mathbb{P} \otimes \lambda)$, that is

$$\mathbb{E}\Big[\int (D_x F)^2\,\lambda(dx)\Big] < \infty. \tag{20.12}$$

Theorem 20.2 *For any $F, G \in L^2_\eta$ such that $DF, DG \in L^2(\mathbb{P} \otimes \lambda)$,*

$$\mathbb{E}[FG] - \mathbb{E}[F]\mathbb{E}[G] = \mathbb{E}\Big[\iint_0^1 (D_x F)(P_t D_x G)\,dt\,\lambda(dx)\Big]. \tag{20.13}$$

Proof Exercise 20.6 shows that the integrand on the right-hand side of (20.13) can be assumed to be measurable. Using the Cauchy–Schwarz inequality and then the contractivity property (20.7) yields

$$\Big(\mathbb{E}\Big[\iint_0^1 |D_x F||P_t D_x G|\,dt\,\lambda(dx)\Big]\Big)^2$$
$$\leq \mathbb{E}\Big[\int (D_x F)^2\,\lambda(dx)\Big]\mathbb{E}\Big[\int (D_x G)^2\,\lambda(dx)\Big], \tag{20.14}$$

which is finite by assumption. Therefore we can use Fubini's theorem and (20.8) to obtain that the right-hand side of (20.13) equals

$$\iint_0^1 t^{-1}\mathbb{E}[(D_x F)(D_x P_t G)]\,dt\,\lambda(dx). \tag{20.15}$$

For $t \in [0,1]$ and λ-a.e. $x \in \mathbb{X}$ we can apply the Fock space isometry (18.15) to D_xF and D_xP_tG. Taking into account Lemma 20.1 this gives

$$\mathbb{E}[(D_xF)(D_xP_tG)] = t\,\mathbb{E}[D_xF]\mathbb{E}[D_xG] + \sum_{n=1}^{\infty} \frac{t^{n+1}}{n!} \int \mathbb{E}[D^{n+1}_{x_1,\dots,x_n,x}F]\,\mathbb{E}[D^{n+1}_{x_1,\dots,x_n,x}G]\,\lambda^n(d(x_1,\dots,x_n)).$$

Inserting this into (20.15), applying Fubini's theorem (to be justified at the end of the proof) and performing the integration over $[0,1]$ shows that the double integral (20.15) equals

$$\int \mathbb{E}[D_xF]\mathbb{E}[D_xG]\,\lambda(dx) + \sum_{n=1}^{\infty} \frac{1}{(n+1)!} \iint \mathbb{E}[D^{n+1}_{x_1,\dots,x_n,x}F]\,\mathbb{E}[D^{n+1}_{x_1,\dots,x_n,x}G]\,\lambda^n(d(x_1,\dots,x_n))\,\lambda(dx)$$
$$= \sum_{m=1}^{\infty} \frac{1}{m!} \int \mathbb{E}[D^m_{x_1,\dots,x_m}F]\,\mathbb{E}[D^m_{x_1,\dots,x_m}G]\,\lambda^m(d(x_1,\dots,x_m)).$$

By (18.15) this equals $\mathbb{E}[FG] - \mathbb{E}[F]\mathbb{E}[G]$, which yields the asserted formula (20.13). By (18.16) and the Cauchy–Schwarz inequality we have that

$$\sum_{m=1}^{\infty} \frac{1}{m!} \int \big|\mathbb{E}[D^m_{x_1,\dots,x_m}F]\big|\,\big|\mathbb{E}[D^m_{x_1,\dots,x_m}G]\big|\,\lambda^m(d(x_1,\dots,x_m)) < \infty,$$

justifying the use of Fubini's theorem. □

Next we prove a symmetric version of Theorem 20.2 that avoids additional integrability assumptions.

Theorem 20.3 *Let $F \in L^2_\eta$ and $G \in L^2_\eta$. Then*

$$\mathbb{E}\left[\iint_0^1 (\mathbb{E}[D_xF \mid \eta_t])^2\,dt\,\lambda(dx)\right] < \infty \tag{20.16}$$

and

$$\mathbb{E}[FG] - \mathbb{E}[F]\mathbb{E}[G] = \mathbb{E}\left[\iint_0^1 \mathbb{E}[D_xF \mid \eta_t]\mathbb{E}[D_xG \mid \eta_t]\,dt\,\lambda(dx)\right]. \tag{20.17}$$

Proof By the thinning theorem (Corollary 5.9), η_t and $\eta - \eta_t$ are independent Poisson processes with intensity measures $t\lambda$ and $(1-t)\lambda$, respectively. Therefore we have for $F \in L^2_\eta$ with representative f and λ-a.e. $x \in \mathbb{X}$ that

$$\mathbb{E}[D_xF \mid \eta_t] = \int D_xf(\eta_t + \mu)\,\Pi_{(1-t)\lambda}(d\mu) \tag{20.18}$$

holds almost surely. By (20.1), the right-hand side of (20.18) is a jointly measurable function of (the suppressed) $\omega \in \Omega$, $x \in \mathbb{X}$ and $t \in [0, 1]$.

Now we take $F, G \in L^2_\eta$ with representatives f and g, respectively. Let us first assume that $DF, DG \in L^2(\mathbb{P} \otimes \lambda)$. Then (20.16) follows from the (conditional) Jensen inequality and the law of the total expectation. The definition (20.2) shows for all $t \in [0, 1]$ and λ-a.e. $x \in \mathbb{X}$ that

$$\mathbb{E}[(D_xF)(P_tD_xG)] = \mathbb{E}\Big[D_xF \int D_xg(\eta_t + \mu)\, \Pi_{(1-t)\lambda}(d\mu)\Big],$$

so that by (20.18)

$$\mathbb{E}[(D_xF)(P_tD_xG)] = \mathbb{E}[D_xF\, \mathbb{E}[D_xG \mid \eta_t]] = \mathbb{E}[\mathbb{E}[D_xF \mid \eta_t]\mathbb{E}[D_xG \mid \eta_t]].$$

Therefore (20.17) is just another version of (20.13).

Now we consider general $F, G \in L^2_\eta$. By Lemma 18.4 there is a sequence F^k, $k \in \mathbb{N}$, of Poisson functionals with representatives in $\mathbf{G}$, such that $\mathbb{E}[(F - F^k)^2] \to 0$ as $k \to \infty$. Equation (18.7) shows for each $k \geq 1$ that $DF^k \in L^2(\mathbb{P} \otimes \lambda)$. By the case of (20.17) already proved we have that

$$\mathbb{V}\mathrm{ar}[F^k - F^l] = \mathbb{E}\Big[\int (\mathbb{E}[D_xF^k \mid \eta_t] - \mathbb{E}[D_xF^l \mid \eta_t])^2\, \lambda^*(d(x,t))\Big]$$

holds for all $k, l \in \mathbb{N}$, where λ^* is the product of λ and Lebesgue measure on $[0, 1]$. Since the space $L^2(\mathbb{P}\otimes\lambda^*)$ is complete, there exists $H \in L^2(\mathbb{P}\otimes\lambda^*)$ satisfying

$$\lim_{k\to\infty} \mathbb{E}\Big[\int (H(x,t) - \mathbb{E}[D_xF^k \mid \eta_t])^2\Big]\lambda^*(d(x,t)) = 0. \tag{20.19}$$

On the other hand, it follows from the triangle inequality for conditional expectations and Lemma 18.5 that, for each $C \in \mathcal{X}$ with $\lambda(C) < \infty$,

$$\int_{C\times[0,1]} \mathbb{E}[|\mathbb{E}[D_xF^k \mid \eta_t] - \mathbb{E}[D_xF \mid \eta_t]|]\, \lambda^*(d(x,t))$$
$$\leq \int_{C\times[0,1]} \mathbb{E}[|D_xF^k - D_xF|]\, \lambda^*(d(x,t)) \to 0, \quad \text{as } k \to \infty.$$

Comparing this with (20.19) shows that $H(\omega, x, t) = \mathbb{E}[D_xF \mid \eta_t](\omega)$ for $\mathbb{P}\otimes\lambda^*$-a.e. $(\omega, x, t) \in \Omega\times C\times[0, 1]$ and hence also for $\mathbb{P}\otimes\lambda^*$-a.e. $(\omega, x, t) \in \Omega \times \mathbb{X} \times [0, 1]$. Therefore since $H \in L^2(\mathbb{P} \otimes \lambda^*)$ we have (20.16). Now let G^k, $k \in \mathbb{N}$, be a sequence approximating G similarly. Then equation (20.17) holds with (F^k, G^k) instead of (F, G). But the right-hand side is just an inner product in $L^2(\mathbb{P} \otimes \lambda^*)$. Taking the limit as $k \to \infty$ and using the L^2-convergence proved above, namely (20.19) with $H(x) = \mathbb{E}[D_xF \mid \eta_t]$ and likewise for G, yields the general result. □

20.3 The Harris–FKG Inequality

As an application of the preceding theorem we obtain a useful correlation inequality for increasing functions of η. Given $B \in \mathcal{X}$, a function $f \in \mathbb{R}(\mathbf{N})$ is said to be *increasing on* B if $f(\mu + \delta_x) \geq f(\mu)$ for all $\mu \in \mathbf{N}$ and all $x \in B$. It is said to be *decreasing on* B if $(-f)$ is increasing on B.

Theorem 20.4 *Suppose $B \in \mathcal{X}$. Let $f, g \in L^2(\mathbb{P}_\eta)$ be increasing on B and decreasing on $\mathbb{X} \setminus B$. Then*

$$\mathbb{E}[f(\eta)g(\eta)] \geq (\mathbb{E}[f(\eta)])(\mathbb{E}[g(\eta)]). \tag{20.20}$$

Proof The assumptions imply that the right-hand side of (20.17) is non-negative. Hence the result follows. □

20.4 Exercises

Exercise 20.1 Suppose that $\mathbb{X}$ is a Borel subspace of a CSMS and that λ is locally finite. Let $U_1, U_2, \dots$ be independent random variables, uniformly distributed on $[0, 1]$, and let η'_{1-t} be a Poisson process with intensity measure $(1-t)\lambda$, independent of the sequence (U_n). Let π_n, $n \in \mathbb{N}$, be as in Proposition 6.3. Given $\mu \in \mathbf{N}$, if μ is locally finite then define

$$\xi_t(\mu) := \eta'_{1-t} + \sum_{n=1}^{k} \mathbf{1}\{U_n \leq t\}\delta_{\pi_n(\mu)}.$$

If $\mu \in \mathbf{N}$ is not locally finite, let $\xi_t(\mu) := 0$. Let $F \in L^1_\eta$ have representative f. Show that $\mu \mapsto \mathbb{E}[f(\xi_t(\mu))]$ is a representative of $P_t F$. (Hint: One needs to show that $f_t(\mu) := \mathbb{E}[f(\xi_t(\mu))]$ is a measurable function of μ and that $\mathbb{E}[g(\eta)f_t(\eta)] = \mathbb{E}[g(\eta)P_t F]$ for all $g \in \mathbb{R}_+(\mathbf{N})$.)

Exercise 20.2 Let $v \in L^1(\lambda)$ and $F := I(v) = \eta(v) - \lambda(v)$; see (12.4). Let $t \in [0, 1]$ and show that $P_t F = tF$, $\mathbb{P}$-a.s. (Hint: Take $f(\mu) := \mu(v) - \lambda(v)$ as a representative of F.)

Exercise 20.3 Let $u \in L^2(\lambda)$ and $F \in L^2_\eta$ such that $DF \in L^2(\mathbb{P} \otimes \lambda)$. Use Theorem 20.2 to show that

$$\mathrm{Cov}[I_1(u), F] = \mathbb{E}\Big[\int u(x) D_x F\, \lambda(dx)\Big].$$

Give an alternative proof using the Mecke equation, making the additional assumption $u \in L^1(\lambda)$.

Exercise 20.4 Let $h \in L^1_s(\lambda^2)$ and $F := I_2(h)$; see (12.9). Show that $D_x F = 2I(h_x)$ for λ-a.e. x, where $h_x(y) := h(x, y)$, $y \in \mathbb{X}$. (Hint: Use Exercise 4.3.)

Exercise 20.5 Let $h \in L^1(\lambda) \cap L^2(\lambda)$ and let $F := I_2(h)$. Use Exercise 20.4 to show that $DF \in L^2(\mathbb{P} \otimes \lambda)$.

Exercise 20.6 Let $f \in \mathbb{R}(\mathbb{X} \times \mathbf{N})$ such that $f_x \in L^1(\mathbb{P}_\eta)$ for each $x \in \mathbb{X}$, where $f_x := f(x, \cdot)$, and let $F_x := f_x(\eta)$. Show that there exists a jointly measurable version of $P_t F_x$, that is an $\tilde{f} \in \mathbb{R}(\mathbb{X} \times [0, 1] \times \Omega)$ satisfying

$$\tilde{f}(x, t, \cdot) = \mathbb{E}\left[\int f_x(\eta_t + \mu) \Pi_{(1-t)\lambda}(d\mu) \,\middle|\, \eta \right], \quad \mathbb{P}\text{-a.s., } x \in \mathbb{X},\ t \in [0, 1].$$

(Hint: Assume first that f does not depend on $x \in \mathbb{X}$. Then (20.11) shows the assertion in the case $f \in \mathbf{G}$, where $\mathbf{G}$ is defined at (18.9). The proof of Lemma 18.4 shows $\sigma(\mathbf{G}) = \mathcal{N}$, so that Theorem A.4 can be used to derive the assertion for a general bounded $f \in \mathbb{R}(\mathbf{N})$.)

Exercise 20.7 Let $h \in L^1_s(\lambda^2) \cap L^2(\lambda^2)$ and $F := I_2(h)$. Show that

$$Z := \iint_0^1 (P_t D_x F)(D_x F)\, dt\, \lambda(dx)$$

is in $L^2(\mathbb{P})$. Prove further that

$$Z = 2 \int I(h_x)^2\, \lambda(dx), \quad \mathbb{P}\text{-a.s.}$$

(with $h_x := h(x, \cdot)$) and

$$D_y Z = 4 \int h(x, y) I(h_x)\, \lambda(dx) + 2 \int h(x, y)^2\, \lambda(dx), \quad \lambda\text{-a.e. } y \in \mathbb{X},\ \mathbb{P}\text{-a.s.}$$

(Hint: Use Exercises 20.2 and 20.4.)

Exercise 20.8 Let Z be a Boolean model as in Definition 17.1 and let $K_1, \dots, K_n$ be compact subsets of $\mathbb{R}^d$. Use Theorem 20.4 to show that

$$\mathbb{P}(Z \cap K_1 \neq \emptyset, \dots, Z \cap K_n \neq \emptyset) \geq \prod_{j=1}^{n} \mathbb{P}(Z \cap K_j \neq \emptyset).$$

Give an alternative proof based on Theorem 17.3.

Exercise 20.9 Let $f \in L^{2+\varepsilon}_\eta$ for some $\varepsilon > 0$ and assume that $\lambda(\mathbb{X}) < \infty$. Show that $Df(\eta) \in L^2(\mathbb{P} \otimes \lambda)$.

21

Normal Approximation

The Wasserstein distance quantifies the distance between two probability distributions. Stein's method is a general tool for obtaining bounds on this distance. Combining this method with the covariance identity of Chapter 20 yields upper bounds on the Wasserstein distance between the distribution of a standardised Poisson functional and the standard normal distribution. In their most explicit form these bounds depend only on the first and second order difference operators.

21.1 Stein's Method

In this chapter we consider a Poisson process η on an arbitrary measurable space $(\mathbb{X}, \mathcal{X})$ with σ-finite intensity measure λ. For a given Poisson functional $F \in L^2_\eta$ (that is, F is a square integrable and $\sigma(\eta)$-measurable almost surely finite random variable) we are interested in the distance between the distribution of F and the standard normal distribution. We use here the *Wasserstein distance* to quantify the discrepancy between the laws of two (almost surely finite) random variables X_0, X_1. This distance is defined by

$$d_1(X_0, X_1) = \sup_{h \in \mathbf{Lip}(1)} |\mathbb{E}[h(X_0)] - \mathbb{E}[h(X_1)]|, \tag{21.1}$$

where $\mathbf{Lip}(1)$ denotes the space of all Lipschitz functions $h\colon \mathbb{R} \to \mathbb{R}$ with a Lipschitz constant less than or equal to one; see (B.3). If a sequence (X_n) of random variables satisfies $\lim_{n\to\infty} d_1(X_n, X_0) = 0$, then Proposition B.9 shows that X_n converges to X_0 in distribution. Here we are interested in the central limit theorem, that is in the case where X_0 has a standard normal distribution.

Let $\mathbf{AC}_{1,2}$ be the set of all differentiable functions $g\colon \mathbb{R} \to \mathbb{R}$ such that the derivative g' is absolutely continuous and satisfies $\sup\{|g'(x)| : x \in \mathbb{R}\} \le 1$ and $\sup\{|g''(x)| : x \in \mathbb{R}\} \le 2$, for some version g'' of the Radon–Nikodým derivative of g'. The next theorem is the key to the results of this chapter. Throughout the chapter we let N denote a standard normal random variable.

Theorem 21.1 (Stein's method) *Let $F \in L^1(\mathbb{P})$. Then*

$$d_1(F, N) \le \sup_{g \in \mathbf{AC}_{1,2}} |\mathbb{E}[g'(F) - Fg(F)]|. \tag{21.2}$$

Proof Let $h \in \mathbf{Lip}(1)$. Proposition B.13 shows that there exists $g \in \mathbf{AC}_{1,2}$ such that

$$h(x) - \mathbb{E}[h(N)] = g'(x) - xg(x), \quad x \in \mathbb{R}. \tag{21.3}$$

It follows that

$$|\mathbb{E}[h(F)] - \mathbb{E}[h(N)]| = |\mathbb{E}[g'(F) - Fg(F)]|.$$

Taking the supremum yields the assertion. □

Next we use the covariance identity of Theorem 20.2 to turn the general bound (21.2) into a result for Poisson functionals.

Theorem 21.2 *Assume that the Poisson process η is proper and suppose that $F \in L^2_\eta$ satisfies $DF \in L^2(\mathbb{P} \otimes \lambda)$ and $\mathbb{E}[F] = 0$. Then*

$$d_1(F, N) \le \mathbb{E}\left[\left|1 - \iint_0^1 (P_t D_x F)(D_x F)\, dt\, \lambda(dx)\right|\right] + \mathbb{E}\left[\iint_0^1 |P_t D_x F|(D_x F)^2\, dt\, \lambda(dx)\right]. \tag{21.4}$$

Proof Let f be a representative of F and let $g \in \mathbf{AC}_{1,2}$. Then we have for λ-a.e. $x \in \mathbb{X}$ and $\mathbb{P}$-a.s. that

$$D_x g(F) = g(f(\eta + \delta_x)) - g(f(\eta)) = g(F + D_x F) - g(F). \tag{21.5}$$

Since g is Lipschitz (by the boundedness of its first derivative) it follows that $|D_x g(F)| \le |D_x F|$ and therefore $Dg(F) \in L^2(\mathbb{P} \otimes \lambda)$. Moreover, since

$$|g(F)| \le |g(F) - g(0)| + |g(0)| \le |F| + |g(0)|,$$

also $g(F) \in L^2_\eta$. Then Theorem 20.2 yields

$$\mathbb{E}[Fg(F)] = \mathbb{E}\left[\iint_0^1 (P_t D_x F)(D_x g(F))\, dt\, \lambda(dx)\right]$$

and it follows that

$$|\mathbb{E}[g'(F) - Fg(F)]| \le \mathbb{E}\left[\left|g'(F) - \iint_0^1 (P_t D_x F)(D_x g(F))\, dt\, \lambda(dx)\right|\right]. \tag{21.6}$$

We assert that there exists a measurable function $R\colon \mathbb{X}\times\mathbf{N}\to\mathbb{R}$ such that for λ-a.e. $x\in\mathbb{X}$ and $\mathbb{P}$-a.s.

$$D_x g(F) = g'(F)D_xF + R(x,\eta)(D_xF)^2, \quad x\in\mathbb{X}. \tag{21.7}$$

Indeed, for $x\in\mathbb{X}$ and $\mu\in\mathbf{N}$ with $D_xf(\mu)\neq 0$, we can define

$$R(x,\mu) := (D_xf(\mu))^{-2}(D_xg(f(\mu)) - g'(f(\mu))D_xf(\mu)).$$

Otherwise we set $R(x,\mu) := 0$. Since $D_xF = 0$ implies that $D_xg(F) = 0$, we obtain (21.7). Using (21.7) in (21.6) gives

$$\begin{aligned}|\mathbb{E}[g'(F) - Fg(F)]| \le \mathbb{E}\Big[|g'(F)|\Big|1 - \iint_0^1 (P_tD_xF)(D_xF)\,dt\,\lambda(dx)\Big|\Big]\\ + \mathbb{E}\Big[\iint_0^1 |P_tD_xF||R(x,\eta)|(D_xF)^2\,dt\,\lambda(dx)\Big].\end{aligned} \tag{21.8}$$

By assumption, $|g'(F)|\le 1$. Moreover, Proposition A.35 and the assumption $|g''(y)|\le 2$ for λ_1-a.e. $y\in\mathbb{R}$ imply for all $(x,\mu)\in\mathbb{X}\times\mathbf{N}$ that

$$|g(f(\mu+\delta_x)) - g(f(\mu)) - g'(f(\mu))D_xf(\mu)| \le (D_xf(\mu))^2,$$

so that $|R(x,\mu)|\le 1$. Using these facts in (21.8) and applying Theorem 21.1 gives the bound (21.4). □

21.2 Normal Approximation via Difference Operators

Since the bound (21.4) involves the operators P_t it is often not easy to apply. The following bound is the main result of this chapter, and involves only the first and second order difference operators. In fact it can be represented in terms of the following three constants:

$$\begin{aligned}\alpha_{F,1} &:= 2\Big[\int (\mathbb{E}[(D_xF)^2(D_yF)^2])^{1/2}\\ &\qquad \times (\mathbb{E}[(D^2_{x,z}F)^2(D^2_{y,z}F)^2])^{1/2}\,\lambda^3(d(x,y,z))\Big]^{1/2},\\ \alpha_{F,2} &:= \Big[\int \mathbb{E}[(D^2_{x,z}F)^2(D^2_{y,z}F)^2]\,\lambda^3(d(x,y,z))\Big]^{1/2},\\ \alpha_{F,3} &:= \int \mathbb{E}[|D_xF|^3]\,\lambda(dx).\end{aligned}$$

Theorem 21.3 *Suppose that $F\in L^2_\eta$ satisfies $DF\in L^2(\mathbb{P}\otimes\lambda)$, $\mathbb{E}[F]=0$ and $\mathbb{V}\mathrm{ar}[F]=1$. Then,*

$$d_1(F,N) \le \alpha_{F,1} + \alpha_{F,2} + \alpha_{F,3}. \tag{21.9}$$

Proof Clearly we can assume that $\alpha_{F,1}, \alpha_{F,2}, \alpha_{F,3}$ are finite.

Since (21.9) concerns only the distribution of η, by Corollary 3.7 it is no restriction of generality to assume that η is a proper point process. Our starting point is the inequality (21.4). By Hölder's inequality (A.2), the second term on the right-hand side can be bounded from above by

$$\iint_0^1 (\mathbb{E}[|P_t D_x F|^3])^{1/3} (\mathbb{E}[|D_x F|^3])^{2/3} \, dt \, \lambda(dx) \le \alpha_{F,3}, \tag{21.10}$$

where the inequality comes from the contractivity property (20.7). Applying Jensen's inequality to the first term on the right-hand side of (21.4), we see that it is enough to show that

$$\left(\mathbb{E}\left[\left(1 - \iint_0^1 (P_t D_x F)(D_x F) \, dt \, \lambda(dx)\right)^2\right]\right)^{1/2} \le \alpha_{F,1} + \alpha_{F,2}. \tag{21.11}$$

Let $Z := \iint_0^1 (P_t D_x F)(D_x F) \, dt \, \lambda(dx)$. Theorem 20.2 and our assumptions on F show that $\mathbb{E}[Z] = 1$. Hence the left-hand side of (21.11) equals $(\mathbb{E}[Z^2]-1)^{1/2}$. By the L^1-version of the Poincaré inequality (Corollary 18.8),

$$\mathbb{E}[Z^2] - 1 \le \mathbb{E}\left[\int (D_y Z)^2 \, \lambda(dy)\right]. \tag{21.12}$$

By Hölder's inequality and (21.10), the random variable

$$W(\eta) := \iint_0^1 |(P_t D_x F)(D_x F)| \, dt \, \lambda(dx)$$

is integrable and therefore $\mathbb{P}$-a.s. finite. Hence, by Exercise 4.1, $W(\eta + \delta_y)$ is also $\mathbb{P}$-a.s. finite for λ-a.e. $y \in \mathbb{X}$. Hence, by the triangle inequality,

$$\iint_0^1 |D_y[(P_t D_x F)(D_x F)]| \, dt \, \lambda(dx) < \infty, \quad \mathbb{P}\text{-a.s., } \lambda\text{-a.e. } y \in \mathbb{X}. \tag{21.13}$$

Therefore

$$D_y Z = \iint_0^1 D_y[(P_t D_x F)(D_x F)] \, dt \, \lambda(dx),$$

again $\mathbb{P}$-a.s. and for λ-a.e. $y \in \mathbb{X}$. Hence we obtain from (21.12) that

$$\mathbb{E}[Z^2] - 1 \le \mathbb{E}\left[\int \left(\iint_0^1 |D_y[(P_t D_x F)(D_x F)]| \, dt \, \lambda(dx)\right)^2 \lambda(dy)\right]. \tag{21.14}$$

Comparison of (21.14) and (21.11) now shows that the inequality

$$\left(\mathbb{E}\left[\int \left(\iint_0^1 |D_y((P_t D_x F)(D_x F))| \, dt \, \lambda(dx)\right)^2 \lambda(dy)\right]\right)^{1/2} \le \alpha_{F,1} + \alpha_{F,2} \tag{21.15}$$

would imply (21.11).

We now verify (21.15). To begin with we apply Exercise 18.2 and the inequality $(a+b+c)^2 \le 3(a^2+b^2+c^2)$ for any $a,b,c \in \mathbb{R}$ (a consequence of Jensen's inequality) to obtain

$$\mathbb{E}\Big[\int\Big(\iint_0^1 |D_y((P_t D_x F)(D_x F))|\, dt\, \lambda(dx)\Big)^2 \lambda(dy)\Big] \le 3(I_1+I_2+I_3), \tag{21.16}$$

where

$$I_1 := \mathbb{E}\Big[\int\Big(\iint_0^1 |D_y P_t D_x F||D_x F|\, dt\, \lambda(dx)\Big)^2 \lambda(dy)\Big],$$

$$I_2 := \mathbb{E}\Big[\int\Big(\iint_0^1 |P_t D_x F||D^2_{x,y} F|\, dt\, \lambda(dx)\Big)^2 \lambda(dy)\Big],$$

$$I_3 := \mathbb{E}\Big[\int\Big(\iint_0^1 |D_y P_t D_x F||D^2_{x,y} F|\, dt\, \lambda(dx)\Big)^2 \lambda(dy)\Big].$$

We shall bound I_1, I_2, I_3 with the help of Lemma 21.4 below.

Since $DF \in L^2(\mathbb{P}\otimes\lambda)$ we have that $D_x F \in L^2(\mathbb{P})$ for λ-a.e. x, so that Mehler's formula (Lemma 20.1) shows that $D_y P_t D_x F = t P_t D^2_{x,y} F$ for λ^2-a.e. $(x,y) \in \mathbb{X}^2$ and $\mathbb{P}$-a.s. Applying Lemma 21.4 with $G(x,y) = D^2_{x,y}F$, $H(x,y) = D_x F$ and $\nu(dt) = 2t\, dt$ gives

$$\begin{aligned} I_1 &\le \frac{1}{4}\int (\mathbb{E}[(D^2_{x_1,y}F)^2(D^2_{x_2,y}F)^2])^{1/2}(\mathbb{E}[(D_{x_1}F)^2(D_{x_2}F)^2])^{1/2}\, \lambda^3(d(x_1,x_2,y)) \\ &\le \frac{1}{16}\alpha^2_{F,1}. \end{aligned} \tag{21.17}$$

Lemma 21.4 with $G(x,y) = D_x F$, $H(x,y) = |D^2_{x,y}F|$ and $\nu(dt) = dt$ gives

$$I_2 \le \frac{1}{4}\alpha^2_{F,1}. \tag{21.18}$$

Finally we apply Lemma 21.4 with $G(x,y) = D^2_{x,y}F$, $H(x,y) = |D^2_{x,y}F|$ and $\nu(dt) = 2t\, dt$ to obtain

$$I_3 \le \frac{1}{4}\int \mathbb{E}[(D^2_{x_1,y}F)^2(D^2_{x_2,y}F)^2]\, \lambda^3(d(x_1,x_2,y)) = \frac{1}{4}\alpha^2_{F,2}. \tag{21.19}$$

Combining the bounds (21.17), (21.18) and (21.19) with the inequality $\sqrt{a+b} \le \sqrt{a} + \sqrt{b}$ (valid for all $a,b \ge 0$) yields

$$\begin{aligned} (3(I_1+I_2+I_3))^{1/2} &\le \sqrt{3}\sqrt{I_1+I_2} + \sqrt{3}\sqrt{I_3} \\ &\le \sqrt{3}\frac{\sqrt{5}}{\sqrt{16}}\alpha_{F,1} + \frac{\sqrt{3}}{2}\alpha_{F,2} \le \alpha_{F,1} + \alpha_{F,2}. \end{aligned}$$

Inserting this into (21.16) yields (21.15), and hence the theorem. □

The following lemma has been used in the preceding proof.

Lemma 21.4 *Let $g \in \mathbb{R}(\mathbb{X}^2 \times \mathbf{N})$ and $h \in \mathbb{R}_+(\mathbb{X}^2 \times \mathbf{N})$. For $(x, y) \in \mathbb{X}^2$ define $G(x, y) := g(x, y, \eta)$ and $H(x, y) := h(x, y, \eta)$. Assume that $\mathbb{E}[|G(x, y)|] < \infty$ for λ^2-a.e. (x, y) and let ν be a probability measure on* $[0, 1]$. *Then*

$$\mathbb{E}\Big[\int \Big(\iint |P_t G(x, y)| H(x, y)\, \nu(dt)\, \lambda(dx)\Big)^2 \lambda(dy)\Big]$$
$$\leq \int (\mathbb{E}[G(x_1, y)^2 G(x_2, y)^2])^{1/2} (\mathbb{E}[H(x_1, y)^2 H(x_2, y)^2])^{1/2}\, \lambda^3(d(x_1, x_2, y)).$$

Proof Given $y \in \mathbb{X}$ define the random variable

$$J(y) := \Big(\iint |P_t G(x, y)| H(x, y)\, \nu(dt)\, \lambda(dx)\Big)^2.$$

By the representation (20.4) of the operator P_t and the triangle inequality for conditional expectations we have almost surely that

$$J(y) \leq \Big(\iiint \mathbb{E}[|g(x, y, \eta_t + \mu)| \mid \eta] H(x, y)\, \Pi_{(1-t)\lambda}(d\mu)\, \nu(dt)\, \lambda(dx)\Big)^2.$$

By the pull out property of conditional expectations, Fubini's theorem and Lemma B.16,

$$J(y) \leq \Big(\iint \mathbb{E}\Big[\int |g(x, y, \eta_t + \mu)| H(x, y)\, \lambda(dx) \,\Big|\, \eta\Big] \Pi_{(1-t)\lambda}(d\mu)\, \nu(dt)\Big)^2.$$

Next we apply Jensen's inequality to the two outer integrations and the conditional expectation to obtain $\mathbb{P}$-a.s. that

$$J(y) \leq \iint \mathbb{E}\Big[\Big(\int |g(x, y, \eta_t + \mu)| H(x, y)\, \lambda(dx)\Big)^2 \,\Big|\, \eta\Big] \Pi_{(1-t)\lambda}(d\mu)\, \nu(dt).$$

By Fubini's theorem (and the pull out property) and Jensen's inequality applied to the conditional expectation and the integration $\Pi_{(1-t)\lambda}(d\mu)\, \nu(dt)$, it follows almost surely that

$$\begin{aligned} J(y) &\leq \int H(x_1, y) H(x_2, y) \iint \mathbb{E}[|g(x_1, y, \eta_t + \mu)||g(x_2, y, \eta_t + \mu)| \mid \eta] \\ &\qquad\qquad \times \Pi_{(1-t)\lambda}(d\mu)\, \nu(dt)\, \lambda^2(d(x_1, x_2)) \\ &\leq \int H(x_1, y) H(x_2, y) \Big(\iint \mathbb{E}[g(x_1, y, \eta_t + \mu)^2 g(x_2, y, \eta_t + \mu)^2 \mid \eta] \\ &\qquad\qquad \times \Pi_{(1-t)\lambda}(d\mu)\, \nu(dt)\Big)^{1/2} \lambda^2(d(x_1, x_2)). \end{aligned}$$

Set $I := \mathbb{E}[\int J(y)\,\lambda(dy)]$, which is the left-hand side of the inequality we seek to prove. By the Cauchy–Schwarz inequality and the law of total expectation it follows that

$$I \le \int \Big(\iint \mathbb{E}[g(x_1, y, \eta_t + \mu)^2 g(x_2, y, \eta_t + \mu)^2]\, \Pi_{(1-t)\lambda}(d\mu)\, \nu(dt) \Big)^{1/2} \\ \times (\mathbb{E}[H(x_1, y)^2 H(x_2, y)^2])^{1/2}\, \lambda^3(d(x_1, x_2, y)).$$

Using the superposition and thinning theorems as in (20.3) we can conclude the proof. □

21.3 Normal Approximation of Linear Functionals

In this section we treat a simple example in the general setting.

Example 21.5 Let $g \in L^1(\lambda) \cap L^3(\lambda)$ such that $\int g^2\, d\lambda = 1$. Let $F := I(g)$ as defined by (12.4), that is $F = \eta(g) - \lambda(g)$. Then we have $\mathbb{E}[F] = 0$, while $\mathbb{V}\mathrm{ar}[F] = 1$ by Proposition 12.4. The definition (12.4) and Proposition 12.1 show for λ^2-a.e. (x_1, x_2) and $\mathbb{P}$-a.s. that $D_{x_1} F = g(x_1)$ and $D^2_{x_1,x_2} F = 0$. Hence Theorem 21.3 implies that

$$d_1(I(g), N) \le \int |g|^3\, d\lambda. \tag{21.20}$$

The bound (21.20) is optimal up to a multiplicative constant:

Proposition 21.6 *Let X_t be a Poisson distributed random variable with parameter $t > 0$ and define $\hat{X}_t := t^{-1/2}(X_t - t)$. Then $d_1(\hat{X}_t, N) \le t^{-1/2}$ and*

$$\liminf_{t\to\infty} \sqrt{t}\, d_1(\hat{X}_t, N) \ge 1/4.$$

Proof The upper bound follows from (21.20) when applied to a homogeneous Poisson process on $\mathbb{R}_+$ of unit intensity with $g(t) := t^{-1/2}\mathbf{1}_{[0,t]}$.

To derive the lower bound we construct a special Lipschitz function. Let $h\colon \mathbb{R} \to \mathbb{R}$ be the continuous 1-periodic function which is zero on the integers and increases (resp. decreases) with rate 1 on the interval $[0, 1/2]$ (resp. $[1/2, 1]$). This function has Lipschitz constant 1. The same is then true for the function $h_t\colon \mathbb{R} \to \mathbb{R}$ defined by $h_t(x) := h(\sqrt{t}x + t)/\sqrt{t}$. By definition, $h_t(\hat{X}_t) \equiv 0$. On the other hand, Exercise 21.1 shows that

$$\lim_{t\to\infty} \sqrt{t}\,\mathbb{E}[h_t(N)] = 1/4; \tag{21.21}$$

since $d_1(\hat{X}_t, N) \ge \mathbb{E}[h_t(N)]$, the desired inequality follows. □

21.4 Exercises

Exercise 21.1 Prove (21.21). Is the result true for other random variables?

Exercise 21.2 Let η be a Poisson process on $\mathbb{R}_+$ with intensity measure ν as in Exercise 7.8. Show the central limit theorem $\nu(t)^{-1/2}(\eta(t) - \nu(t)) \xrightarrow{d} N$ as $t \to \infty$. (Hint: Use Example 21.5.)

Exercise 21.3 Let $u \in L^1(\lambda)$, $v \in \mathbb{R}_+(\mathbb{X})$ and $F := \eta(u) + \exp[-\eta(v)]$. Show that $\mathbb{P}$-a.s. and for λ^2-a.e. (x, y) we have that $|D_x F| \le |u(x)| + v(x)\exp[-\eta(v)]$ and $|D^2_{x,y}F| \le v(x)v(y)\exp[-\eta(v)]$. (Hint: Use Lemma 18.2.)

Exercise 21.4 Let $u, v \in L^1(\lambda) \cap L^2(\lambda)$. Assume that $\lambda(u^2) > 0$ and $v \ge 0$. For $t > 0$ let η_t be a Poisson process with intensity measure $t\lambda$. Define $F_t := \eta_t(u) + \exp[-\eta_t(v)]$. Show that there exists $a > 0$ such that $\mathbb{V}\mathrm{ar}[F_t] \ge at$ for all $t \ge 1$. (Hint: Use Exercise 18.8.)

Exercise 21.5 Let $u, v \in L^1(\lambda) \cap L^2(\lambda)$ and assume that $v \ge 0$. Let $F := \eta(u) + \exp[-\eta(v)]$ and show that

$$\int \left(\mathbb{E}[(D_x F)^2 (D_y F)^2]\right)^{1/2} \left(\mathbb{E}[(D^2_{x,z}F)^2 (D^2_{y,z}F)^2]\right)^{1/2} \lambda^3(d(x,y,z)) \\ \le \lambda(v^2)\lambda((|u| + v)v)^2 \exp\left[-2^{-1}\lambda(1 - e^{-4v})\right].$$

Assume in addition that $v \in L^4(\lambda)$ and show that

$$\int \mathbb{E}[(D^2_{x,z}F)^2 (D^2_{y,z}F)^2]\, \lambda^3(d(x,y,z)) \le \lambda(v^2)^2 \lambda(v^4) \exp\left[-\lambda(1 - e^{-4v})\right].$$

Exercise 21.6 Let $u, v \in L^1(\lambda) \cap L^3(\lambda)$ and assume that $v \ge 0$. Let $F := \eta(u) + \exp[-\eta(v)]$ and show that

$$\int \mathbb{E}[|D_x F|^3]\, \lambda(dx) \le \lambda(|u|^3) + 7(\lambda(|u|^3) + \lambda(v^3)) \exp\left[-\lambda(1 - e^{-v})\right].$$

Exercise 21.7 Let $u \in L^1(\lambda) \cap L^3(\lambda)$ and $v \in L^1(\lambda) \cap L^4(\lambda)$. Assume that $v \ge 0$. For $t > 0$ let η_t be a Poisson process with intensity measure $t\lambda$. Define $F_t := \eta_t(u) + \exp[-\eta_t(v)]$ and assume that $\sigma_t := (\mathbb{V}\mathrm{ar}[F_t])^{1/2} > 0$. Let $\hat{F}_t := \sigma_t^{-1}(F_t - \mathbb{E}[F_t])$. Show that $d_1(\hat{F}_t, N) \le c_1\sigma_t^{-2} t^{3/2} + c_2\sigma_t^{-3} t$, $t \ge 1$, where $c_1, c_2 > 0$ depend on u, v and λ, but not on t. Assume in addition that $\lambda(u^2) > 0$ and show that then $d_1(\hat{F}_t, N) \le c_3 t^{-1/2}$, $t \ge 1$, for some $c_3 > 0$. (Hint: Combine Theorem 21.3 with Exercises 21.4–21.6.)

22

Normal Approximation in the Boolean Model

The intersection of a Boolean model Z having convex grains with a convex observation window W is a finite union of convex sets and hence amenable to additive translation invariant functionals φ, such as the intrinsic volumes. The general results of Chapter 21 yield bounds on the Wasserstein distance between the distribution of the standardised random variable $\varphi(Z \cap W)$ and the standard normal. These bounds depend on the variance of $\varphi(Z \cap W)$ and are of the presumably optimal order $\lambda_d(W)^{-1/2}$ whenever this variance grows like the volume $\lambda_d(W)$ of W.

22.1 Normal Approximation of the Volume

As in Definition 17.1, let $d \in \mathbb{N}$, let $\mathbb{Q}$ be a probability measure on the space $C^{(d)}$ of non-empty compact sets in $\mathbb{R}^d$, let ξ be an independent $\mathbb{Q}$-marking of a stationary Poisson process η in $\mathbb{R}^d$ with intensity $\gamma \in (0, \infty)$ and let Z be the Boolean model induced by ξ. Recall from Theorem 5.6 that ξ is a Poisson process on $\mathbb{R}^d \times C^{(d)}$ with intensity measure $\lambda = \gamma\lambda_d \otimes \mathbb{Q}$. Throughout this chapter we assume that the integrability condition (17.10) holds. Proposition 17.5 then shows that Z is a random element of $\mathcal{F}^d$.

In the whole chapter we fix a Borel set (observation window) $W \subset \mathbb{R}^d$ satisfying $\lambda_d(W) \in (0, \infty)$. In Section 22.2 we shall assume W to be convex. As in Section 19.3 we study Poisson functionals of the form $\varphi(Z \cap W)$, where φ is a (geometric) function defined on compact sets. Under certain assumptions on φ and $\mathbb{Q}$ we shall derive from Theorem 21.3 that $\varphi(Z \cap rW)$ is approximately normal for large $r > 0$. We start with the volume $\varphi = \lambda_d$.

By Proposition 17.4 and Fubini's theorem,

$$F_W := \lambda_d(Z \cap W) \tag{22.1}$$

is a (bounded) random variable and hence a Poisson functional. Recall from Proposition 17.4 that $\mathbb{E}[F_W] = p\lambda_d(W)$, where $p < 1$ is given at (17.8). The variance of F_W is given as follows.

Proposition 22.1 *We have that*

$$\mathbb{V}\mathrm{ar}[F_W] = (1-p)^2 \int \lambda_d(W \cap (W+x))(e^{\gamma\beta_d(x)} - 1)\,dx, \tag{22.2}$$

where $\beta_d(x) = \int \lambda_d(K \cap (K+x))\,\mathbb{Q}(dK)$, *as in* (17.13). *If* $\lambda_d(\partial W) = 0$ *and* $\int \lambda_d(K)^2\,\mathbb{Q}(dK) < \infty$, *then*

$$\lim_{r\to\infty} \lambda_d(rW)^{-1}\,\mathbb{V}\mathrm{ar}[F_{rW}] = (1-p)^2 \int (e^{\gamma\beta_d(x)} - 1)\,dx. \tag{22.3}$$

Proof By Fubini's theorem

$$\begin{aligned}\mathbb{E}[F_W^2] &= \mathbb{E}\Big[\iint \mathbf{1}\{x \in W, y \in W\}\mathbf{1}\{x \in Z, y \in Z\}\,dy\,dx\Big] \\ &= \iint \mathbf{1}\{x \in W, y \in W\}\mathbb{P}(x \in Z, y \in Z)\,dy\,dx.\end{aligned}$$

By Theorem 17.8 this equals

$$p^2\lambda_d(W)^2 + (1-p)^2 \iint \mathbf{1}\{x \in W, y \in W\}(e^{\gamma\beta_d(x-y)} - 1)\,dy\,dx.$$

Changing variables and using the equation $\mathbb{E}[\lambda_d(Z \cap W)] = p\lambda_d(W)$, (22.2) follows.

The second assertion follows from (22.2) upon combining Exercises 17.10 and 17.11 with dominated convergence. □

In the next theorem we need to assume that $\phi_{d,3} < \infty$, where

$$\phi_{d,k} := \int (\lambda_d(K))^k\,\mathbb{Q}(dK), \quad k \in \mathbb{N}. \tag{22.4}$$

In particular, $\phi_d = \phi_{d,1}$ as given by (17.11). It follows from Exercise 22.2 that $\mathbb{V}\mathrm{ar}[F_W] > 0$ provided that $\phi_d > 0$. In this case we use the notation

$$c_W := \lambda_d(W)(1-p)^{-2}\Big[\int \lambda_d(W \cap (W+x))(e^{\gamma\beta_d(x)} - 1)\,dx\Big]^{-1}. \tag{22.5}$$

Recall from (21.1) the definition of the Wasserstein distance d_1 and let N be a standard normal random variable.

Theorem 22.2 *Define the random variable* F_W *by* (22.1). *Assume that* $\phi_{d,3} < \infty$ *and also that* $\phi_d > 0$. *Define* $\hat{F}_W := (\mathbb{V}\mathrm{ar}[F_W])^{-1/2}(F_W - \mathbb{E}[F_W])$. *Then*

$$d_1(\hat{F}_W, N) \le [2(\gamma\phi_{d,2})^{3/2}c_W + \gamma^{3/2}\phi_{d,2}c_W + \gamma\phi_{d,3}(c_W)^{3/2}](\lambda_d(W))^{-1/2}. \tag{22.6}$$

Proof We apply Theorem 21.3. To simplify notation, we assume $\gamma = 1$ and leave the general case to the reader. Let $\sigma := (\mathbb{V}\text{ar}[F_W])^{1/2}$. We shall use the notation $K_x := K + x$ for $(x, K) \in \mathbb{R}^d \times \mathcal{C}^{(d)}$.

By the additivity of Lebesgue measure, we have almost surely and for $\lambda_d \otimes \mathbb{Q}$-a.e. $(x, K) \in \mathbb{R}^d \times \mathcal{C}^{(d)}$ (similarly to Example 18.1) that

$$D_{(x,K)}F_W = \lambda_d(K_x \cap W) - \lambda_d(Z \cap K_x \cap W). \tag{22.7}$$

(We leave it to the reader to construct a suitable representative of F_W.) Iterating this identity (or using (18.3)) yields $\mathbb{P}$-a.s. and for $(\lambda_d \otimes \mathbb{Q})^2$-a.e. $((x, K), (y, L))$ that

$$D^2_{(x,K),(y,L)}F_W = \lambda_d(Z \cap K_x \cap L_y \cap W) - \lambda_d(K_x \cap L_y \cap W). \tag{22.8}$$

In particular,

$$|D_{(x,K)}F_W| \le \lambda_d((K + x) \cap W), \tag{22.9}$$

$$|D^2_{(x,K),(y,L)}F_W| \le \lambda_d((K + x) \cap (L + y) \cap W). \tag{22.10}$$

The following calculations rely on the monotonicity of Lebesgue measure and the following direct consequence of Fubini's theorem:

$$\int \lambda_d(A \cap (B + x))\, dx = \lambda_d(A)\lambda_d(B), \quad A, B \in \mathcal{B}^d. \tag{22.11}$$

By (22.9),

$$\mathbb{E}\Big[\int (D_{(x,K)}F_W)^2\, \lambda(d(x, K))\Big] \le \iint \lambda_d(K_x \cap W)\lambda_d(W)\, dx\, \mathbb{Q}(dK)$$
$$= (\lambda_d(W))^2 \int \lambda_d(K)\, \mathbb{Q}(dK).$$

Hence $DF_W \in L^2(\mathbb{P} \otimes \lambda)$ and Theorem 21.3 applies. Let $\hat{F} := \hat{F}_W$. Using the bounds (22.9) and (22.10) in the definition of $\alpha_{\hat{F},1}$ yields

$$(\alpha_{\hat{F},1})^2 \le \frac{4}{\sigma^4} \iiiint \lambda_d(K_x \cap W)\lambda_d(L_y \cap W)\lambda_d(K_x \cap M_z \cap W)$$
$$\times \lambda_d(L_y \cap M_z \cap W)\, dx\, dy\, dz\, \mathbb{Q}^3(d(K, L, M)). \tag{22.12}$$

Since $\lambda_d(K_x) = \lambda_d(K)$ we obtain

$$(\alpha_{\hat{F},1})^2 \le \frac{4}{\sigma^4} \iiiint \lambda_d(K)\lambda_d(L)\lambda_d(K_x \cap M_z \cap W)\lambda_d(L_y \cap M_z \cap W)$$
$$\times\, dx\, dy\, dz\, \mathbb{Q}^3(d(K, L, M)). \tag{22.13}$$

Therefore, by (22.11),

$$(\alpha_{\hat{F},1})^2 \le \frac{4}{\sigma^4} \iiint \lambda_d(K)^2 \lambda_d(L) \lambda_d(M_z \cap W) \lambda_d(L_y \cap M_z \cap W) \times dy\, dz\, \mathbb{Q}^3(d(K, L, M)). \tag{22.14}$$

Since $\lambda_d(M_z \cap W) \le \lambda_d(M)$, applying (22.11) twice gives

$$(\alpha_{\hat{F},1})^2 \le \frac{4}{\sigma^4} \int \lambda_d(K)^2 \lambda_d(L)^2 \lambda_d(M)^2 \lambda_d(W)\, \mathbb{Q}^3(d(K, L, M)), \tag{22.15}$$

that is

$$\alpha_{\hat{F},1} \le 2(\phi_{d,2})^{3/2} \frac{(\lambda_d(W))^{1/2}}{\sigma^2} = 2(\phi_{d,2})^{3/2} c_W (\lambda_d(W))^{-1/2},$$

where we have used that $\sigma^2 = \lambda_d(W)/c_W$; see (22.5) and (22.2). We leave it to the reader to prove similarly that

$$\alpha_{\hat{F},2} \le \phi_{d,2} \frac{(\lambda_d(W))^{1/2}}{\sigma^2} = \phi_{d,2} c_W (\lambda_d(W))^{-1/2}$$

and

$$\alpha_{\hat{F},3} \le \phi_{d,3} \frac{\lambda_d(W)}{\sigma^3} = \phi_{d,3} (c_W)^{3/2} (\lambda_d(W))^{-1/2}.$$

Inserting these bounds into (21.9) gives the result (22.6). □

As a corollary we obtain a central limit theorem for the volume.

Corollary 22.3 *Assume that* $\lambda_d(\partial W) = 0$. *Assume also that* $\phi_d > 0$ *and* $\phi_{d,3} < \infty$. *For* $r > 0$ *let* $W_r := r^{1/d} W$. *Then* $\hat{F}_{W_r,d} \xrightarrow{d} N$ *as* $r \to \infty$.

Proof By Proposition 22.1, $c_{W_r} = \lambda_d(W_r) \mathbb{V}\mathrm{ar}[F_{W_r}]^{-1}$ tends, as $r \to \infty$, to the inverse of $(1-p)^2 \int (e^{\gamma \beta_d(x)} - 1)\, dx$, which is finite by Exercise 22.1 and our assumption $\phi_d > 0$. Hence the result follows from Theorem 22.2. □

The rate of convergence (with respect to the Wasserstein distance) in Corollary 22.3 is $r^{-1/2}$. Proposition 21.6 suggests that this rate is presumably optimal. Indeed, if $\mathbb{Q}$ is concentrated on a single grain with small volume $v > 0$ then $\lambda_d(Z \cap W)$ approximately equals $v\eta(W)$.

22.2 Normal Approximation of Additive Functionals

In the remainder of this chapter we assume that $W \in \mathcal{K}^{(d)}$ is a compact, convex set with $\lambda_d(W) > 0$. In particular, the boundary of W has Lebesgue measure 0. Let $W_r := r^{1/d} W$ for $r > 0$, so that $\lambda_d(W_r) = r\lambda_d(W)$. We also assume that $\mathbb{Q}$ is concentrated on the system $\mathcal{K}^{(d)}$ of all convex $K \in \mathcal{C}^{(d)}$. As in Section 19.3 we can then assume that $Z \cap K$ is for each $K \in \mathcal{K}^d$ a random element of the convex ring $\mathcal{R}^d$. By Lemma A.30, $(\omega, K) \mapsto Z(\omega) \cap K$ is a measurable mapping from $\Omega \times \mathcal{K}^d$ to $\mathcal{R}^d$.

A function $\varphi\colon \mathcal{C}^d \to \mathbb{R}$ is said to be *translation invariant* if, for all $(x, K) \in \mathbb{R}^d \times \mathcal{C}^d$, $\varphi(K + x) = \varphi(K)$. We say that a measurable function $\varphi\colon \mathcal{R}^d \to \mathbb{R}$ is *geometric* if it is translation invariant, additive and satisfies

$$M(\varphi) := \sup\{|\varphi(K)| : K \in \mathcal{K}^{(d)}, K \subset Q_0\} < \infty, \tag{22.16}$$

where $Q_0 := [-1/2, 1/2]^d$ denotes the unit cube centred at the origin. Fundamental examples of geometric functions are the intrinsic volumes $V_0, \dots, V_d$; see Section A.3. Given a geometric function φ, we wish to apply Theorem 21.3 to the Poisson functional

$$F_{W,\varphi} := \varphi(Z \cap W). \tag{22.17}$$

Recalling that B^d denotes the unit ball in $\mathbb{R}^d$, we define $\overline{V}\colon \mathcal{K}^d \to \mathbb{R}$ by

$$\overline{V}(K) := \lambda_d(K \oplus B^d), \quad K \in \mathcal{K}^d. \tag{22.18}$$

Clearly $\overline{V}$ is translation invariant. By the Steiner formula (A.22), $\overline{V}$ is a linear combination of the intrinsic volumes. Therefore $\overline{V}$ is continuous on $\mathcal{K}^{(d)}$ and hence measurable on $\mathcal{K}^d$. Throughout this section we strengthen (17.10) by assuming that

$$\int \overline{V}(K)^3 \, \mathbb{Q}(dK) < \infty; \tag{22.19}$$

see also Exercise 17.2.

Before stating the main result of this section, we provide two preliminary results. For a given geometric function φ it is convenient to write

$$\varphi_Z(K) := |\varphi(K)| + |\varphi(Z \cap K)|, \quad K \in \mathcal{R}^d. \tag{22.20}$$

The constants $c, c_1, c_2, \dots$ appearing in the following are allowed to depend on d, $\mathbb{Q}$, γ and φ, but not on anything else.

Proposition 22.4 *Let φ be a geometric function. Then there is a constant $c > 0$ such that for any $K, L \in \mathcal{K}^d$,*

$$\mathbb{E}[\varphi_Z(K)^2 \varphi_Z(L)^2] \le c\overline{V}(K)^2 \overline{V}(L)^2, \tag{22.21}$$

$$\mathbb{E}[\varphi_Z(K)^2] \le c\overline{V}(K)^2, \quad \mathbb{E}[\varphi_Z(K)^3] \le c\overline{V}(K)^3. \tag{22.22}$$

Proof For $z \in \mathbb{R}^d$ we set $Q_z := Q_0 + z$. Let $K \in \mathcal{K}^{(d)}$ and define

$$I(K) := \{z \in \mathbb{Z}^d : Q_z \cap K \neq \emptyset\}.$$

By the inclusion–exclusion principle (A.30) we have

$$|\varphi(Z \cap K)| = \Big|\varphi\Big(Z \cap K \cap \bigcup_{z \in I(K)} Q_z\Big)\Big| \leq \sum_{I \subset I(K): I \neq \emptyset} \Big|\varphi\Big(Z \cap K \cap \bigcap_{z \in I} Q_z\Big)\Big|.$$

For each compact set $C \subset \mathbb{R}^d$ let

$$N(C) := \int \mathbf{1}\{(M + x) \cap C \neq \emptyset\}\, \xi(d(x, M)) \tag{22.23}$$

denote the number of grains in $\{M + x : (x, M) \in \xi\}$ hitting C. For each non-empty $I \subset I(K)$, fix some $z(I) \in I$ and let $Z_1, \dots, Z_{N(Q_{z(I)})}$ denote the grains hitting $Q_{z(I)}$. Then, for $\emptyset \neq J \subset \{1, \dots, N(Q_{z(I)})\}$, assumption (22.16) and the translation invariance of φ yield that

$$\Big|\varphi\Big(\bigcap_{j \in J} Z_j \cap K \cap \bigcap_{z \in I} Q_z\Big)\Big| \leq M(\varphi).$$

Using the inclusion–exclusion formula again and taking into account the fact that $\varphi(\emptyset) = 0$, we obtain

$$\begin{aligned}
|\varphi(Z \cap K)| &\leq \sum_{I \subset I(K): I \neq \emptyset} \Big|\varphi\Big(\bigcup_{j=1}^{N(Q_{z(I)})} Z_j \cap K \cap \bigcap_{Q \in I} Q\Big)\Big| \\
&\leq \sum_{I \subset I(K): I \neq \emptyset} \ \sum_{J \subset \{1, \dots, N(Q_{z(I)})\}: J \neq \emptyset} \Big|\varphi\Big(\bigcap_{j \in J} Z_j \cap K \cap \bigcap_{z \in I} Q_z\Big)\Big| \\
&\leq \sum_{I \subset I(K): I \neq \emptyset} \mathbf{1}\Big\{\bigcap_{z \in I} Q_z \neq \emptyset\Big\} 2^{N(Q_{z(I)})} M(\varphi).
\end{aligned}$$

Taking into account a similar (but simpler) bound for $|\varphi(K)|$ we obtain

$$|\varphi_Z(K)| \leq \sum_{I \subset I(K): I \neq \emptyset} \mathbf{1}\Big\{\bigcap_{z \in I} Q_z \neq \emptyset\Big\} (2^{N(Q_{z(I)})} + 1) M(\varphi). \tag{22.24}$$

Exercise 22.3 shows that the expectations

$$\mathbb{E}[(2^{N(Q_x)} + 1)(2^{N(Q_y)} + 1)(2^{N(Q_z)} + 1)(2^{N(Q_w)} + 1)]$$

are uniformly bounded in $x, y, z, w \in \mathbb{Z}^d$. Therefore we obtain from (22.24) for each $L \in \mathcal{K}^{(d)}$ that

$$\mathbb{E}[\varphi_Z(K)^2 \varphi_Z(L)^2] \leq c_2 \Big(\sum_{I \subset I(K): I \neq \emptyset} \mathbf{1}\Big\{\bigcap_{z \in I} Q_z \neq \emptyset\Big\}\Big)^2 \Big(\sum_{I \subset I(L): I \neq \emptyset} \mathbf{1}\Big\{\bigcap_{z \in I} Q_z \neq \emptyset\Big\}\Big)^2, \tag{22.25}$$

for some $c_2 > 0$, not depending on (K, L).

A combinatorial argument (left to the reader) shows that

$$\operatorname{card}\Big\{I \subset I(K) : I \neq \emptyset, \bigcap_{z \in I} Q_z \neq \emptyset\Big\} \leq 2^{2^d} \operatorname{card} I(K). \tag{22.26}$$

Since $\operatorname{card} I(K) \leq V_d(K + B(0, \sqrt{d}))$, Steiner's formula (A.22) yields

$$\operatorname{card} I(K) \leq \sum_{i=0}^{d} \kappa_{d-i} d^{(d-i)/2} V_i(K) \leq c_3 \overline{V}(K)$$

for some $c_3 > 0$ depending only on d. Using this bound, together with (22.26) in (22.25), yields inequality (22.21). The inequalities (22.22) follow in the same way. □

Proposition 22.5 *Let $K, L \in \mathcal{K}^d$. Then*

$$\int \overline{V}(K \cap (L + x))\, dx \leq \overline{V}(K)\overline{V}(L). \tag{22.27}$$

Proof For each $x \in \mathbb{R}^d$ we have the inclusion

$$(K \cap (L + x)) \oplus B^d \subset (K \oplus B^d) \cap ((L + x) \oplus B^d).$$

Since $(L + x) \oplus B^d = (L \oplus B^d) + x$, the result follows from (22.11). □

Lemma 22.6 *Suppose that φ is a geometric function and that $K \in \mathcal{K}^d$. Define $F_{K,\varphi} := \varphi(Z \cap K)$ and let $n \in \mathbb{N}$. Then we have $\mathbb{P}$-a.s. and for $(\lambda_d \otimes \mathbb{Q})^n$-a.e. $((x_1, K_1), \dots, (x_n, K_n))$ that*

$$\begin{aligned} D^n_{(x_1,K_1),\dots,(x_n,K_n)} F_{K,\varphi} = (-1)^n[&\varphi(Z \cap (K_1 + x_1) \cap \cdots \cap (K_n + x_n) \cap K) \\ &- \varphi((K_1 + x_1) \cap \cdots \cap (K_n + x_n) \cap K)]. \end{aligned} \tag{22.28}$$

Proof Let

$$\mathbb{X} := \{(x, L) \in \mathbb{R}^d \times \mathcal{K}^{(d)} : (L + x) \cap K \neq \emptyset\}.$$

For $\mu \in \mathbf{N}(\mathbb{R}^d \times \mathcal{K}^{(d)})$ we define $Z(\mu) := \bigcup_{(x,K)\in\mu}(K + x)$ if $\mu(\mathbb{X}) < \infty$ and $Z(\mu) := \emptyset$ otherwise. Then $f(\mu) := \varphi(Z(\mu) \cap K)$ defines a representative f of $F_{K,\varphi}$; see the proof of Proposition 19.5. For each $(x, L) \in \mathbb{R}^d \times \mathcal{K}^{(d)}$ we recall that $L_x := L + x$. By additivity of φ,

$$\begin{aligned} f(\mu + \delta_{(x,L)}) &= \varphi((Z(\mu) \cap K) \cup (L_x \cap K)) \\ &= \varphi(Z(\mu) \cap K) + \varphi(L_x \cap K) - \varphi(Z(\mu) \cap L_x \cap K), \end{aligned}$$

so that

$$D_{(x,L)} f(\mu) = \varphi(L_x \cap K) - \varphi(Z(\mu) \cap L_x \cap K).$$

This can be iterated to yield for each $(y, M) \in \mathbb{R}^d \times \mathcal{K}^d$ that

$$D^2_{(x,L),(y,M)} f(\mu) = \varphi(Z(\mu) \cap L_x \cap M_y \cap K) - \varphi(L_x \cap M_y \cap K).$$

Hence the assertion follows by induction. □

We are now ready to present the main result of this section.

Theorem 22.7 *Suppose that* φ *is a geometric function and that* (22.19) *holds. Assume that* $\sigma_{W,\varphi} := (\mathbb{V}\mathrm{ar}[F_{W,\varphi}])^{1/2} > 0$, *where* $F_{W,\varphi}$ *is given by* (22.17). *Let* $\hat{F}_{W,\varphi} := (\sigma_{W,\varphi})^{-1}(F_{W,\varphi} - \mathbb{E}[F_{W,\varphi}])$. *Then*

$$d_1(\hat{F}_{W,\varphi}, N) \leq c_1 \sigma^{-2}_{W,\varphi} \overline{V}(W)^{1/2} + c_2 \sigma^{-3}_{W,\varphi} \overline{V}(W), \tag{22.29}$$

where c_1, c_2 *do not depend on* W.

Proof We intend to apply Theorem 21.3. Proposition 22.4 shows that $\mathbb{E}[F^2_{W,\varphi}] < \infty$. Recall the definition (22.20) of φ_Z. By Lemma 22.6 we have for $(\lambda_d \otimes \mathbb{Q})^2$-a.e. $((x, K), (y, L))$ and $\mathbb{P}$-a.s. that

$$|D_{(x,K)} F_{W,\varphi}| \leq \varphi_Z(K_x \cap W), \tag{22.30}$$

$$|D^2_{(x,K),(y,L)} F_{W,\varphi}| \leq \varphi_Z(K_x \cap L_y \cap W), \tag{22.31}$$

where we recall the notation $K_x := K + x$. By (22.30), (22.22) and Proposition 22.5,

$$\begin{aligned}\mathbb{E}\Big[\iint (D_{(x,K)} F_{W,\varphi})^2 \, dx \, \mathbb{Q}(dK)\Big] &\leq c\overline{V}(W) \iint \overline{V}(K_x \cap W) \, dx \, \mathbb{Q}(dK) \\ &\leq c\overline{V}(W)^2 \int \overline{V}(K) \, \mathbb{Q}(dK),\end{aligned}$$

which is finite by assumption (22.19). Therefore Theorem 21.3 applies.

Let $\sigma := \sigma_{W,\varphi}$ and $F := \hat{F}_{W,\varphi}$. Using (22.30), (22.31) and Proposition 22.4 yields (similarly to (22.12))

$$\begin{aligned}\alpha^2_{F,1} \leq \frac{4c\gamma^3}{\sigma^4} \iiiint & \overline{V}(K_x \cap W)\overline{V}(L_y \cap W)\overline{V}(K_x \cap M_z \cap W) \\ & \times \overline{V}(L_y \cap M_z \cap W) \, dx \, dy \, dz \, \mathbb{Q}^3(d(K, L, M)).\end{aligned}$$

Since the function $\overline{V}$ is monotone and translation invariant we can use Proposition 22.5 to conclude as at (22.13), (22.14) and (22.15) that

$$\alpha^2_{F,1} \leq \frac{4c\gamma^3}{\sigma^4} \overline{V}(W) \int \overline{V}(K)^2 \overline{V}(L)^2 \overline{V}(M)^2 \, \mathbb{Q}^3(d(K, L, M)).$$

By assumption (22.19), the preceding integral is finite. The constants $\alpha_{F,2}$ can be treated in the same way. For $\alpha_{F,3}$ we obtain

$$\begin{aligned}\alpha_{F,3} &\leq \frac{\gamma}{\sigma^3}\iint \mathbb{E}[\varphi_Z(K_x \cap W)^3]\, dx\, \mathbb{Q}(dK)\\ &\leq \frac{c\gamma}{\sigma^3}\iint \overline{V}(K_x \cap W)^3\, dx\, \mathbb{Q}(dK)\\ &\leq \frac{c\gamma}{\sigma^3}\iint \overline{V}(K)^2\overline{V}(K_x \cap W)\, dx\, \mathbb{Q}(dK) \leq \frac{c\gamma\overline{V}(W)}{\sigma^3}\int \overline{V}(K)^3\, \mathbb{Q}(dK),\end{aligned}$$

which is finite by assumption (22.19). □

22.3 Central Limit Theorems

Recall that W is a convex compact set such that $\lambda_d(W) > 0$. As before we define $W_r := r^{1/d}W$ for $r > 0$. Then Theorem 22.7 yields a central limit theorem, provided that

$$\liminf_{r\to\infty} r^{-1}\sigma^2_{W_r,\varphi} > 0. \tag{22.32}$$

Theorem 22.8 *Suppose that the assumptions of Theorem 22.7 hold and, in addition, that* (22.32) *holds. Then there exists* $\bar{c} > 0$ *such that*

$$d_1(\hat{F}_{W_r,\varphi}, N) \leq \bar{c}r^{-1/2}, \quad r \geq 1. \tag{22.33}$$

In particular, $\hat{F}_{W_r,\varphi} \xrightarrow{d} N$ *as* $r \to \infty$.

Proof By the scaling property of λ_d we have $\overline{V}(W_r) = r\lambda_d(W \oplus r^{-1/d}B^d)$. Dominated convergence shows that $r^{-1}\overline{V}(W_r) \to \lambda_d(W)$ as $r \to \infty$. Therefore (22.33) is a consequence of Theorem 22.7 and assumption (22.32). □

Proposition 22.1 and Exercise 22.1 show that (22.32) holds for the volume $\varphi = V_d$ provided that $\phi_d > 0$. Finally in this chapter we prove this result in the general case.

Theorem 22.9 *Assume that* (22.19) *holds and suppose that* φ *is a geometric function satisfying* $\mathbb{Q}(\{K \in \mathcal{K}^{(d)} : \varphi(K) \neq 0\}) > 0$. *Then* (22.32) *holds.*

Proof Define a measurable function $\varphi^* : \mathcal{K}^d \to \mathbb{R}$ by

$$\varphi^*(K) := \mathbb{E}[\varphi(Z \cap K)] - \varphi(K), \quad K \in \mathcal{K}^d.$$

The stationarity of Z (Proposition 17.6) and translation invariance of φ

show that φ^* is translation invariant. From Theorem 18.6 and Lemma 22.6 we have for each $K \in \mathcal{K}^{(d)}$ that

$$\begin{aligned}
\mathbb{V}\text{ar}[\varphi(Z \cap K)] &= \sum_{n=1}^{\infty} \frac{\gamma^n}{n!} \iint (\mathbb{E}[D^n_{(x_1,K_1),\dots,(x_n,K_n)} F_{K,\varphi}])^2 \\
&\qquad \times d(x_1,\dots,x_n)\, \mathbb{Q}^n(d(K_1,\dots,K_n)) \\
&= \sum_{n=1}^{\infty} \frac{\gamma^n}{n!} \iint [\varphi^*((K_1+x_1)\cap\cdots\cap(K_n+x_n)\cap K)]^2 \\
&\qquad \times d(x_1,\dots,x_n)\, \mathbb{Q}^n(d(K_1,\dots,K_n)).
\end{aligned} \tag{22.34}$$

Therefore we obtain for each $r > 0$ that

$$\begin{aligned}
\mathbb{V}\text{ar}[F_{W_r,\varphi}] &\geq \sum_{n=1}^{\infty} \frac{\gamma^n}{n!} \iint [\varphi^*((K_1+x_1)\cap\cdots\cap(K_n+x_n))]^2 \\
&\qquad \times \mathbf{1}\{K_1 + x_1 \subset W_r\}\, d(x_1,\dots,x_n)\, \mathbb{Q}^n(d(K_1,\dots,K_n)) \\
&= \sum_{n=1}^{\infty} \frac{\gamma^n}{n!} \iint [\varphi^*(K_1\cap(K_2+y_2)\cap\cdots\cap(K_n+y_n))]^2 \\
&\qquad \times \mathbf{1}\{K_1 + y_1 \subset W_r\}\, d(y_1,\dots,y_n)\, \mathbb{Q}^n(d(K_1,\dots,K_n)),
\end{aligned}$$

where we have used the translation invariance of φ^* and a change of variables. A change of variables yields for each fixed $K_1 \in \mathcal{K}^{(d)}$ that

$$r^{-1} \int \mathbf{1}\{K_1 + y_1 \subset W_r\}\, dy_1 = \int \mathbf{1}\{r^{-1/d} K_1 + y \subset W\}\, dy.$$

Note that for each y in the interior of W the inclusion $r^{-1/d}K_1 + y \subset W$ holds for all sufficiently large r. Fatou's lemma (Lemma A.7) shows that

$$\begin{aligned}
\liminf_{r\to\infty} r^{-1}\, \mathbb{V}\text{ar}[F_{W_r,\varphi}] &\geq \lambda_d(W) \int [\varphi^*(K_1)]^2\, \mathbb{Q}(dK_1) \\
&\quad + \lambda_d(W) \sum_{n=2}^{\infty} \frac{\gamma^n}{n!} \iint [\varphi^*(K_1\cap(K_2+y_2)\cap\cdots\cap(K_n+y_n))]^2 \\
&\qquad \times d(y_2,\dots,y_n)\, \mathbb{Q}^n(d(K_1,\dots,K_n)),
\end{aligned} \tag{22.35}$$

where we have used the fact that the boundary of a convex set has Lebesgue measure 0. We now assume that the left-hand side of (22.35) vanishes. Then we obtain, for all $m \in \mathbb{N}$, $(\lambda_d \otimes \mathbb{Q})^m$-a.e. $((y_1, K_1), \dots, (y_m, K_m))$ and for $\mathbb{Q}$-a.e. K, that $\varphi^*(K) = 0$ and

$$\varphi^*(K \cap (K_1 + y_1) \cap \cdots \cap (K_m + y_m)) = 0. \tag{22.36}$$

Hence we obtain from (22.34) that $\mathbb{V}\mathrm{ar}[\varphi(Z \cap K)] = 0$ for $\mathbb{Q}$-a.e. K, that is

$$\varphi(Z \cap K) = \mathbb{E}[\varphi(Z \cap K)], \quad \mathbb{P}\text{-a.s., } \mathbb{Q}\text{-a.e. } K.$$

Since $\varphi^*(K) = 0$ we have $\mathbb{E}[\varphi(Z \cap K)] = \varphi(K)$ for $\mathbb{Q}$-a.e. K. Moreover, by Theorem 17.3 and assumption (17.10), $\mathbb{P}(Z \cap K = \emptyset) > 0$ for each $K \in \mathcal{K}^{(d)}$. Therefore $\varphi(K) = \varphi(\emptyset) = 0$ for $\mathbb{Q}$-a.e. K, as asserted. □

22.4 Exercises

Exercise 22.1 Show that $\int (e^{\gamma\beta_d(x)} - 1)\, dx > 0$ if $\int \lambda_d(K)\, \mathbb{Q}(dK) > 0$.

Exercise 22.2 Show that

$$\int \lambda_d(W \cap (W + x))(e^{\gamma\beta_d(x)} - 1)\, dx \geq \gamma \int \lambda_d(W \cap (K + x))^2\, dx\, \mathbb{Q}(dK).$$

Use this and Proposition 22.1 (or Exercise 18.8) to show that $\mathbb{V}\mathrm{ar}[F_W] > 0$ provided that $\phi_d > 0$.

Exercise 22.3 Let ξ be a Poisson process on $\mathbb{R}^d \times C^{(d)}$ with intensity measure $\lambda = \gamma\lambda_d \otimes \mathbb{Q}$ and suppose that (17.10) holds. Let $C \subset \mathbb{R}^d$ be compact and let $C_i := C + x_i$ for $i \in \{1, \dots, m\}$, where $m \in \mathbb{N}$ and $x_1, \dots, x_m \in \mathbb{R}^d$. For $I \subset [m]$ let N_I denote the number of points $(x, K) \in \xi$ such that $(K + x) \cap C_i \neq \emptyset$ for $i \in I$ and $(K + x) \cap C_i = \emptyset$ for $i \notin I$. Show that the N_I are independent Poisson random variables. Use this fact to show that

$$\mathbb{E}[2^{N(C_1)} \cdots 2^{N(C_m)}] \leq \exp\left[2^m(2^m - 1)\gamma \int \lambda_d(K \oplus C^*)\, \mathbb{Q}(dK)\right],$$

where $C^* := (-1)C$ and the random variables $N(C_i)$ are defined by (22.23). (Hint: Use that $N(C_1) + \cdots + N(C_m) \leq m \sum_{I \neq \emptyset} N_I$.)

Exercise 22.4 Assume that (22.19) holds and suppose that φ is a geometric function such that

$$\iint |\varphi((K_1 + x_1) \cap \cdots \cap (K_n + x_n))|\, d(x_1, \dots, x_n)\, \mathbb{Q}^n(d(K_1, \dots, K_n)) > 0$$

for some $n \in \mathbb{N}$. Use the final part of the proof of Theorem 22.9 to show that (22.32) holds.

Exercise 22.5 Consider the point process η_1 of isolated nodes in the Gilbert graph with deterministic radius $s/2$ based on a stationary Poisson process η with intensity $\gamma > 0$; see Corollary 16.12. Let $W \subset \mathbb{R}^d$ and set $F := \eta_1(W)$. Show for $x, y \in \mathbb{R}^d$ that

$$|D_x F| \le \eta(B(x,s) \cap W) + \mathbf{1}\{x \in W\}$$

and

$$|D^2_{x,y} F| \le \eta(B(x,s) \cap B(y,s) \cap W) \\ + 2 \cdot \mathbf{1}\{\{x,y\} \cap W \ne \emptyset\}\mathbf{1}\{\|x-y\| \le s\}.$$

Exercise 22.6 Let $W \subset \mathbb{R}^d$ be a Borel set with $0 < \lambda_d(W) < \infty$ such that the boundary of W has Lebesgue measure 0. Let the point process η be given as in Exercise 8.9. Show that

$$\lim_{r\to\infty} \lambda_d(rW)^{-1} \mathbb{V}\mathrm{ar}[\eta(rW)] = \gamma^2 \int (\rho_2(x) - 1)\, dx + \gamma.$$

(Hint: Use Exercises 8.9 and 17.11.)

Exercise 22.7 Let η_1 be as in Exercise 22.5 and let $W \in \mathcal{B}^d$ be such that $\lambda_d(W) < \infty$. Show that

$$\mathbb{V}\mathrm{ar}[\eta_1(W)] \ge \lambda_d(W)(\gamma e^{-\gamma\kappa_d s^d} - \gamma^2\kappa_d s^d e^{-2\gamma\kappa_d s^d})$$

and that $\gamma e^{-\gamma\kappa_d s^d} - \gamma^2\kappa_d s^d e^{-2\gamma\kappa_d s^d} > 0$. Assume now that $\lambda_d(W) > 0$ and that the boundary of W has Lebesgue measure 0. Show that

$$\lim_{r\to\infty} \lambda_d(rW)^{-1} \mathbb{V}\mathrm{ar}[\eta_1(rW)] = \gamma e^{-\gamma\kappa_d s^d} - \gamma^2\kappa_d s^d e^{-2\gamma\kappa_d s^d} \\ + \gamma^2 e^{-2\gamma\kappa_d s^d} \int \mathbf{1}\{s < \|x\| \le 2s\}(\exp[\gamma\lambda_d(B(0,s) \cap B(x,s))] - 1)\, dx.$$

(Hint: Combine Exercises 8.9 and 22.6 with Corollary 16.12.)

Exercise 22.8 Let η_1 be as in Exercise 22.5 and let $W \subset \mathbb{R}^d$ be a compact set with $\lambda_d(W) > 0$. Let $\hat{\eta}_1(W) := \mathbb{V}\mathrm{ar}[\eta_1(W)]^{-1}(\eta_1(W) - \mathbb{E}[\eta_1(W)])$. Show that

$$d_1(\hat{\eta}_1(W), N) \le c_1 \lambda_d(W_{\oplus s})^{1/2} \lambda_d(W)^{-1} + c_2 \lambda_d(W_{\oplus s}) \lambda_d(W)^{-3/2},$$

where $W_{\oplus s} := W \oplus B(0,s)$ and $c_1, c_2 > 0$ do not depend on W. (Hint: Use Theorem 21.3, Exercise 22.5 and Exercise 22.7.)

Exercise 22.9 Let η_1 be as in Exercise 22.5 and let $W \subset \mathbb{R}^d$ be a compact convex set with $\lambda_d(W) > 0$. Show that $\hat{\eta}_1(rW) \xrightarrow{d} N$ as $r \to \infty$, where $\hat{\eta}_1(rW)$ is defined as in Exercise 22.8. (Hint: Use Exercise 22.8 and the Steiner formula (A.22).)

Appendix A

Some Measure Theory

A.1 General Measure Theory

We assume that the reader is familiar with measure theory but provide here the basic concepts and results. More detail can be found in [13, 16, 30, 63].

Given a function (mapping) f from a set $\mathbb{X}$ to a set $\mathbb{Y}$, we write $f: \mathbb{X} \to \mathbb{Y}$ and denote by $f(x)$ the value of f at x. Let f and g be functions from $\mathbb{X}$ to the extended real line $\overline{\mathbb{R}} := [-\infty, +\infty]$. We often write $\{f \leq g\} := \{x \in \mathbb{X} : f(x) \leq g(x)\}$. Similarly we define $\{f \leq a, g \leq b\}$ (for $a, b \in \overline{\mathbb{R}}$) and other sets of this type. Using the convention $0\infty = \infty 0 = 0(-\infty) = (-\infty)0 = 0$ we may define the product fg pointwise by $(fg)(x) := f(x)g(x)$. Similarly, we define the function $f + g$, whenever the sets $\{f = -\infty, g = \infty\}$ and $\{f = \infty, g = -\infty\}$ are empty. Here we use the common rules for calculating with ∞ and $-\infty$. Let $\mathbf{1}_A$ denote the indicator function of A on $\mathbb{X}$ taking the value one on A and zero on $\mathbb{X} \setminus A$. Given $f: A \to \overline{\mathbb{R}}$, we do not hesitate to interpret $\mathbf{1}_A f$ as a function on $\mathbb{X}$ with the obvious definition. Then the equality $\mathbf{1}_A f = g$ means that f and g agree on A (i.e. $f(x) = g(x)$ for all $x \in A$) and g vanishes outside A. Sometimes it is convenient to write $\mathbf{1}\{x : x \in A\}$ instead of $\mathbf{1}_A$ and $\mathbf{1}\{x \in A\}$ instead of $\mathbf{1}_A(x)$.

In what follows all sets under consideration will be subsets of a fixed set Ω. A class $\mathcal{H}$ of sets is said to be *closed* with respect to finite intersections if $A \cap B \in \mathcal{H}$ whenever $A, B \in \mathcal{H}$. In this case one also says that $\mathcal{H}$ is a π-*system.* One defines similarly the notion of $\mathcal{H}$ being closed with respect to countable intersections, or closed with respect to finite unions, and so on. A class $\mathcal{A}$ of sets is called a *field* (on Ω) if, firstly, $\Omega \in \mathcal{A}$ and, secondly, $A, B \in \mathcal{A}$ implies that $A \setminus B \in \mathcal{A}$ and $A \cup B \in \mathcal{A}$, that is $\mathcal{A}$ is closed with respect to finite unions and set differences. A field $\mathcal{A}$ that is closed with respect to countable unions (or, equivalently, countable intersections) is called a σ-*field.* The symbol $\sigma(\mathcal{H})$ denotes the σ-field generated by $\mathcal{H}$, i.e. the smallest σ-field containing $\mathcal{H}$. In this case $\mathcal{H}$ is called a *generator* of $\sigma(\mathcal{H})$. A class $\mathcal{D}$ of sets is called a *monotone system* if it is closed with respect to countable increasing unions and with respect to countable decreasing intersections, i.e. $A_n \subset A_{n+1}$, $A_n \in \mathcal{D}$, implies $\bigcup_{n=1}^{\infty} A_n \in \mathcal{D}$ and $A_n \supset A_{n+1}$, $A_n \in \mathcal{D}$, implies $\bigcap_{n=1}^{\infty} A_n \in \mathcal{D}$. Thus, a field $\mathcal{D}$ is a σ-field if it is monotone. A class $\mathcal{D}$ of sets is called a *Dynkin system* (also known as a λ-system) if $\Omega \in \mathcal{D}$ and if it is closed with respect to countable increasing unions and it is closed with respect to proper differences, i.e. if $A, B \in \mathcal{D}$ with $A \subset B$ implies $B \setminus A \in \mathcal{D}$. In this case $\mathcal{D}$ is a monotone system. The following theorem is a well-known version of a so-called *monotone class theorem.* If nothing else is stated then all definitions and theorems in this chapter can be found in [63], which is our basic reference for measure and probability theory.

Theorem A.1 (Monotone class theorem) *Let $\mathcal{H}$ and $\mathcal{D}$ be classes of subsets of Ω satisfying $\mathcal{H} \subset \mathcal{D}$. Suppose that $\mathcal{H}$ is a π-system and that $\mathcal{D}$ is a Dynkin system. Then $\sigma(\mathcal{H}) \subset \mathcal{D}$.*

Later in this appendix we use the following version of the monotone class theorem (see e.g. Th. 4.4.2 in [30]).

Theorem A.2 (Monotone class theorem) *Let $\mathcal{A}$ and $\mathcal{M}$ be classes of subsets of Ω with $\mathcal{A} \subset \mathcal{M}$. Suppose that $\mathcal{A}$ is a field and that $\mathcal{M}$ is a monotone system. Then $\sigma(\mathcal{A}) \subset \mathcal{M}$.*

Let $n \in \mathbb{N}$ and let $B_1, \ldots, B_n \subset \Omega$. Define $\mathcal{A} := \sigma(\{B_1, \ldots, B_n\})$. An *atom* of $\mathcal{A}$ is a non-empty set in the field $\mathcal{A}$ having no non-empty proper subset in the field. The atoms of the field are the non-empty sets of the form $B_1^{i_1} \cap \cdots \cap B_n^{i_n}$, where $i_1, \ldots, i_n \in \{0, 1\}$ and, for $B \subset \mathbb{X}$, $B^1 := B$ and $B^0 := \mathbb{X} \setminus B$. Every non-empty set in $\mathcal{A}$ is a union of some atoms.

A *measurable space* is a pair $(\mathbb{X}, \mathcal{X})$, where $\mathbb{X}$ is a set and $\mathcal{X}$ is a σ-field of subsets of $\mathbb{X}$. Let $(\mathbb{X}, \mathcal{X})$ be a measurable space and let f be a mapping from Ω into $\mathbb{X}$. If $\mathcal{F}$ is a σ-field on Ω, then f is said to be $\mathcal{F}$-$\mathcal{X}$-measurable if $f^{-1}(\mathcal{X}) \subset \mathcal{F}$, where $f^{-1}(\mathcal{X}) := \{f^{-1}(B) : B \in \mathcal{X}\}$. If there is no risk of ambiguity, we will also speak of $\mathcal{F}$-measurability or, simply, of measurability. The σ-field $\sigma(f)$ generated by f is the smallest σ-field $\mathcal{G}$ such that f is $\mathcal{G}$-$\mathcal{X}$-measurable; it is given by $f^{-1}(\mathcal{X})$. If $\mathbb{X} = \overline{\mathbb{R}}$ and nothing else is said, then $\mathcal{X}$ will always be given by the σ-field $\mathcal{B}(\overline{\mathbb{R}})$ on $\overline{\mathbb{R}}$, which is generated by the system of open sets in $\mathbb{R}$ along with $\{-\infty\}$ and $\{+\infty\}$. This is the *Borel σ-field* on $\overline{\mathbb{R}}$; see Section A.2. More generally, if $\mathbb{X} \in \mathcal{B}(\overline{\mathbb{R}})$, then we shall take $\mathcal{X}$ as the *trace σ-field* $\{B \cap \mathbb{X} : B \in \mathcal{B}(\overline{\mathbb{R}})\}$; see Section A.2. Now let $\mathcal{F}$ be fixed. Then we denote by $\overline{\mathbb{R}}(\Omega)$ the set of all measurable functions from Ω to $\overline{\mathbb{R}}$. The symbols $\overline{\mathbb{R}}_+(\Omega)$ and $\mathbb{R}_+(\Omega)$ denote the set of $[0, \infty]$-valued (resp. $[0, \infty)$-valued) functions in $\overline{\mathbb{R}}(\Omega)$.

Theorem A.3 *Let f be a mapping from Ω into a measurable space $(\mathbb{X}, \mathcal{X})$ and let g be an $\overline{\mathbb{R}}$-valued function on Ω. Then g is $\sigma(f)$-measurable if and only if there exists an $\overline{\mathbb{R}}$-valued measurable function h from $\mathbb{X}$ into $\overline{\mathbb{R}}$ such that $g = h \circ f$.*

If $\mathbf{G}$ is a class of functions from Ω into $\mathbb{X}$, then we denote by $\sigma(\mathbf{G})$ the smallest σ-field $\mathcal{A}$ on Ω such that f is $\mathcal{A}$-$\mathcal{X}$-measurable for all $f \in \mathbf{G}$. It is given by $\sigma\{f^{-1}(B) : f \in \mathbf{G}, B \in \mathcal{X}\}$. The next theorem is a functional version of the monotone class theorem; see Th. 2.12.9 in [16].

Theorem A.4 *Let $\mathbf{W}$ be a vector space of $\mathbb{R}$-valued bounded functions on Ω that contains the constant functions. Further, suppose that, for every increasing sequence of non-negative functions $f_n \in \mathbf{W}$, $n \in \mathbb{N}$, satisfying $\sup\{|f_n(\omega)| : n \in \mathbb{N}, \omega \in \Omega\} < \infty$, the function $f = \lim_{n\to\infty} f_n$ belongs to $\mathbf{W}$. Assume also that $\mathbf{W}$ is closed under uniform convergence. Let $\mathbf{G}$ be a subset of $\mathbf{W}$ that is closed with respect to multiplication. Then $\mathbf{W}$ contains all bounded $\sigma(\mathbf{G})$-measurable functions on Ω.*

It is possible to show that the final assumption made on $\mathbf{W}$ in Theorem A.4 can be dropped.

Let $(\mathbb{X}, \mathcal{X})$ and $(\mathbb{Y}, \mathcal{Y})$ be two measurable spaces. When $\mathcal{X}$ and $\mathcal{Y}$ are fixed and nothing else is said, measurability on $\mathbb{X} \times \mathbb{Y}$ always refers to the *product σ-field* $\mathcal{X} \otimes \mathcal{Y}$ generated by all sets of the form $A \times B$ with $A \in \mathcal{X}$ and $B \in \mathcal{Y}$. The measurable space $(\mathbb{X} \times \mathbb{Y}, \mathcal{X} \otimes \mathcal{Y})$ is called the *product* of $(\mathbb{X}, \mathcal{X})$ and $(\mathbb{Y}, \mathcal{Y})$. Given a finite number $(\mathbb{X}_1, \mathcal{X}_1), \dots, (\mathbb{X}_n, \mathcal{X}_n)$ of measurable spaces we can define the product $(\mathbb{X}_1 \times \cdots \times \mathbb{X}_n, \mathcal{X}_1 \otimes \cdots \otimes \mathcal{X}_n)$ in a similar way. In the case where $(\mathbb{X}_i, \mathcal{X}_i) = (\mathbb{X}, \mathcal{X})$ for every $i \in \{1, \dots, n\}$ we abbreviate this product as $(\mathbb{X}^n, \mathcal{X}^n)$ and refer to it as the n-th power of $(\mathbb{X}, \mathcal{X})$. Let $(\mathbb{X}_i, \mathcal{X}_i)$, $i \in \mathbb{N}$, be a countable collection of measurable spaces. The infinite product $\otimes_{i=1}^{\infty} \mathcal{X}_i$ is the σ-field on $\times_{i=1}^{\infty} \mathbb{X}_i$ generated by the sets

$$B_1 \times \cdots \times B_n \times \mathop{\times}_{i=n+1}^{\infty} \mathbb{X}_i, \tag{A.1}$$

where $B_i \in \mathcal{X}_i$ for $i \in \{1, \dots, n\}$ and $n \in \mathbb{N}$.

Let $(\mathbb{X}, \mathcal{X})$ be a measurable space. A function $\lambda \colon \mathcal{X} \to [0, \infty]$ is said to be *additive* if $\lambda(B \cup B') = \lambda(B) + \lambda(B')$ for all disjoint $B, B' \in \mathcal{X}$. In this case λ is finitely additive in the obvious sense. A *measure* on a measurable space $(\mathbb{X}, \mathcal{X})$ is a function $\lambda \colon \mathcal{X} \to [0, \infty]$ such that $\lambda(\emptyset) = 0$ and such that λ is σ-additive (countably additive), that is

$$\lambda\Big(\bigcup_{n=1}^{\infty} B_n\Big) = \sum_{n=1}^{\infty} \lambda(B_n),$$

whenever $B_1, B_2, \dots$ are pairwise disjoint sets in $\mathcal{X}$. In this case the triple $(\mathbb{X}, \mathcal{X}, \lambda)$ is called a *measure space*. For simplicity we sometimes speak of a measure on $\mathbb{X}$. A measure λ on $(\mathbb{X}, \mathcal{X})$ is said to be *σ-finite* if there are sets $B_n \in \mathcal{X}$, $n \in \mathbb{N}$, such that $\cup_{n=1}^{\infty} B_n = \mathbb{X}$ and $\lambda(B_n) < \infty$ for all $n \in \mathbb{N}$. In this case we say that $(\mathbb{X}, \mathcal{X}, \lambda)$ is a σ-finite measure space. The *counting measure* on $(\mathbb{X}, \mathcal{X})$ *supported* by a set $A \subset \mathbb{X}$ is the measure $B \mapsto \mathrm{card}(A \cap B)$, where $\mathrm{card}\, B$ denotes the number of elements of a set B. If, for instance, $(\mathbb{X}, \mathcal{X}) = (\mathbb{R}, \mathcal{B}(\mathbb{R}))$ with $\mathcal{B}(\mathbb{R})$ generated by the open sets, then this measure is σ-finite if and only if A is finite or countably infinite.

The following result can easily be proved using Theorem A.1.

Theorem A.5 *Let μ, ν be measures on $(\mathbb{X}, \mathcal{X})$. Assume that μ and ν agree on a π-system $\mathcal{H}$ with $\sigma(\mathcal{H}) = \mathcal{X}$. Assume moreover that there is an increasing sequence $B_n \in \mathcal{H}$, $n \in \mathbb{N}$, such that $\mu(B_n) < \infty$ for all $n \in \mathbb{N}$ and $\cup_{n=1}^{\infty} B_n = \mathbb{X}$. Then $\mu = \nu$.*

Let $(\mathbb{X}, \mathcal{X}, \lambda)$ be a measure space. The *integral* $\int f\, d\lambda$ of $f \in \overline{\mathbb{R}}_+(\mathbb{X})$ with respect to λ is defined as follows. If f is *simple*, that is of the form

$$f = \sum_{i=1}^{m} c_i \mathbf{1}_{B_i}$$

for some $m \in \mathbb{N}$, $c_1, \dots, c_m \in \mathbb{R}_+$ and $B_1, \dots, B_m \in \mathcal{X}$, then

$$\int f\, d\lambda := \sum_{i=1}^{m} c_i \lambda(B_i).$$

Any $f \in \overline{\mathbb{R}}_+(\mathbb{X})$ is the limit of simple functions f_n given, for $n \in \mathbb{N}$, by

$$f_n(x) := n\mathbf{1}\{n \le f(x)\} + \sum_{j=1}^{n2^n - 1} j2^{-n} \mathbf{1}\{j2^{-n} \le f(x) < (j+1)2^{-n}\},$$

and one defines $\int f\, d\lambda$ as the finite or infinite limit of $\int f_n\, d\lambda$. To extend the integral to $f \in \overline{\mathbb{R}}(\mathbb{X})$ we define

$$\int f\, d\lambda = \int f^+\, d\lambda - \int f^-\, d\lambda$$

whenever one of the integrals on the right-hand side is finite. Here

$$f^+(x) := f(x) \vee 0, \quad f^-(x) := -(f(x) \wedge 0),$$

and $a \vee b$ (resp. $a \wedge b$) denotes the maximum (minimum) of two numbers $a, b \in \mathbb{R}$. For definiteness we put $\int f\, d\lambda := 0$ in the case $\int f^+\, d\lambda =$

$\int f^- \, d\lambda = \infty$. Sometimes we abbreviate $\lambda(f) := \int f \, d\lambda$. For $B \in \mathcal{X}$ one writes $\int_B f \, d\lambda := \lambda(\mathbf{1}_B f)$.

Given measurable functions $f, g \in \overline{\mathbb{R}}(\mathbb{X})$, we write $f \leq g$, λ-almost everywhere (short: λ-a.e.) if $\lambda(\{f > g\}) = 0$. We also write $f(x) \leq g(x)$, λ-a.e. $x \in \mathbb{X}$. Similar notation is used for other measurable relationships.

Given $p > 0$ let $L^p(\lambda) = \{f \in \mathbb{R}(\mathbb{X}) : \lambda(|f|^p) < \infty\}$. The mapping $f \mapsto \lambda(f)$ is linear on $L^1(\lambda)$ and satisfies the *triangle inequality* $|\lambda(f)| \leq \lambda(|f|)$. If $f \geq 0$, λ-a.e. (that is, $\lambda(\{f < 0\}) = 0$) then $\lambda(f) = 0$ implies that $f = 0$, λ-a.e.

The next results show that the integral has nice continuity properties.

Theorem A.6 (Monotone convergence) *Let $f_n \in \overline{\mathbb{R}}_+(\mathbb{X})$, $n \in \mathbb{N}$, be such that $f_n \uparrow f$ (pointwise) for some $f \in \overline{\mathbb{R}}_+(\mathbb{X})$. Then $\lambda(f_n) \uparrow \lambda(f)$.*

Lemma A.7 (Fatou's lemma) *Let $f_n \in \overline{\mathbb{R}}_+(\mathbb{X})$, $n \in \mathbb{N}$. Then*

$$\liminf_{n\to\infty} \lambda(f_n) \geq \lambda(\liminf_{n\to\infty} f_n).$$

Theorem A.8 (Dominated convergence) *Let $f_n \in \overline{\mathbb{R}}_+(\mathbb{X})$, $n \in \mathbb{N}$, be such that $f_n \to f$ (pointwise) for some $f \in \overline{\mathbb{R}}_+(\mathbb{X})$ and $|f_n| \leq g$ (pointwise) for some $g \in L^1(\lambda)$. Then $\lambda(f_n) \to \lambda(f)$.*

Suppose that $p, q > 1$ with $1/p + 1/q = 1$. Let $f \in L^p(\lambda)$ and $g \in L^q(\lambda)$. *Hölder's inequality* says that then

$$\int |fg| \, d\lambda \leq \left(\int |f|^p \, d\lambda\right)^{1/p} \left(\int |f|^p \, d\lambda\right)^{1/q}. \tag{A.2}$$

In the special case $p = q = 2$ this is known as the *Cauchy–Schwarz inequality*. Hölder's inequality can be generalised to

$$\int |f_1| \cdots |f_m| \, d\lambda \leq \prod_{i=1}^{m} \left(\int |f_i|^{p_i} \, d\lambda\right)^{1/p_i} \tag{A.3}$$

whenever $m \in \mathbb{N}$, $p_1, \dots, p_m$ are positive numbers with $1/p_1 + \cdots + 1/p_m = 1$ and $f_i \in L^{p_i}(\lambda)$ for $i \in \{1, \dots, m\}$.

Let $p > 1$. A quick consequence of (A.2) is the *Minkowski inequality*

$$\left(\int |f + g|^p \, d\lambda\right)^{1/p} \leq \left(\int |f|^p \, d\lambda\right)^{1/p} + \left(\int |g|^p \, d\lambda\right)^{1/p}, \quad f, g \in L^p(\lambda).$$

Identifying $f, \tilde{f} \in L^p(\lambda)$ whenever $\lambda(\{f \neq \tilde{f}\}) = 0$, and giving f the norm $\left(\int |f|^p \, d\lambda\right)^{1/p}$, $L^p(\lambda)$ becomes a normed vector space. A remarkable feature of this space is its *completeness*. This means that if (f_n) is a Cauchy sequence in $L^p(\lambda)$, that is $\lim_{m,n\to\infty} \int |f_m - f_n|^p \, d\lambda = 0$, then there exists $f \in L^p(\lambda)$ such that $\lim_{n\to\infty} f_n = f$ in $L^p(\lambda)$, that is $\lim_{n\to\infty} \int |f_n - f|^p \, d\lambda = 0$.

Let λ, ν be measures on $(\mathbb{X}, \mathcal{X})$. If $\nu(B) = 0$ for all $B \in \mathcal{X}$ with $\lambda(B) = 0$, then ν is said to be *absolutely continuous* with respect to λ and one writes $\nu \ll \lambda$. The two measures ν and λ are said to be *mutually singular* if there exists some $A \in \mathcal{X}$ such that $\nu(A) = \lambda(\mathbb{X} \setminus A) = 0$. A *finite signed measure* (on $\mathbb{X}$ or $(\mathbb{X}, \mathcal{X})$) is a σ-additive bounded function $\nu \colon \mathcal{X} \to \mathbb{R}$.

Theorem A.9 (Hahn–Jordan decomposition) *Let ν be a finite signed measure on $\mathbb{X}$. Then there exist uniquely determined mutually singular finite measures ν_+ and ν_- such that $\nu = \nu_+ - \nu_-$.*

Theorem A.10 (Radon–Nikodým theorem) *Let λ, ν be σ-finite measures on $(\mathbb{X}, \mathcal{X})$ such that $\nu \ll \lambda$. Then there exists $f \in \mathbb{R}_+(\mathbb{X})$ such that*

$$\nu(B) = \int_B f \, d\lambda, \quad B \in \mathcal{X}. \tag{A.4}$$

The function f in (A.4) is called the *Radon–Nikodým derivative* (or *density*) of ν with respect to λ. We write $\nu = f\lambda$. If g is another such function then $f = g$, λ-a.e., that is $\lambda(\{f \neq g\}) = 0$.

We need to integrate with respect to a finite signed measure ν on $(\mathbb{X}, \mathcal{X})$. For $f \in \overline{\mathbb{R}}(\mathbb{X})$ we define

$$\int f \, d\nu := \int f \, d\nu_+ - \int f \, d\nu_-$$

whenever this expression is well defined. This can be written as an ordinary integral as follows. Let ρ be a finite measure such that $\nu_- \ll \rho$ and $\nu_+ \ll \rho$; let h_- and h_+ denote the corresponding Radon-Nikodým derivatives. Then

$$\int f \, d\nu = \int f(h_+ - h_-) \, d\rho, \tag{A.5}$$

where the values $-\infty$ and ∞ are allowed. A natural choice is $\rho = \nu_+ + \nu_-$. This is called the *total variation measure* of ν.

Any countable sum of measures is a measure. A measure λ is said to be *s-finite* if

$$\lambda = \sum_{n=1}^{\infty} \lambda_n \tag{A.6}$$

is a countable sum of *finite* measures λ_n. Given $f \in \overline{\mathbb{R}}_+(\mathbb{X})$ we then have

$$\int f \, d\lambda = \sum_{n=1}^{\infty} \int f \, d\lambda_n. \tag{A.7}$$

This remains true for $f \in L^1(\lambda)$. Any σ-finite measure is s-finite. The converse is not true. If μ is a measure on $(\mathbb{X}, \mathcal{X})$ such that $\mu(\mathbb{X}) < \infty$ we can define a measure ν by multiplying μ by infinity. (Recall that $0 \cdot \infty = 0$.) Then $\nu(B) = 0$ if $\mu(B) = 0$ and $\nu(B) = \infty$ otherwise. If $\mu(\mathbb{X}) > 0$ then ν is s-finite but not σ-finite. If the measure ν is of this form, i.e. if $\nu = \infty \cdot \mu$ for some finite measure μ on $\mathbb{X}$ with $\mu(X) > 0$, then we say that ν is *totally infinite*. The sum of a σ-finite and a totally infinite measure is s-finite. The converse is also true:

Theorem A.11 *Let λ be an s-finite measure on $\mathbb{X}$. Then there exist a σ-finite measure λ' and a measure λ'' such that λ' and λ'' are mutually singular, $\lambda = \lambda' + \lambda''$ and λ'' is either totally infinite or the zero measure.*

Proof Assume that λ is given as in (A.6) and let ν be a finite measure such that $\lambda_n \ll \nu$ for all $n \in \mathbb{N}$. (The construction of ν is left as an exercise.) By Theorem A.10 there are $f_n \in \mathbb{R}_+(\mathbb{X})$ such that $\lambda_n = f_n\nu$, i.e. $\lambda_n(B) = \nu(\mathbf{1}_B f_n)$ for all $B \in \mathcal{X}$. Define $f := \sum_{n=1}^{\infty} f_n$. Then f is a measurable function from $\mathbb{X}$ to $[0, \infty]$ and, by Theorem A.6 (monotone convergence), $\lambda = f\nu$. It is easy to see that the restriction of λ to $A := \{f < \infty\}$ (defined by $B \mapsto \lambda(A \cap B)$) is σ-finite. Moreover, if $\nu(\mathbb{X} \setminus A) > 0$, then by the definition of integrals the restriction of λ to $\mathbb{X} \setminus A$ is totally infinite. □

Suppose λ is an s-finite measure, given by (A.6). Then $\nu_n := \sum_{j=1}^{n} \lambda_j \uparrow \lambda$, in the sense that $\nu_n(B) \le \nu_{n+1}(B)$ for all $n \in \mathbb{N}$ and all $B \in \mathcal{X}$ and $\nu_n(B) \to \lambda(B)$ as $n \to \infty$. We use this notation also for general measures. Theorem A.6 on monotone convergence can be generalised as follows.

Theorem A.12 *Let ν_n, $n \in \mathbb{N}$, be measures on $\mathbb{X}$ such that $\nu_n \uparrow \nu$ for some measure ν. Assume also that $f_n \in \overline{\mathbb{R}}(\mathbb{X})$, $n \in \mathbb{N}$, satisfy $\nu_n(\{f_n < 0\}) = 0$, $n \in \mathbb{N}$, and $f_n \uparrow f$ for some $f \in \overline{\mathbb{R}}(\mathbb{X})$. Then $\nu_n(f_n) \uparrow \nu(f)$.*

Proof For all $n \in \mathbb{N}$ we have $\nu_n(\{f < 0\}) \le \nu_n(\{f_n < 0\}) = 0$ and hence $\nu(\{f < 0\}) = 0$. Assume $\nu(\{f > 0\}) > 0$. (Else we have for each $n \in \mathbb{N}$ that $\nu(\{f \neq 0\}) = \nu(\{f_n \neq 0\}) = 0$ and there is nothing to prove.) Let $c \in (0, \nu(f))$. Then there exists a simple $g \in \mathbb{R}_+(\mathbb{X})$ with $g \le f^+$ such that $\nu(g) > c$. Next we can pick $n \in \mathbb{N}$ with $\nu_n(g) > c$ and then $m_0 \ge n$ with $\nu_n(f_m) > c$ for all $m \ge m_0$. Then we obtain for all $m \ge m_0$ that $\nu_m(f_m) \ge \nu_n(f_m) > c$, and the result follows. □

Let λ be an s-finite measure on $(\mathbb{X}, \mathcal{X})$. Then $(\mathbb{X}, \mathcal{X}, \lambda)$ is said to be an s-finite measure space. Let $(\mathbb{Y}, \mathcal{Y})$ be an additional measurable space. If $f \in \overline{\mathbb{R}}(\mathbb{X} \times \mathbb{Y})$, then $y \mapsto \int f(x, y)\,\lambda(dx)$ is a measurable function on $\mathbb{Y}$. Hence, if ν is a measure on $(\mathbb{Y}, \mathcal{Y})$ we can form the double integral

$\iint f(x,y)\,\lambda(dx)\,\nu(dy)$. In particular, we can define the *product measure* $\lambda\otimes\nu$ as the measure on $(\mathbb{X}\times\mathbb{Y},\mathcal{X}\otimes\mathcal{Y})$ given by

$$(\lambda\otimes\nu)(A) := \iint \mathbf{1}_A(x,y)\,\lambda(dx)\,\nu(dy), \quad A\in\mathcal{X}\otimes\mathcal{Y}. \tag{A.8}$$

If ν is also s-finite, and given as the sum of finite measures ν_m, $m\in\mathbb{N}$, then (A.7) and monotone convergence (Theorem A.6) show that

$$\lambda\otimes\nu = \sum_{n,m\in\mathbb{N}} \lambda_n\otimes\nu_m.$$

In particular, $\lambda\otimes\nu$ is s-finite. The product is linear with respect to countable sums and therefore also associative.

Theorem A.13 (Fubini's theorem) *Let $(\mathbb{X},\mathcal{X},\lambda)$ and $(\mathbb{Y},\mathcal{Y},\nu)$ be two s-finite measure spaces and let $f\in\mathbb{R}_+(\mathbb{X}\times\mathbb{Y})$. Then*

$$\iint f(x,y)\,\lambda(dx)\,\nu(dy) = \iint f(x,y)\,\nu(dy)\,\lambda(dx) \tag{A.9}$$

and both integrals coincide with $(\lambda\otimes\nu)(f)$. These assertions remain true for all $f\in L^1(\lambda\otimes\nu)$.

If λ and ν are σ-finite, then $\lambda\otimes\nu$ is σ-finite and uniquely determined by

$$(\lambda\otimes\nu)(B\times C) = \lambda(B)\nu(C), \quad B\in\mathcal{X}, C\in\mathcal{Y}. \tag{A.10}$$

In this case the proof of Fubini's theorem can be found in the textbooks. The s-finite case can be derived by using the formula (A.7) for both λ and ν and then applying Fubini's theorem in the case of two finite measures.

Let us now consider s-finite measure spaces $(\mathbb{X}_i,\mathcal{X}_i,\lambda_i)$, $i\in\{1,\dots,n\}$, for some $n\geq 2$. Then the product $\otimes_{i=1}^n\lambda_i$ of $\lambda_1,\dots,\lambda_n$ is an s-finite measure on $(\times_{i=1}^n\mathbb{X}_i,\otimes_{i=1}^n\mathcal{X}_i)$, defined inductively in the obvious way. Of particular importance is the case $(\mathbb{X}_i,\mathcal{X}_i,\lambda_i) = (\mathbb{X},\mathcal{X},\lambda)$ for all $i\in\{1,\dots,n\}$. Then we write $\lambda^n := \otimes_{i=1}^n\lambda_i$ and call this the n-th *power* of λ. The power ν^n can also be defined for a finite *signed* measure ν. Similarly to (A.5) we have for $f\in\mathbb{R}(\mathbb{X}^n)$ that

$$\int f\,d\nu^n = \int f(h_+ - h_-)^{\otimes n}\,d|\nu|, \tag{A.11}$$

where the tensor product $(h_+ - h_-)^{\otimes n}$ is defined by (18.6).

A *kernel* from $(\mathbb{X},\mathcal{X})$ to $(\mathbb{Y},\mathcal{Y})$ (or abbreviated: from $\mathbb{X}$ to $\mathbb{Y}$) is a mapping K from $\mathbb{X}\times\mathcal{Y}$ to $\overline{\mathbb{R}}_+$ such that $K(\cdot,A)$ is $\mathcal{X}$-measurable for all $A\in\mathcal{Y}$ and such that $K(x,\cdot)$ is a measure on $\mathbb{Y}$ for all $x\in\mathbb{X}$. It is called a *probability kernel* (resp. *sub-probability kernel*) if $K(x,\mathbb{Y}) = 1$ (≤ 1) for all

$x \in \mathbb{X}$. A countable sum of kernels is again a kernel. A countable sum of sub-probability kernels is called an *s-finite kernel*. If K is an s-finite kernel and $f \in \overline{\mathbb{R}}(\mathbb{X} \times \mathbb{Y})$ then $x \mapsto \int f(x, y)\, K(x, dy)$ is a measurable function. If, in addition, λ is an s-finite measure on $(\mathbb{X}, \mathcal{X})$, then

$$(\lambda \otimes K)(A) := \iint \mathbf{1}_A(x, y)\, K(x, dy)\, \lambda(dx), \quad A \in \mathcal{X} \otimes \mathcal{Y},$$

defines an s-finite measure $\lambda \otimes K$ on $(\mathbb{X} \times \mathbb{Y}, \mathcal{X} \otimes \mathcal{Y})$.

For the next result we recall from Definition 6.1 the concept of a Borel space.

Theorem A.14 (Disintegration theorem) *Suppose that $(\mathbb{X}, \mathcal{X})$ is a measurable space and that $(\mathbb{Y}, \mathcal{Y})$ is a Borel space. Let ν be a measure on $(\mathbb{X} \times \mathbb{Y}, \mathcal{X} \otimes \mathcal{Y})$ such that $\lambda := \nu(\cdot \times \mathbb{Y})$ is σ-finite. Then there exists a probability kernel K from $\mathbb{X}$ to $\mathbb{Y}$ such that $\nu = \lambda \otimes K$.*

In the remainder of this section we prove Proposition 4.3. To this end we need some notation and auxiliary results. Let $\mathbf{N}_{<\infty}$ denote the set of all $\mu \in \mathbf{N} := \mathbf{N}(\mathbb{X})$ with $\mu(\mathbb{X}) < \infty$. For $\mu \in \mathbf{N}_{<\infty}$ the recursion (4.9) is solved by

$$\mu^{(m)} = \int \cdots \int \mathbf{1}\{(x_1, \ldots, x_m) \in \cdot\} \Big(\mu - \sum_{j=1}^{m-1} \delta_{x_j}\Big)(dx_m) \cdots \mu(dx_1), \quad \text{(A.12)}$$

where the integrations are with respect to finite signed measures. Note that $\mu^{(m)}$ is a signed measure such that $\mu^{(m)}(C) \in \mathbb{Z}$ for all $C \in \mathcal{X}^m$. At this stage it might not be obvious that $\mu^{(m)}(C) \geq 0$. If, however, μ is given by (4.3) with $k \in \mathbb{N}$, then Lemma 4.2 shows that (A.12) coincides with (4.4). Hence $\mu^{(m)}$ is a measure in this case. For each $\mu \in \mathbf{N}_{<\infty}$ we denote by $\mu^{(m)}$ the signed measure (A.12). This is in accordance with the recursion (4.9). The next lemma is our main tool for proving Proposition 4.3.

Lemma A.15 *Let $n \in \mathbb{N}$ and $B_1, \ldots, B_n \in \mathcal{X}$. Let $\mathcal{A}$ be the field generated by these sets and let $\mu \in \mathbf{N}_{<\infty}$. Then there exists a finite sum μ' of Dirac measures such that $\mu^{(m)}(B) = (\mu')^{(m)}(B)$ for each $m \in \mathbb{N}$ and each $B \in \mathcal{A}^m$.*

Proof Let $\mathcal{A}_0$ denote the set of all atoms of $\mathcal{A}$. For $A \in \mathcal{A}_0$ we take $x_A \in A$ and set

$$\mu' := \sum_{A \in \mathcal{A}_0} \mu(A) \delta_{x_A}.$$

Since $\mu(\mathbb{X}) < \infty$ and $\mu(A) \in \mathbb{N}_0$ for each $A \in \mathcal{A}_0$, this is a finite sum of Dirac measures. By definition, μ and μ' coincide on $\mathcal{A}_0$ and hence by additivity also on $\mathcal{A}$. To prove that $\mu^{(m)} = (\mu')^{(m)}$ on $\mathcal{A}^m$ for $m \in \mathbb{N}$ it is by additivity sufficient to show that

$$\mu^{(m)}(A_1 \times \cdots \times A_m) = (\mu')^{(m)}(A_1 \times \cdots \times A_m) \tag{A.13}$$

holds for all $m \in \mathbb{N}$ and all $A_1, \dots, A_m \in \mathcal{A}$. By the recursion (4.9) we have for each $m \in \mathbb{N}$ and all $A_1, \dots, A_{m+1} \in \mathcal{A}$ that

$$\mu^{(m+1)}(A_1 \times \cdots \times A_{m+1}) = \mu(A_{m+1})\mu^{(m)}(A_1 \times \cdots \times A_m) - \sum_{j=1}^{m} \mu^{(m)}(A_1 \times \cdots \times A_j \cap A_{m+1} \times \cdots \times A_m),$$

so that (A.13) follows by induction. □

Now we can show that $\mu^{(m)}$ is a measure for $\mu \in \mathbf{N}_{<\infty}$ and $m \in \mathbb{N}$.

Lemma A.16 *Let $\mu \in \mathbf{N}_{<\infty}$ and $m \in \mathbb{N}$. Then $\mu^{(m)}(C) \geq 0$ for all $C \in \mathcal{X}^m$.*

Proof Given $B_1, \dots, B_m \in \mathcal{X}$ we assert that $\mu^{(m)}(B_1 \times \cdots \times B_m) \geq 0$. To see this, we apply Lemma A.15 in the case $n = m$. By (4.4) (applied to μ') we have $(\mu')^{(m)}(B_1 \times \cdots \times B_m) \geq 0$ and hence the assertion.

Let $\mathcal{A}_m$ be the system of all finite disjoint unions of sets of the form $C_1 \times \cdots \times C_m$, with $C_1, \dots, C_m \in \mathcal{X}$. This is a field; see Prop. 3.2.3 in [30]. From the first step of the proof and additivity of $\mu^{(m)}$ we deduce that $\mu^{(m)}(A) \geq 0$ holds for all $A \in \mathcal{A}_m$. The system $\mathcal{M}$ of all sets $A \in \mathcal{X}^m$ with the property that $\mu^{(m)}(A) \geq 0$ is closed with respect to (countable) monotone unions and intersections. Hence Theorem A.2 implies that $\mathcal{M} = \mathcal{X}^m$. Therefore $\mu^{(m)}$ is non-negative. □

Lemma A.17 *Let $\mu, \nu \in \mathbf{N}_{<\infty}$ with $\mu \leq \nu$. Let $m \in \mathbb{N}$. Then $\mu^{(m)} \leq \nu^{(m)}$.*

Proof By Theorem A.2 it suffices to show that

$$\mu^{(m)}(B_1 \times \cdots \times B_m) \leq \nu^{(m)}(B_1 \times \cdots \times B_m) \tag{A.14}$$

for all $B_1, \dots, B_m \in \mathcal{X}$. Fixing the latter sets we apply Lemma A.15 to both μ and ν to obtain finite sums μ' and ν' of Dirac measures with the stated properties. Since $\mu \leq \nu$ we have $\mu' \leq \nu'$. Therefore (4.4) (applied to μ' and ν') yields $(\mu')^{(m)} \leq (\nu')^{(m)}$ and hence the asserted inequality (A.14). □

We are now in a position to prove a slightly more detailed version of Proposition 4.3.

Proposition A.18 *For each $\mu \in \mathbf{N}$ there is a sequence $\mu^{(m)}$, $m \in \mathbb{N}$, of* *measures on* $(\mathbb{X}^m, \mathcal{X}^m)$ *satisfying* $\mu^{(1)} := \mu$ *and the recursion* (4.9). *Moreover, the mapping* $\mu \mapsto \mu^{(m)}$ *is measurable. Finally, if* $\mu_n \uparrow \mu$ *for a sequence* (μ_n) *of finite measures in* $\mathbf{N}$, *then* $(\mu_n)^{(m)} \uparrow \mu^{(m)}$.

Proof For $\mu \in \mathbf{N}_{<\infty}$ the functions defined by (A.12) satisfy the recursion (4.9) and are measures by Lemma A.16.

For general $\mu \in \mathbf{N}$ we proceed by induction. For $m = 1$ we have $\mu^{(1)} = \mu$ and there is nothing to prove. Assume now that $m \geq 1$ and that the measures $\mu^{(1)}, \dots, \mu^{(m)}$ satisfy the first $m-1$ recursions and have the properties stated in the proposition. Then (4.9) forces the definition

$$\mu^{(m+1)}(C) := \int K(x_1, \dots, x_m, \mu, C)\, \mu^{(m)}(d(x_1, \dots, x_m)) \tag{A.15}$$

for $C \in \mathcal{X}^{m+1}$, where

$$K(x_1, \dots, x_m, \mu, C)$$
$$:= \int \mathbf{1}\{(x_1, \dots, x_{m+1}) \in C\}\, \mu(dx_{m+1}) - \sum_{j=1}^{m} \mathbf{1}\{(x_1, \dots, x_m, x_j) \in C\}.$$

The function $K \colon \mathbb{X}^m \times \mathbf{N} \times \mathcal{X}^m \to (-\infty, \infty]$ is a *signed kernel* in the following sense. The mapping $(x_1, \dots, x_m, \mu) \mapsto K(x_1, \dots, x_m, \mu, C)$ is measurable for all $C \in \mathcal{X}^{m+1}$, while $K(x_1, \dots, x_m, \mu, \cdot)$ is σ-additive for all $(x_1, \dots, x_m, \mu) \in \mathbb{X}^m \times \mathbf{N}$. Hence it follows from (A.15) and the measurability properties of $\mu^{(m)}$ (which are part of the induction hypothesis) that $\mu^{(m+1)}(C)$ is a measurable function of μ.

Next we show that

$$K(x_1, \dots, x_m, \mu, C) \geq 0, \quad \mu^{(m)}\text{-a.e. } (x_1, \dots, x_m) \in \mathbb{X}^m \tag{A.16}$$

holds for all $\mu \in \mathbf{N}$ and all $C \in \mathcal{X}^{m+1}$. Since $\mu^{(m)}$ is a measure (by the induction hypothesis), (A.15), (A.16) and monotone convergence then imply that $\mu^{(m+1)}$ is a measure. Fix $\mu \in \mathbf{N}$. By definition of $\mathbf{N}$ we can choose a sequence (μ_n) of finite measures in $\mathbf{N}$ such that $\mu_n \uparrow \mu$. Lemma A.16 (applied to μ_n and $m+1$) shows that

$$K(x_1, \dots, x_m, \mu_n, C) \geq 0, \quad (\mu_n)^{(m)}\text{-a.e. } (x_1, \dots, x_m) \in \mathbb{X}^m,\ n \in \mathbb{N}.$$

Indeed, we have for all $B \in \mathcal{X}^m$ that

$$\int_B K(x_1, \dots, x_m, \mu_n, C)\, (\mu_n)^{(m)}(d(x_1, \dots, x_m)) = (\mu_n)^{(m+1)}((B \times \mathbb{X}) \cap C) \geq 0.$$

Since $K(x_1, \dots, x_m, \cdot, C)$ is increasing, this implies that

$$K(x_1, \dots, x_m, \mu, C) \geq 0, \quad (\mu_n)^{(m)}\text{-a.e. } (x_1, \dots, x_m) \in \mathbb{X}^m,\ n \in \mathbb{N}.$$

By the induction hypothesis we have that $(\mu_n)^{(m)} \uparrow \mu^{(m)}$ so that (A.16) follows.

To finish the induction we take $\mu \in \mathbf{N}$ and $\mu_n \in \mathbf{N}_{<\infty}$, $n \in \mathbb{N}$, as above. We need to show that $(\mu_n)^{(m+1)}(C) \uparrow \mu^{(m+1)}(C)$ for each $C \in \mathcal{X}^{n+1}$. For each $n \in \mathbb{N}$, let us define a measurable function $f_n \colon \mathbb{X}^m \to (-\infty, \infty]$ by $f_n(x_1, \dots, x_m) := K(x_1, \dots, x_m, \mu_n, C)$. Then $f_n \uparrow f$, where the function f is given by $f(x_1, \dots, x_m) := K(x_1, \dots, x_m, \mu, C)$. Hence we can apply Theorem A.12 (and (A.15)) to obtain

$$(\mu_n)^{(m+1)}(C) = (\mu_n)^{(m)}(f_n) \uparrow \mu^{(m)}(f) = \mu^{(m+1)}(C).$$

This finishes the proof. □

For each $\mu \in \mathbf{N}$ and each $m \in \mathbb{N}$ the measure $\mu^{(m)}$ is *symmetric*, that is

$$\int f(x_1, \dots, x_m)\, \mu^{(m)}(d(x_1, \dots, x_m)) = \int f(x_{\pi(1)}, \dots, x_{\pi(m)})\, \mu^{(m)}(d(x_1, \dots, x_m)) \tag{A.17}$$

for each $f \in \mathbb{R}_+(\mathbb{X}^m)$ and all bijective mappings π from $[m] := \{1, \dots, m\}$ to $[m]$. To see this, we may first assume that $\mu(\mathbb{X}) < \infty$. If f is the product of indicator functions, then (A.17) is implied by Lemma A.15 and (4.4). The case of a general $f \in \mathbb{R}_+(\mathbb{X}^m)$ follows by a monotone class argument. For a general $\mu \in \mathbf{N}$ we can use the final assertion of Proposition A.18. Product measures λ^m yield other examples of measures satisfying (A.17).

A.2 Metric Spaces

A *metric* on a set $\mathbb{X}$ is a symmetric function $\rho \colon \mathbb{X} \times \mathbb{X} \to \mathbb{R}_+$ satisfying $\rho(x, y) = 0$ if and only if $x = y$ and the *triangle inequality*

$$\rho(x, y) \le \rho(x, z) + \rho(z, y), \quad x, y, z \in \mathbb{X}.$$

Then the pair $(\mathbb{X}, \rho)$ is called a *metric space*. A sequence $x_n \in \mathbb{X}$, $n \in \mathbb{N}$, *converges* to $x \in \mathbb{X}$ if $\lim_{n \to \infty} \rho(x_n, x) = 0$. The closed *ball* with centre $x_0 \in \mathbb{X}$ and radius $r \ge 0$ is defined by

$$B(x_0, r) := \{x \in \mathbb{X} : \rho(x, x_0) \le r\}. \tag{A.18}$$

A set $U \subset \mathbb{X}$ is said to be *open* if for each $x_0 \in U$ there exists $\varepsilon > 0$ such that $B(x_0, \varepsilon) \subset U$. A set $F \subset \mathbb{X}$ is said to be *closed* if its complement $\mathbb{X} \setminus F$ is open. The *closure* of a set $B \subset \mathbb{X}$ is the smallest closed set containing B. The *interior* $\operatorname{int} B$ of $B \subset \mathbb{X}$ is the largest open subset of B. The *boundary* ∂B of B is the set theoretic difference of its closure and its interior.

The *Borel σ-field* $\mathcal{B}(\mathbb{X})$ on a metric space $\mathbb{X}$ is the σ-field generated by the open sets; see [30]. Another generator of $\mathcal{B}(\mathbb{X})$ is the system of closed sets. If $C \subset \mathbb{X}$ then we can restrict the metric to $C \times C$ to obtain a *subspace* of $\mathbb{X}$. With respect to this restricted metric the open (resp. closed) sets are of the form $C \cap U$, where U is open (resp. closed) in $\mathbb{X}$. Therefore the σ-field $\mathcal{B}(C)$ generated by the open sets is given by $\mathcal{B}(C) = \{B \cap C : B \in \mathcal{B}(\mathbb{X})\}$. If $C \in \mathcal{B}(\mathbb{X})$, then we call $(C, \mathcal{B}(C))$ a *Borel subspace* of $\mathbb{X}$.

Let $(\mathbb{X}, \rho)$ be a metric space. A sequence $x_n \in \mathbb{X}$, $n \in \mathbb{N}$, is called a *Cauchy sequence* if $\lim_{m,n\to\infty} \rho(x_m, x_n) = 0$. A subset of $\mathbb{X}$ is said to be *dense* if its closure equals $\mathbb{X}$. A metric space is said to be *complete* if every Cauchy sequence converges in $\mathbb{X}$, and *separable* if it has a countable dense subset. A complete separable metric space is abbreviated as CSMS. The following result on Borel spaces (see Definition 6.1) is quite useful. We refer to [63, Th. A1.2] and [65, Th. 1.1].

Theorem A.19 *Let $(C, \mathcal{B}(C))$ be a Borel subspace of a CSMS $\mathbb{X}$. Then $(C, \mathcal{B}(C))$ is a Borel space.*

A metric space is said to be *σ-compact* if it is a countable union of compact sets. A metric space is said to be *locally compact* if every $x \in \mathbb{X}$ has a compact neighbourhood U, that is, a compact set containing x in its interior. It is easy to see that any σ-compact metric space is separable. Here is a partial converse of this assertion:

Lemma A.20 *Let $\mathbb{X}$ be a locally compact separable metric space. Then $\mathbb{X}$ is σ-compact.*

Proof Let $C \subset \mathbb{X}$ be an at most countable dense subset of $\mathbb{X}$ and let $\mathcal{U}$ be the collection of all open sets $\{z \in \mathbb{X} : \rho(y, z) < 1/n\}$, where $y \in C$ and $n \in \mathbb{N}$. For each $x \in \mathbb{X}$ there exists an open set U_x with compact closure and with $x \in U_x$. There exists $V_x \in \mathcal{U}$ such that $x \in V_x \subset U_x$. The closures of the sets V_x, $x \in \mathbb{X}$, are compact and cover $\mathbb{X}$. □

The next fact is easy to prove.

Lemma A.21 *Each closed subset of a locally compact metric space is locally compact.*

Lemma A.22 *Let $\mathbb{X}$ be a separable metric space and let B be a subspace of $\mathbb{X}$. Then B is also separable.*

Proof Let $C \subset \mathbb{X}$ be an at most countable dense subset of $\mathbb{X}$. For each $x \in C$ and each $n \in \mathbb{N}$ we take a point $y(x, n) \in B(x, 1/n) \cap B$, provided this intersection is not empty. The set of all such points $y(x, n)$ is dense in B. □

Let ν be a measure on a metric space $\mathbb{X}$. The *support* $\operatorname{supp} \nu$ of ν is the intersection of all closed sets $F \subset \mathbb{X}$ such that $\nu(\mathbb{X} \setminus F) = 0$.

Lemma A.23 *Let ν be a measure on a separable metric space $\mathbb{X}$. Then $\nu(\mathbb{X} \setminus \operatorname{supp} \nu) = 0$.*

Proof By definition, the set $\mathbb{X} \setminus \operatorname{supp} \nu$ is the union of all open sets $U \subset \mathbb{X}$ with $\nu(U) = 0$. Any such U is the union of some closed balls $B(x, q)$, where x is in a given at most countable dense subset of $\mathbb{X}$ and q is a positive rational number. (The proof of this fact is left to the reader.) Hence the result follows from the sub-additivity of ν. □

If $(\mathbb{X}, \rho)$ and $(\mathbb{X}', \rho')$ are metric spaces then $\mathbb{X} \times \mathbb{X}'$ can be made into a metric space in its own right. One natural choice of metric is

$$((x, x'), (y, y')) \mapsto (\rho(x, y)^2 + \rho(x', y')^2)^{1/2}.$$

Let $\mathcal{B}(\mathbb{X} \times \mathbb{Y})$ be the Borel σ-field on $\mathbb{X} \times \mathbb{Y}$ based on this metric.

Lemma A.24 *Suppose that $\mathbb{X}$ and $\mathbb{Y}$ are separable metric spaces. Then so is $\mathbb{X} \times \mathbb{Y}$ and $\mathcal{B}(\mathbb{X} \times \mathbb{Y}) = \mathcal{B}(\mathbb{X}) \otimes \mathcal{B}(\mathbb{Y})$. If, moreover, $\mathbb{X}$ and $\mathbb{Y}$ are complete, then so is $\mathbb{X} \times \mathbb{Y}$.*

A *topology* on a given set $\mathbb{X}$ is a system $\mathcal{O}$ of subsets of $\mathbb{X}$ containing $\emptyset$ and $\mathbb{X}$ and being closed under finite intersections and arbitrary unions. The sets in $\mathcal{O}$ are said to be *open* and the pair $(\mathbb{X}, \mathcal{O})$ is called a *topological space*. An example is a metric space with $\mathcal{O}$ given as the system of open sets. Let $(\mathbb{X}, \mathcal{O})$ be a topological space. A sequence $(x_n)_{n \geq 1}$ of points in $\mathbb{X}$ is said to *converge* to $x \in \mathbb{X}$ if for every $U \in \mathcal{O}$ with $x \in U$ there is an $n_0 \in \mathbb{N}$ such that $x_n \in U$ for all $n \geq n_0$.

A.3 Hausdorff Measures and Additive Functionals

In this section we fix a number $d \in \mathbb{N}$ and consider the Euclidean space $\mathbb{R}^d$ with scalar product $\langle \cdot, \cdot \rangle$, norm $\| \cdot \|$ and Borel σ-field $\mathcal{B}^d := \mathcal{B}(\mathbb{R}^d)$. We shall discuss a few basic properties of Lebesgue and Hausdorff measure, referring to [63] for more detail on the first and to [35] for more information on the second. We shall also introduce the intrinsic volumes of convex bodies, referring to [147] and [146] for further detail.

The *diameter* of a non-empty set $B \subset \mathbb{R}^d$ is the possibly infinite number $d(B) := \sup\{\|x - y\| : x, y \in B\}$. The *Lebesgue measure* (or *volume* function)

λ_d on $(\mathbb{R}^d, \mathcal{B}^d)$ is the unique measure satisfying $\lambda_d([0,1]^d) = 1$ and the *translation invariance* $\lambda_d(B) = \lambda_d(B+x)$ for all $(B,x) \in \mathcal{B}^d \times \mathbb{R}^d$, where $B + x := \{y + x : y \in B\}$. In particular, λ_d is *locally finite*, that is $\lambda_d(B) < \infty$ for all bounded Borel sets $B \subset \mathbb{R}^d$. We also have $\lambda_d = (\lambda_1)^d$ and therefore $\lambda_d(rB) = r^d\lambda_d(B)$ for all $r \geq 0$ and $B \in \mathcal{B}^d$, where $rB := \{rx : x \in B\}$. The Lebesgue measure is also invariant under rotations, that is we have $\lambda_d(\rho B) = \lambda_d(B)$ for all $B \in \mathcal{B}^d$ and all rotations $\rho\colon \mathbb{R}^d \to \mathbb{R}^d$. (Here we write $\rho B := \{\rho x : x \in B\}$.) Recall that a *rotation* is a linear isometry (called *proper* if it preserves the orientation, that is has determinant 1). For $f \in \overline{\mathbb{R}}(\mathbb{R}^d)$ one usually writes $\int f(x)\,dx$ instead of $\int f(x)\,\lambda_d(dx)$.

The volume of the *unit ball* $B^d := \{x \in \mathbb{R}^d : \|x\| \leq 1\}$ is denoted by $\kappa_d := \lambda_d(B^d)$. This volume can be expressed with the help of the Gamma function; see Exercise 7.16. We mention the special cases $\kappa_1 = 2$, $\kappa_2 = \pi$, and $\kappa_3 = (4\pi)/3$. It is convenient to define $\kappa_0 := 1$. Note that the ball

$$B(x,r) := rB^d + x = \{y \in \mathbb{R}^d : \|y - x\| \leq r\}$$

centred at $x \in \mathbb{R}^d$ with radius $r \geq 0$ has volume $\kappa_d r^d$.

For $k \in \{0, \dots, d\}$ and $\delta > 0$ we set

$$\mathcal{H}_{k,\delta}(B) := \frac{\kappa_k}{2^k} \inf\left\{\sum_{j=1}^{\infty} d(B_j)^k : B \subset \bigcup_{j=1}^{\infty} B_j, d(B_j) \leq \delta\right\}, \quad B \subset \mathbb{R}^d, \quad (A.19)$$

where the infimum is taken over all countable collections $B_1, B_2, \dots$ of subsets of $\mathbb{R}^d$ and where $d(\emptyset) := 0$. Note that $\mathcal{H}_{k,\delta}(B) = \infty$ is possible for $k < d$ even for bounded sets B. Define

$$\mathcal{H}_k(B) := \lim_{\delta \downarrow 0} \mathcal{H}_{k,\delta}(B). \quad (A.20)$$

The restriction of $\mathcal{H}_k$ to $\mathcal{B}^d$ is a measure, the k-dimensional *Hausdorff measure*. For $k = d$ we obtain the Lebesgue measure, while $\mathcal{H}_0$ is the counting measure supported by $\mathbb{R}^d$. If $B \subset \mathbb{R}^d$ is a k-dimensional smooth manifold then $\mathcal{H}_k(B)$ coincides with its differential geometric volume measure.

For $K, L \subset \mathbb{R}^d$ we define the *Minkowski sum* $K \oplus L$ by

$$K \oplus L := \{x + y : x \in K, y \in L\}.$$

Note that $K \oplus L = \emptyset$ if $K = \emptyset$. The Minkowski sum of K and the ball $B(0,r)$ centred at the origin with radius r is called the *parallel set* of K at distance r. If $K \subset \mathbb{R}^d$ is closed, then

$$K \oplus rB^d = \{x \in \mathbb{R}^d : d(x,K) \leq r\} = \{x \in \mathbb{R}^d : B(x,r) \cap K \neq \emptyset\},$$

where

$$d(x, B) := \inf\{\|x - y\| : y \in B\} \tag{A.21}$$

is the distance of x from a set $B \subset \mathbb{R}^d$ and $\inf \emptyset := \infty$.

A set $C \subset \mathbb{R}^d$ is said to be *convex* if for all $x, y \in C$ and all $t \in [0, 1]$ the point $tx + (1 - t)y$ belongs to C. A non-empty, compact convex subset of $\mathbb{R}^d$ is called a *convex body* for short. The system of all convex bodies is denoted by $\mathcal{K}^{(d)}$. We let $\mathcal{K}^d := \mathcal{K}^{(d)} \cup \{\emptyset\}$. It turns out that the volume of the parallel set of a convex body is a polynomial of degree d:

$$\lambda_d(K \oplus rB^d) = \sum_{j=0}^{d} r^{d-j}\kappa_{d-j}V_j(K), \quad K \in \mathcal{K}^d. \tag{A.22}$$

This is known as the *Steiner formula* and determines the *intrinsic volumes* $V_0(K), \dots, V_d(K)$ of K. Clearly $V_i(\emptyset) = 0$ for all $i \in \{0, \dots, d\}$. Taking $r \to 0$ and taking $r \to \infty$ in (A.22) shows, respectively, that $V_d(K) = \lambda_d(K)$ and $V_0(K) = 1$ if $K \neq \emptyset$. More generally, if the dimension of the affine hull of K equals j, then $V_j(K)$ equals the j-dimensional Hausdorff measure $\mathcal{H}_j(K)$ of K. If K has non-empty interior, then

$$V_{d-1}(K) = \frac{1}{2}\mathcal{H}_{d-1}(\partial K), \tag{A.23}$$

where ∂K denotes the boundary of K. If the interior of K is empty, then $V_{d-1}(K) = \mathcal{H}_{d-1}(\partial K) = \mathcal{H}_{d-1}(K)$. These facts are suggested by the following consequence of (A.22):

$$2V_{d-1}(K) = \lim_{r \downarrow 0} r^{-1}(\lambda_d(K \oplus rB^d) - \lambda_d(K)).$$

Together with Fubini's theorem they can be used to show that

$$\int V_{d-1}(A \cap (B + x))\, dx = V_d(A)V_{d-1}(B) + V_{d-1}(A)V_d(B). \tag{A.24}$$

Taking $K = B^d$ in (A.22), and comparing the coefficients in the resulting identity between polynomials, yields

$$V_i(B^d) = \binom{d}{i}\frac{\kappa_d}{\kappa_{d-i}}, \quad i = 0, \dots, d. \tag{A.25}$$

The intrinsic volumes inherit from Lebesgue measure the properties of invariance under translations and rotations. Moreover, the scaling property of Lebesgue measure implies for any $i \in \{0, \dots, d\}$ that the function V_i is *homogeneous* of degree i, that is

$$V_i(rK) = r^i V_i(K), \quad K \in \mathcal{K}^d,\ r \geq 0. \tag{A.26}$$

A less obvious property of the intrinsic volumes is that they are monotone increasing with respect to set inclusion. In particular, since $V_i(\emptyset) = 0$, the intrinsic volumes are non-negative. The restrictions of the intrinsic volumes to $\mathcal{K}^{(d)}$ are continuous with respect to the *Hausdorff distance*, defined by

$$\delta(K, L) := \min\{\varepsilon \geq 0 : K \subset L \oplus \varepsilon B^d, L \subset K \oplus \varepsilon B^d\}, \quad K, L \in \mathcal{K}^{(d)}. \tag{A.27}$$

The intrinsic volumes have the important property of *additivity*, that is

$$V_i(K \cup L) = V_i(K) + V_i(L) - V_i(K \cap L), \quad i = 0, \dots, d, \tag{A.28}$$

whenever $K, L, (K \cup L) \in \mathcal{K}^d$. The following result highlights the relevance of intrinsic volumes for convex geometry.

Theorem A.25 (Hadwiger's characterisation) *Suppose that $\varphi\colon \mathcal{K}^d \to \mathbb{R}$ is additive, continuous on $\mathcal{K}^d \setminus \{\emptyset\}$ and invariant under translations and proper rotations. Then there exist $c_0, \dots, c_d \in \mathbb{R}$ such that*

$$\varphi(K) = \sum_{i=0}^{d} c_i V_i(K), \quad K \in \mathcal{K}^d.$$

For applications in stochastic geometry it is necessary to extend the intrinsic volumes to the *convex ring* $\mathcal{R}^d$. A set $K \subset \mathbb{R}^d$ belongs to $\mathcal{R}^d$ if it can be represented as a finite (possibly empty) union of compact convex sets. (Note that $\emptyset \in \mathcal{R}^d$.) The space $\mathcal{R}^d \setminus \{\emptyset\}$ is a subset of the space $\mathcal{C}^{(d)}$ of all non-empty compact subsets of $\mathbb{R}^d$. The latter can be equipped with the Hausdorff distance (defined again by (A.27)) and the associated Borel σ-field. For the following result we refer to [146, Th. 1.8.4] and [147, Th. 2.4.2]; see also Exercise 17.3.

Theorem A.26 *The space $\mathcal{C}^{(d)}$ is a CSMS, $\mathcal{K}^{(d)}$ is a closed subset of $\mathcal{C}^{(d)}$ and $\mathcal{R}^d \setminus \{\emptyset\}$ is a measurable subset of $\mathcal{C}^{(d)}$.*

Upon extending the Borel σ-field from $\mathcal{C}^{(d)}$ to $\mathcal{C}^d$ in the usual minimal way (all elements of $\mathcal{B}(\mathcal{C}^{(d)})$ and the singleton $\{\emptyset\}$ should be measurable), $\mathcal{R}^d$ is a measurable subset of $\mathcal{C}^d$. The σ-fields on these spaces are denoted by $\mathcal{B}(\mathcal{C}^d)$ and $\mathcal{B}(\mathcal{R}^d)$, respectively.

A function $\varphi\colon \mathcal{R}^d \to \mathbb{R}$ is said to be *additive* if $\varphi(\emptyset) = 0$ and

$$\varphi(K \cup L) = \varphi(K) + \varphi(L) - \varphi(K \cap L), \quad K, L \in \mathcal{R}^d. \tag{A.29}$$

Such an additive function satisfies the *inclusion–exclusion principle*

$$\varphi(K_1 \cup \dots \cup K_m) = \sum_{n=1}^{m} (-1)^{n-1} \sum_{1 \leq i_1 < \dots < i_n \leq m} \varphi(K_{i_1} \cap \dots \cap K_{i_n}) \tag{A.30}$$

for all $K_1, \dots, K_m \in \mathcal{R}^d$ and all $m \in \mathbb{N}$. The intrinsic volumes V_i can be extended from $\mathcal{K}^d$ to $\mathcal{R}^d$ such that this extended function (still denoted by V_i) is additive. By (A.30) this extension must be unique. It is the existence that requires a (non-trivial) proof. Then V_d is still the volume, while (A.23) holds whenever $K \in \mathcal{R}^d$ is the closure of its interior. Moreover, $V_{d-1}(K) \geq 0$ for all $K \in \mathcal{R}^d$. The function V_0 is known as the *Euler characteristic* and takes on integer values. In particular, $V_0(K) = 1$ for all $K \in \mathcal{K}^{(d)}$. When $d = 2$ the number $V_0(K)$ can be interpreted as the number of connected components minus the number of holes of $K \in \mathcal{R}^2$. The intrinsic volumes are measurable functions on $\mathcal{R}^d$.

Let $\mathcal{F}^d$ denote the space of all closed subsets of $\mathbb{R}^d$. The *Fell topology* on this space is the smallest topology such that the sets $\{F \in \mathcal{F}^d : F \cap G \neq \emptyset\}$ and $\{F \in \mathcal{F}^d : F \cap K = \emptyset\}$ are open for all open sets $G \subset \mathbb{R}^d$ and all compact sets $K \subset \mathbb{R}^d$. It can be shown that $F_n \to F$ in $\mathcal{F}^d$ if and only if $d(x, F_n) \to d(x, F)$ for each x in a dense subset of $\mathbb{R}^d$; see [63, Th. A2.5]. This shows that the Fell topology is *second countable*, that is there is a countable family of open subsets of $\mathcal{F}^d$ such that every open set is the union of some sets in the family; see again [63, Th. A2.5].

In this book we use the following properties of the Fell topology. Further information can be found in [101, 112, 147].

Lemma A.27 *The mapping $(F, x) \mapsto F + x$ from $\mathcal{F}^d \times \mathbb{R}^d$ to $\mathcal{F}^d$ is continuous.*

Proof Suppose that $x_n \to x$ in $\mathbb{R}^d$ and $F_n \to F$ in $\mathcal{F}^d$. For each $y \in \mathbb{R}^d$ we need to show that $d(y, F_n + x_n) \to d(y, F + x)$. Assuming (for simplicity) that $F_n \neq \emptyset$ for all $n \in \mathbb{N}$, this follows from the identities

$$d(y, F_n + x_n) = d(y - x_n, F_n) = (d(y - x_n, F_n) - d(y - x, F_n)) + d(y - x, F_n)$$

and the Lipschitz property from Exercise 2.8. □

We denote by $\mathcal{B}(\mathcal{F}^d)$ the σ-field generated by the open subsets of $\mathcal{F}^d$. Then $(\mathcal{F}^d, \mathcal{B}(\mathcal{F}^d))$ is a measurable space.

Lemma A.28 *The sets $\{F \in \mathcal{F}^d : F \cap K = \emptyset\}$, $K \in \mathcal{C}^d$, form a π-system generating $\mathcal{B}(\mathcal{F}^d)$.*

Proof The first assertion follows from the fact that for given $K, L \in \mathcal{C}^d$ the equations $F \cap K = \emptyset$ and $F \cap L = \emptyset$ are equivalent to $F \cap (K \cup L) = \emptyset$. Let $G \subset \mathbb{R}^d$ be open. Then there exists a sequence $K_n \in \mathcal{C}^d$, $n \in \mathbb{N}$, such that $G = \cup_n K_n$. For $F \in \mathcal{F}^d$ we then have that $F \cap G \neq \emptyset$ if and only if there exists $n \in \mathbb{N}$ such that $F \cap K_n \neq \emptyset$. This shows the second assertion. □

Lemma A.29 *We have $C^d \in \mathcal{B}(\mathcal{F}^d)$ and $\mathcal{B}(C^d) = \{C^d \cap A : A \in \mathcal{B}(\mathcal{F}^d)\}$.*

Proof To prove that $C^d \in \mathcal{B}(\mathcal{F}^d)$ it is sufficient to note that a set $F \in \mathcal{F}^d$ is compact if and only if there exists $n \in \mathbb{N}$ such that $F \cap (\mathbb{R}^d \setminus B(0, n)) = \emptyset$. The second assertion follows from Lemma A.28 and Lemma 17.2. □

Lemma A.30 *The mappings $(F, F') \mapsto F \cap F'$ from $\mathcal{F}^d \times \mathcal{F}^d$ to $\mathcal{F}^d$ and $(F, K) \mapsto F \cap K$ from $\mathcal{F}^d \times C^d$ to C^d are measurable.*

Proof We sketch the proof, leaving some of the details to the reader. By Lemma A.29 we only need to prove the first assertion. Let $C \in C^{(d)}$. We shall show that $\mathcal{H} := \{(F, F') \in \mathcal{F}^d \times \mathcal{F}^d : F \cap F' \cap C = \emptyset\}$ is open in the *product topology* on $\mathcal{F}^d \times \mathcal{F}^d$, which is the smallest topology containing the sets $\mathcal{G} \times \mathcal{G}'$ for all open $\mathcal{G}, \mathcal{G}' \subset \mathcal{F}^d$. Assume that $\mathcal{H}$ is not open. Then there exists $(F, F') \in \mathcal{H}$ such that every open neighbourhood of (F, F') (an open set containing (F, F')) has a non-empty intersection with the complement of $\mathcal{H}$. Since the Fell topology is second countable, there exist sequences (F_n) and (F'_n) of closed sets such that $F_n \cap F'_n \cap C \neq \emptyset$ for all $n \in \mathbb{N}$ and $(F_n, F'_n) \to (F, F')$ as $n \to \infty$. For each $n \in \mathbb{N}$ choose $x_n \in F_n \cap F'_n \cap C$. Since C is compact, there exists $x \in C$ such that $x_n \to x$ along a subsequence. Since $F_n \to F$ we have $x \in F$. (Otherwise there is a compact neighbourhood of x, not intersecting F, contradicting the definition of the Fell topology.) Similarly, we have $x \in F'$. This shows that $x \in F \cap F' \cap C$, a contradiction. Hence $\mathcal{H}$ is open, and since the Fell topology is second countable it can be shown that $\mathcal{H}$ is a member of the product σ-field $\mathcal{B}(\mathcal{F}^d) \otimes \mathcal{B}(\mathcal{F}^d)$; see [30, Prop. 4.1.7]. Since $C \in C^{(d)}$ was arbitrarily chosen, Lemma A.28 implies the assertion. □

A.4 Measures on the Real Half-Line

In this section we consider a locally finite measure ν on $\mathbb{R}_+ = [0, \infty)$ with the Borel σ-field. We abbreviate $\nu(t) := \nu([0, t])$, $t \in \mathbb{R}_+$, and note that ν can be identified with the (right-continuous) mapping $t \mapsto \nu(t)$. We define a function $\nu^{\leftarrow} : \mathbb{R}_+ \to [0, \infty]$ by

$$\nu^{\leftarrow}(t) := \inf\{s \geq 0 : \nu(s) \geq t\}, \quad t \geq 0, \tag{A.31}$$

where $\inf \emptyset := \infty$. This function is increasing, left-continuous and thus measurable; see e.g. [139]. We also define $\nu(\infty) := \lim_{t\to\infty} \nu(t)$.

Proposition A.31 *Let $f \in \mathbb{R}_+(\mathbb{R}_+)$. Then*

$$\int f(t)\,\nu(dt) = \int_0^{\nu(\infty)} f(\nu^{\leftarrow}(t))\,dt. \tag{A.32}$$

If ν is diffuse, then we have for all $g \in \mathbb{R}_+(\mathbb{R}_+)$ that

$$\int g(\nu(t))\,\nu(dt) = \int_0^{\nu(\infty)} g(t)\,dt. \tag{A.33}$$

Proof For all $s, t \in \mathbb{R}_+$ the inequalities $\nu^{\leftarrow}(t) \le s$ and $t \le \nu(s)$ are equivalent; see [139]. For $0 \le a < b < \infty$ and $f := \mathbf{1}_{(a,b]}$ we therefore obtain

$$\begin{aligned}\int_0^{\nu(\infty)} f(\nu^{\leftarrow}(t))\,dt &= \int_0^{\nu(\infty)} \mathbf{1}\{a < \nu^{\leftarrow}(t) \le b\}\,dt \\ &= \int_0^{\nu(\infty)} \mathbf{1}\{\nu(a) < t \le \nu(b)\}\,dt = \nu(b) - \nu(a),\end{aligned}$$

so that (A.32) follows for this choice of f. Also (A.32) holds for $f = \mathbf{1}_{\{0\}}$, since $\nu^{\leftarrow}(t) = 0$ if and only if $t \le \nu(\{0\})$. We leave it to the reader to prove the case of a general f using the tools from measure theory presented in Section A.1.

Assume now that ν is diffuse and let $g \in \mathbb{R}_+(\mathbb{R}_+)$. Applying (A.32) with $f(t) := g(\nu(t))$ yields

$$\int g(\nu(t))\,\nu(dt) = \int_0^{\nu(\infty)} g(\nu(\nu^{\leftarrow}(t)))\,dt = \int_0^{\nu(\infty)} g(t)\,dt,$$

since $\nu(\nu^{\leftarrow}(t)) = t$; see [139]. □

Assume now that ν is a measure on $\mathbb{R}_+$ with $\nu(\mathbb{R}_+) \le 1$. With the definition $\nu(\{\infty\}) := 1 - \nu(\mathbb{R}_+)$ we may then interpret ν as a probability measure on $\overline{\mathbb{R}}$. The *hazard measure* of ν is the measure R_ν on $\mathbb{R}_+$ given by

$$R_\nu(dt) := (\nu[t, \infty))^{\oplus}\nu(dt), \tag{A.34}$$

where $a^{\oplus} := \mathbf{1}\{a \ne 0\}a^{-1}$ is the generalised inverse of $a \in \mathbb{R}$. The following result is a consequence of the exponential formula of Lebesgue–Stieltjes calculus (see [18, 88]) and a special case of Th. A5.10 in [88].

Proposition A.32 *If ν is a diffuse measure on $\mathbb{R}_+$ with $\nu(\mathbb{R}_+) \le 1$ and hazard measure R_ν, then $\nu((t, \infty]) = \exp[-R_\nu([0, t])]$ for all $t \in \mathbb{R}_+$.*

A.5 Absolutely Continuous Functions

Let $I \subset \mathbb{R}$ be a non-empty *interval*. This means that the relations $a, b \in I$ and $a < b$ imply that $[a, b] \subset I$. A function $f: I \to \mathbb{R}$ is said to be *absolutely continuous* if for every $\epsilon > 0$ there exists $\delta > 0$ such that

$$\sum_{i=1}^{n} |f(y_i) - f(x_i)| \leq \varepsilon$$

whenever $n \in \mathbb{N}$ and $x_1, \dots, x_n, y_1, \dots, y_n \in I$ satisfy $x_i \leq y_i$ for all $i \in \{1, \dots, n\}$, $y_i < x_{i+1}$ for all $i \in \{1, \dots, n-1\}$ and $\sum_{i=1}^{n} |y_i - x_i| \leq \delta$.

Recall that λ_1 denotes the Lebesgue measure on $\mathbb{R}$. For $f \in \mathbb{R}(I)$ and $a, b \in I$ we write

$$\int_a^b f(t)\,dt := \int_{[a,b]} f(t)\,\lambda_1(dt)$$

if $a \leq b$ and

$$\int_a^b f(t)\,dt := -\int_{[b,a]} f(t)\,\lambda_1(dt)$$

if $a > b$. Absolutely continuous functions can be characterised as follows.

Theorem A.33 *Let $a < b$ and suppose that $f: [a, b] \to \mathbb{R}$ is a function. Then f is absolutely continuous if and only if there is a function $f' \in L^1((\lambda_1)_{[a,b]})$ such that*

$$f(x) = f(a) + \int_a^x f'(t)\,dt, \quad x \in [a, b]. \tag{A.35}$$

The function f' in (A.35) is called the *Radon–Nikodým derivative* of f. It is uniquely determined almost everywhere with respect to Lebesgue measure on $[a, b]$.

Proposition A.34 (Product rule) *Suppose that $f, g \in \mathbb{R}(I)$ are absolutely continuous with Radon–Nikodým derivatives f', g'. Then the product fg is absolutely continuous with Radon–Nikodým derivative $f'g + fg'$.*

Proof Let $x \in [a, b]$. By Theorem A.33 and Fubini's theorem,

$$\begin{aligned}
&(f(x) - f(a))(g(x) - g(a)) \\
&= \int_a^x \int_a^x \mathbf{1}\{s > t\} f'(s) g'(t)\,ds\,dt + \int_a^x \int_a^x \mathbf{1}\{t \geq s\} f'(s) g'(t)\,ds\,dt \\
&= \int_a^x f'(s)(g(s) - g(a))\,ds + \int_a^x (f(t) - f(a)) g'(t)\,dt.
\end{aligned}$$

Again by Theorem A.33 this can be simplified to

$$f(x)g(x) = f(a)g(a) + \int_a^x f'(s)g(s)\,ds + \int_a^x f(t)g'(t)\,dt.$$

Thus, by Theorem A.33 the proposition is true. □

Let $\mathbf{AC}^2$ be the space of functions $f: \mathbb{R} \to \mathbb{R}$ such that f is differentiable with an absolutely continuous derivative f'. The following result can be proved using the preceding product rule (Proposition A.34).

Proposition A.35 *Let $f \in \mathbf{AC}^2$. Then for all $x \in \mathbb{R}$ we have*

$$f(x) = f(a) + f'(a)(x-a) + \int_a^x f''(t)(x-t)\,dt, \tag{A.36}$$

where f'' is a Radon–Nikodým derivative of f'.

Proof We claim that it suffices to prove

$$f(x) = f(a) - af'(a) + xf'(x) - \int_a^x f''(t)t\,dt \tag{A.37}$$

for all $x \in \mathbb{R}$. Indeed, by Theorem A.33 (applied with f' instead of f) the right-hand side of (A.36) equals that of (A.37). Both sides of (A.37) agree for $x = a$. By Theorem A.33 it is hence enough to show that both sides have the same Radon–Nikodým derivative. This follows from Proposition A.34 applied to $xf'(x)$ and Theorem A.33. □

Appendix B

Some Probability Theory

B.1 Fundamentals

For the reader's convenience we here provide terminology and some basic results of measure-theoretic probability theory. More detail can be found, for instance, in [13, 30] or in the first chapters of [63].

A *probability space* is a measure space $(\Omega, \mathcal{F}, \mathbb{P})$ with $\mathbb{P}(\Omega) = 1$. Then $\mathbb{P}$ is called a *probability measure* (sometimes also a *distribution*), the sets $A \in \mathcal{F}$ are called *events*, while $\mathbb{P}(A)$ is known as the *probability* of the event A. In this book the probability space $(\Omega, \mathcal{F}, \mathbb{P})$ will be fixed.

Let $(\mathbb{X}, \mathcal{X})$ be a measurable space. A *random element* of $\mathbb{X}$ (or of $(\mathbb{X}, \mathcal{X})$) is a measurable mapping $X: \Omega \to \mathbb{X}$. The *distribution* $\mathbb{P}_X$ of X is the image of $\mathbb{P}$ under X, that is $\mathbb{P}_X := \mathbb{P} \circ X^{-1}$ or, written more explicitly, $\mathbb{P}_X(A) = \mathbb{P}(X \in A)$, $A \in \mathcal{X}$. Here we use the common abbreviation

$$\mathbb{P}(X \in A) := \mathbb{P}(\{\omega \in \Omega : X(\omega) \in A\}).$$

We write $X \overset{d}{=} Y$ to express the fact that two random elements X, Y of $\mathbb{X}$ have the same distribution.

Of particular importance is the case $(\mathbb{X}, \mathcal{X}) = (\overline{\mathbb{R}}, \mathcal{B}(\overline{\mathbb{R}}))$. A random element X of this space is called a *random variable* while the integral $\int X \, d\mathbb{P}$ is called the *expectation* (or *mean*) $\mathbb{E}[X]$ of X. If Y is a random element of a measurable space $(\mathbb{Y}, \mathcal{Y})$ and $f \in \overline{\mathbb{R}}(\mathbb{Y})$, then $f(Y)$ is a random variable, and it is easy to prove that $\mathbb{E}[f(Y)] = \int f \, d\mathbb{P}_Y$. If X is a random variable with $\mathbb{E}[|X|^a] < \infty$ for some $a > 0$ then $\mathbb{E}[|X|^b] < \infty$ for all $b \in [0, a]$. In the case $a = 1$ we say that X is *integrable*, while in the case $a = 2$ we say that X is *square integrable*. In the latter case the *variance* of X is defined as

$$\mathbb{V}\mathrm{ar}[X] := \mathbb{E}[(X - \mathbb{E}[X])^2] = \mathbb{E}[X^2] - (\mathbb{E}[X])^2.$$

We have $\mathbb{P}(X = \mathbb{E}[X]) = 1$ if and only if $\mathbb{V}\mathrm{ar}[X] = 0$. For random variables X, Y we write $X \le Y$, $\mathbb{P}$-almost surely (shorter: $\mathbb{P}$-a.s. or just a.s.) if $\mathbb{P}(X \le Y) = 1$. For $A \in \mathcal{F}$ we write $X \le Y$, $\mathbb{P}$-a.s. on A if $\mathbb{P}(A \setminus \{X \le Y\}) = 0$.

The *covariance* between two square integrable random variables X and Y is defined by

$$\mathbb{C}\text{ov}[X, Y] := \mathbb{E}[(X - \mathbb{E}[X])(Y - \mathbb{E}[Y])] = \mathbb{E}[XY] - (\mathbb{E}[X])(\mathbb{E}[Y]).$$

The Cauchy–Schwarz inequality (see (A.2)) says that

$$|\mathbb{E}[XY]| \leq (\mathbb{E}[X^2])^{1/2}(\mathbb{E}[Y^2])^{1/2},$$

or $|\mathbb{C}\text{ov}[X, Y]| \leq (\mathbb{V}\text{ar}[X])^{1/2}(\mathbb{V}\text{ar}[Y])^{1/2}$. Here is another useful inequality for the expectation of convex functions of a *random vector* X in $\mathbb{R}^d$, that is of a random element of $\mathbb{R}^d$.

Proposition B.1 (Jensen's inequality) *Let $X = (X_1, \dots, X_d)$ be a random vector in $\mathbb{R}^d$ whose components are in $L^1(\mathbb{P})$ and let $f : \mathbb{R}^d \to \mathbb{R}$ be convex such that $\mathbb{E}[|f(X)|] < \infty$. Then $\mathbb{E}[f(X)] \geq f(\mathbb{E}[X_1], \dots, \mathbb{E}[X_d])$.*

Jensen's inequality $(\mathbb{E}[X])^2 \leq \mathbb{E}[X^2]$ and the Cauchy–Schwarz inequality $\mathbb{E}[XY] \leq (\mathbb{E}[X^2])^{1/2}(\mathbb{E}[Y^2])^{1/2}$ hold for all $\overline{\mathbb{R}}_+$-valued random variables. To see this, we can apply these inequalities with $X \wedge n$ and $Y \wedge n$, $n \in \mathbb{N}$, in place of X (resp. Y) and then let $n \to \infty$.

Let $T \neq \emptyset$ be an (index) set. A family $\{\mathcal{F}_t : t \in T\}$ of σ-fields contained in $\mathcal{F}$ is said to be *independent* if

$$\mathbb{P}(A_{t_1} \cap \cdots \cap A_{t_k}) = \mathbb{P}(A_{t_1}) \cdots \mathbb{P}(A_{t_k})$$

for any distinct $t_1, \dots, t_k \in T$ and any $A_{t_1} \in \mathcal{F}_{t_1}, \dots, A_{t_k} \in \mathcal{F}_{t_k}$. A family $\{X_t : t \in T\}$ of random variables with values in measurable spaces $(\mathbb{X}_t, \mathcal{X}_t)$ is said to be *independent* if the family $\{\sigma(X_t) : t \in T\}$ of generated σ-fields is independent.

The following result guarantees the existence of infinite sequences of independent random variables in a general setting.

Theorem B.2 *Let $(\Omega_n, \mathcal{F}_n, \mathbb{Q}_n)$, $n \in \mathbb{N}$, be probability spaces. Then there exists a unique probability measure $\mathbb{Q}$ on the space $(\times_{n=1}^{\infty}\Omega_n, \otimes_{n=1}^{\infty}\mathcal{F}_n)$ such that $\mathbb{Q}(A \times \times_{m=n+1}^{\infty}\Omega_m) = \otimes_{i=1}^{m}\mathbb{Q}_i(A)$ for all $n \in \mathbb{N}$ and $A \in \otimes_{m=1}^{n}\mathcal{F}_m$.*

Under a Borel assumption, Theorem B.2 extends to general probability measures on infinite products.

Theorem B.3 *Let $(\Omega_n, \mathcal{F}_n)$, $n \in \mathbb{N}$, be a sequence of Borel spaces and let $\mathbb{Q}_n$ be probability measures on $(\Omega_1 \times \cdots \times \Omega_n, \mathcal{F}_1 \otimes \cdots \otimes \mathcal{F}_n)$ such that $\mathbb{Q}_{n+1}(\cdot \times \Omega_{n+1}) = \mathbb{Q}_n$ for all $n \in \mathbb{N}$. Then there is a unique probability measure $\mathbb{Q}$ on $(\times_{n=1}^{\infty}\Omega_n, \otimes_{n=1}^{\infty}\mathcal{F}_n)$ such that $\mathbb{Q}(A \times \times_{m=n+1}^{\infty}\Omega_m) = \mathbb{Q}_n(A)$ for all $n \in \mathbb{N}$ and $A \in \otimes_{m=1}^{n}\mathcal{F}_m$.*

The *characteristic function* of a random vector $X = (X_1, \dots, X_d)$ in $\mathbb{R}^d$ is the function $\varphi_X : \mathbb{R}^d \to \mathbb{C}$ defined by

$$\varphi_X(t) := \mathbb{E}[\exp[-\mathbf{i}\langle X, t\rangle]], \quad t = (t_1, \dots, t_d) \in \mathbb{R}^d, \tag{B.1}$$

where $\mathbb{C}$ denotes the complex numbers and $\mathbf{i} := \sqrt{-1}$ is the imaginary unit. The *Laplace transform* of a random vector $X = (X_1, \dots, X_d)$ in $\mathbb{R}^d_+$ (a random element of $\mathbb{R}^d_+$) is the function L_X on $\mathbb{R}^d_+$ defined by

$$L_X(t) := \mathbb{E}[\exp(-\langle X, t\rangle)], \quad t = (t_1, \dots, t_d) \in \mathbb{R}^d_+. \tag{B.2}$$

Proposition B.4 (Uniqueness theorem) *Two random vectors in $\mathbb{R}^d$ (resp. in $\mathbb{R}^d_+$) have the same distribution if and only if their characteristic functions (resp. Laplace transforms) coincide.*

The Laplace transform of an $\mathbb{R}_+$-valued random variable is analytic on $(0, \infty)$. Therefore it is determined by its values on any open (non-empty) interval $I \subset (0, \infty)$ (see [78]), and Proposition B.4 yields:

Proposition B.5 *Two $\mathbb{R}_+$-valued random variables have the same distribution if and only if their Laplace transforms coincide on a non-empty open interval.*

A sequence (X_n) of finite random variables is said to *converge $\mathbb{P}$-almost surely* (shorter: $\mathbb{P}$-a.s. or just a.s.) to a random variable X if the event

$$\left\{\lim_{n\to\infty} X_n = X\right\} := \left\{\omega \in \Omega : \lim_{n\to\infty} X_n(\omega) = X(\omega)\right\}$$

has probability 1. A similar notation is used for infinite series of random variables.

Theorem B.6 (Law of large numbers) *Let $X_1, X_2, \dots$ be independent and identically distributed random variables such that $\mathbb{E}[|X_1|] < \infty$. Then the sequence $n^{-1}(X_1 + \cdots + X_n)$, $n \in \mathbb{N}$, converges almost surely to $\mathbb{E}[X_1]$.*

The following criterion for the convergence of a series with independent summands is useful.

Proposition B.7 *Let $X_n \in L^2(\mathbb{P})$, $n \in \mathbb{N}$, be independent random variables satisfying $\sum_{n=1}^{\infty} \mathbb{V}\mathrm{ar}[X_n] < \infty$. Then the series $\sum_{n=1}^{\infty}(X_n - \mathbb{E}[X_n])$ converges $\mathbb{P}$-a.s. and in $L^2(\mathbb{P})$.*

A sequence (X_n) of random variables *converges in probability* to a random variable X if $\mathbb{P}(|X_n - X| \geq \varepsilon) \to 0$ as $n \to \infty$ for each $\varepsilon > 0$. Each almost surely converging sequence converges in probability. *Markov's inequality* says that every non-negative random variable Z satisfies

$$\mathbb{P}(Z \geq \varepsilon) \leq \frac{\mathbb{E}[Z]}{\varepsilon}, \quad \varepsilon > 0,$$

and implies the following fact.

Proposition B.8 *Let $p \geq 1$ and suppose that the random variables X_n, $n \in \mathbb{N}$, converge in $L^p(\mathbb{P})$ to X. Then $X_n \to X$ in probability.*

Let $X, X_1, X_2, \ldots$ be random elements of a metric space $\mathbb{X}$ (equipped with its Borel σ-field). The sequence (X_n) is said to *converge in distribution* to X if $\lim_{n\to\infty} \mathbb{E}[f(X_n)] = \mathbb{E}[f(X)]$ for every bounded continuous function $f : \mathbb{X} \to \mathbb{R}$. One writes $X_n \xrightarrow{d} X$ as $n \to \infty$. Let ρ denote the metric on $\mathbb{X}$. A function $f : \mathbb{X} \to \mathbb{R}$ is said to be *Lipschitz* if there exists $c \geq 0$ such that

$$|f(x) - f(y)| \leq c\rho(x, y), \quad x, y \in \mathbb{X}. \tag{B.3}$$

The smallest of such c is the *Lipschitz constant* of f. The following result is proved (but not stated) in [12].

Proposition B.9 *A sequence (X_n) of random elements of a metric space $\mathbb{X}$ converges in distribution to X if and only if $\lim_{n\to\infty} \mathbb{E}[f(X_n)] = \mathbb{E}[f(X)]$ for every bounded Lipschitz function $f : \mathbb{X} \to \mathbb{R}$.*

Proposition B.10 *A sequence $(X_n)_{n\geq 1}$ of random vectors in $\mathbb{R}^d$ converges in distribution to a random vector X if and only if $\lim_{n\to\infty} \varphi_{X_n}(t) = \varphi_X(t)$ for all $t \in \mathbb{R}^d$. A sequence $(X_n)_{n\geq 1}$ of random vectors in $\mathbb{R}^d_+$ converges in distribution to a random vector X if and only if $\lim_{n\to\infty} L_{X_n}(t) = L_X(t)$ for all $t \in \mathbb{R}^d_+$.*

B.2 Mean Ergodic Theorem

Random variables $X_1, X_2, X_3 \ldots$ are said to form a *stationary sequence* if $(X_1, \ldots, X_k) \overset{d}{=} (X_2, \ldots, X_{k+1})$ for all $k \in \mathbb{N}$. The following result is well known; see e.g. [30, 63]. For completeness we provide here a simple proof, which was inspired by [67].

Theorem B.11 (Mean ergodic theorem) *Suppose $(X_n)_{n\geq 1}$ is a stationary sequence of integrable random variables. For $n \in \mathbb{N}$, set $S_n := \sum_{i=1}^n X_i$ and $A_n := S_n/n$. Then there exists a random variable Y with $A_n \to Y$ in $L^1(\mathbb{P})$ as $n \to \infty$.*

Proof In the first part of the proof we assume that there exists $c \in \mathbb{R}$ such that $|X_i| \leq c$ for all $i \in \mathbb{N}$. We show that there exists a random variable L with $A_n \to L$ a.s. It suffices to prove this in the case where $c = 1$ and $\mathbb{E}[X_1] = 0$, so assume this. Define the random variable $L := \limsup_{n\to\infty} A_n$. It is enough to show that $\mathbb{E}[L] \leq 0$, since then the same argument shows that $\mathbb{E}[\limsup_{n\to\infty}(-A_n)] \leq 0$ so that $\mathbb{E}[\liminf_{n\to\infty} A_n] \geq 0$, and hence

$$\mathbb{E}[\limsup_{n\to\infty} A_n - \liminf_{n\to\infty} A_n] \leq 0,$$

so that $\limsup_{n\to\infty} A_n = \liminf_{n\to\infty} A_n$ almost surely.

Let $\varepsilon \in (0, 1/4)$. Setting $T := \min\{n : A_n > L - \varepsilon\}$, we can and do choose $k \in \mathbb{N}$ such that $\mathbb{P}(T > k) < \varepsilon$. For $n, m \in \mathbb{N}$, define

$$A_{n,m} := m^{-1}(X_n + \cdots + X_{n+m-1})$$

and $T_n := \min\{m : A_{n,m} > L - \varepsilon\}$. In particular, $T_1 = T$. Also, T_n has the same distribution as T for all n, because $L = \limsup_{m\to\infty} A_{n,m}$ for all n. Now set

$$X_n^* := X_n + 2\mathbf{1}\{T_n > k\},$$

and note that $\mathbb{E}[X_n^*] = \mathbb{E}[X_1^*]$. Set $S_n^* := \sum_{i=1}^n X_i^*$ and $A_n^* := S_n^*/n$. Set $T^* := \min\{n : A_n^* > L-\varepsilon\}$. Then $T^* \leq T$ and if $T > k$ then $S_1^* = X_1+2 \geq 1 > L-\varepsilon$, so that $T^* = 1$. It follows that $\mathbb{P}(T^* \leq k) = 1$.

Set $M_0 := 0$ and $M_1 := T^*$. Then $M_1 \in (0, k]$ with $A_{M_1}^* > L-\varepsilon$. Repeating the argument, there exists $M_2 \in (M_1, M_1 + k]$ such that

$$(M_2 - M_1)^{-1}(S_{M_2}^* - S_{M_1}^*) > L - \varepsilon.$$

Continuing in this way we have a strictly increasing sequence of random variables $M_0, M_1, M_2, \ldots$ such that the average of X_n^* over each interval $(M_{i-1}, M_i]$ exceeds $L - \varepsilon$ for each $n \in \mathbb{N}$. Then for $m \in \mathbb{N}$ the average over $(0, M_m]$ satisfies the same inequality: indeed, setting $S_0^* = 0$ we have

$$S_{M_m}^* = \sum_{i=1}^{m} (S_{M_i}^* - S_{M_{i-1}}^*) \geq (L - \varepsilon) \sum_{i=1}^{m} (M_i - M_{i-1}) = (L - \varepsilon)M_m. \quad \text{(B.4)}$$

Given $n \in \mathbb{N}$ with $n \geq k$, at least one of the times M_i (denoted M') lies in $(n-k, n]$. Since $X_i^* \geq -1$ for all i we have $S_n^* \geq S_{M'}^* - k$ and since $A_{M'}^* \geq L-\varepsilon$ by (B.4) and $L - \varepsilon \leq 1$, we obtain

$$S_n^* \geq (L - \varepsilon)n - (L - \varepsilon)(n - M') - k \geq (L - \varepsilon)n - 2k$$

so that, for large enough n, $\mathbb{E}[A_n^*] \geq \mathbb{E}[L] - 2\varepsilon$. However, for all n we have $\mathbb{E}[A_n^*] = \mathbb{E}[X_1^*] \leq 2\varepsilon$; hence $\mathbb{E}[L] \leq 4\varepsilon$, and hence $\mathbb{E}[L] \leq 0$, as required.

Now we turn to the second part of the proof, dropping the assumption that X_1 is bounded. By taking positive and negative parts, it suffices to treat the case where $X_1 \geq 0$, so assume this. For $k, n \in \mathbb{N}$, set $X_{n,k} := \min\{X_n, k\}$ and $A_{n,k} := n^{-1} \sum_{i=1}^{n} X_{i,k}$. By the first part of the proof and dominated convergence, there exists a random variable L_k such that $A_{n,k} \to L_k$ almost surely and in L^1. Then $L_k \geq 0$, L_k is non-decreasing in k and $\mathbb{E}[L_k] \leq \mathbb{E}[X_1]$ for all k. Hence there is a limit variable $L := \lim_{k\to\infty} L_k$.

Let $\varepsilon > 0$ and choose $k > 0$ such that $\mathbb{E}[X_{1,k}] \geq \mathbb{E}[X_1] - \varepsilon$, and such that moreover $\mathbb{E}[|L - L_k|] < \varepsilon$. Then for large enough n we have

$$\mathbb{E}[|A_n - L|] \leq \mathbb{E}[|A_n - A_{n,k}|] + \mathbb{E}[|A_{n,k} - L_k|] + \mathbb{E}[|L_k - L|] < 3\varepsilon,$$

which yields the result. □

B.3 The Central Limit Theorem and Stein's Equation

A random variable N is said to be *standard normal* if its distribution has density $x \mapsto (2\pi)^{-1/2} \exp(-x^2/2)$ with respect to Lebesgue measure on $\mathbb{R}$. Its characteristic function is given by $t \mapsto \exp(-t^2/2)$ while its moments are given by

$$\mathbb{E}[N^k] = \begin{cases} (k-1)!!, & \text{if } k \text{ is even,} \\ 0, & \text{otherwise,} \end{cases} \tag{B.5}$$

where for an even integer $k \geq 2$ we define the *double factorial* of $k - 1$ by

$$(k-1)!! := (k-1) \cdot (k-3) \cdots 3 \cdot 1. \tag{B.6}$$

Note that this is the same as the number of *matchings* of $[k] := \{1, ..., k\}$ (a matching of $[k]$ is a partition of $[k]$ into disjoint blocks of size 2). Indeed, it can be easily checked by induction that

$$\operatorname{card} M(k) = \begin{cases} (k-1)!!, & \text{if } k \text{ is even,} \\ 0, & \text{otherwise,} \end{cases} \tag{B.7}$$

where $M(k)$ denotes the set of matchings of $[k]$. The moment formula (B.5) can be proved by partial integration or by writing the characteristic function of N as a power series. Taking $c > 0$ and using a change of variables we can derive from (B.5) that

$$\mathbb{E}[\exp(-cN^2)N^{2m}] = (1 + 2c)^{-m-1/2}(2m-1)!!, \quad m \in \mathbb{N}_0, \tag{B.8}$$

where $(-1)!! := 1$. A random variable X is said to have a *normal distribution* with mean $a \in \mathbb{R}$ and variance $b \geq 0$, if $X \overset{d}{=} bN + a$, where N is standard normal. A sequence $(X_n)_{n\geq 1}$ of random variables is said to satisfy the *central limit theorem* if $X_n \xrightarrow{d} N$ as $n \to \infty$.

A random vector $X = (X_1, \dots, X_d)$ is said to have a *multivariate normal distribution* if $\langle X, t\rangle$ has a normal distribution for all $t \in \mathbb{R}^d$. In this case the distribution of X is determined by the means $\mathbb{E}[X_i]$ and covariances $\mathbb{E}[X_i X_j]$, $i, j \in \{1, \dots, d\}$. Moreover, if a sequence $(X^{(n)})_{n\geq 1}$ of random vectors with a multivariate normal distribution converges in distribution to a random vector X, then X has a multivariate normal distribution.

Proposition B.12 *Let $X, X_1, X_2, \dots$ be random variables and assume that $\mathbb{E}[|X|^k] < \infty$ for all $k \in \mathbb{N}$. Suppose that*

$$\lim_{n\to\infty} \mathbb{E}[X_n^k] = \mathbb{E}[X^k], \quad k \in \mathbb{N},$$

and that the distribution of X is uniquely determined by the moments $\mathbb{E}[X^k]$, $k \in \mathbb{N}$. Then $X_n \xrightarrow{d} X$ as $n \to \infty$.

Let $\mathbf{Lip}(1)$ denote the space of Lipschitz functions $h\colon \mathbb{X} \to \mathbb{R}$ with Lipschitz constant less than or equal to 1. For a given $h \in \mathbf{Lip}(1)$ a function $g\colon \mathbb{R} \to \mathbb{R}$ is said to satisfy *Stein's equation* for h if

$$h(x) - \mathbb{E}[h(N)] = g'(x) - xg(x), \quad x \in \mathbb{R}, \tag{B.9}$$

where N is a standard normal random variable.

Proposition B.13 (Stein's equation) *Suppose that $h \in \mathbf{Lip}(1)$. Then there exists a differentiable solution g of* (B.9) *such that g' is absolutely continuous and such that $g'(x) \leq \sqrt{2/\pi}$ and $g''(x) \leq 2$ for λ_1-a.e. $x \in \mathbb{R}$, where g'' is a Radon–Nikodým derivative of g'.*

Proof We assert that the function

$$g(x) := e^{x^2/2} \int_{-\infty}^{x} e^{-y^2/2}(h(y) - \mathbb{E}[h(N)])\, dy, \quad x \in \mathbb{R},$$

is a solution. Indeed, the product rule (Proposition A.34) implies that g is absolutely continuous. Moreover, one version of the Radon–Nikodým derivative is given by

$$\begin{aligned} g'(x) &= xe^{x^2/2} \int_{-\infty}^{x} e^{-y^2/2}(h(y) - \mathbb{E}[h(N)])\, dy + e^{x^2/2}e^{-x^2/2}(h(x) - \mathbb{E}[h(N)]) \\ &= h(x) - \mathbb{E}[h(N)] + xg(x). \end{aligned}$$

Hence (B.9) holds. Since a Lipschitz function is absolutely continuous (this can be checked directly) it follows from the product rule that g' is absolutely continuous. The bounds for g' and g'' follow from some lines of calculus which we omit; see [20, Lem. 4.2] for the details. □

B.4 Conditional Expectations

Let X be a random variable and let $\mathcal{G} \subset \mathcal{F}$ be a σ-field. If there exists a $\mathcal{G}$-measurable random variable Y such that

$$\mathbb{E}[\mathbf{1}_C X] = \mathbb{E}[\mathbf{1}_C Y], \quad C \in \mathcal{G}, \tag{B.10}$$

then Y is said to be a version of the *conditional expectation* of X given $\mathcal{G}$. If Y' is another version, then it follows that $Y = Y'$, $\mathbb{P}$-a.s. If, on the other hand, Y is a version of the conditional expectation of X given $\mathcal{G}$, and Y' is another $\mathcal{G}$-measurable random variable satisfying $Y = Y'$, $\mathbb{P}$-a.s., then Y' is also a version of the conditional expectation of X, given $\mathcal{G}$. We use the notation $Y = \mathbb{E}[X \mid \mathcal{G}]$ to denote one fixed version of the conditional expectation, if it exists. If the σ-field $\mathcal{G}$ is generated by an at most countable family of pairwise disjoint sets $A_1, A_2, \dots$ of $\mathcal{F}$, whose union is Ω, then

$$\mathbb{E}[X \mid \mathcal{G}] = \sum_{n \geq 1} \mathbf{1}_{A_n} \mathbb{E}[X \mid A_n], \quad \mathbb{P}\text{-a.s.}$$

Here we use the conditional expectation $\mathbb{E}[X \mid A] := \mathbb{P}(A)^{-1}\mathbb{E}[\mathbf{1}_A X]$ of X with respect to an event $A \in \mathcal{F}$, where $0/0 := 0$.

In the general case one has the following result, which can be proved with the aid of the Radon–Nikodým theorem (Theorem A.10).

Proposition B.14 *Let X be a random variable and let $\mathcal{G} \subset \mathcal{F}$ be a σ-field.*

(i) *If X is non-negative, then $\mathbb{E}[X \mid \mathcal{G}]$ exists and has an almost surely finite version if and only if the measure $C \mapsto \mathbb{E}[\mathbf{1}_C X]$ is σ-finite on $\mathcal{G}$.*
(ii) *If $X \in L^1(\mathbb{P})$, then $\mathbb{E}[X \mid \mathcal{G}]$ exists and has an almost surely finite version.*

For $A \in \mathcal{F}$ the random variable

$$\mathbb{P}(A \mid \mathcal{G}) := \mathbb{E}[\mathbf{1}_A \mid \mathcal{G}]$$

is called (a version of the) *conditional probability* of A given $\mathcal{G}$. Let Y be a random element of a measurable space $(\mathbb{Y}, \mathcal{Y})$ and let X be a random variable. We write $\mathbb{E}[X \mid Y] := \mathbb{E}[X \mid \sigma(Y)]$ if the latter expression is defined. Further, we write $\mathbb{P}(A \mid Y) := \mathbb{E}[\mathbf{1}_A \mid Y]$ for $A \in \mathcal{F}$. If X is a random element of the space $(\mathbb{X}, \mathcal{X})$ then the mapping $(\omega, B) \mapsto \mathbb{P}(\{X \in B\} \mid Y)(\omega)$

from $\Omega\times\mathcal{X}$ to $\mathbb{R}_+$ is called the *conditional distribution* of X given Y. If this mapping can be chosen as a probability kernel from Ω to $\mathbb{X}$, it is called a *regular version* of this conditional distribution.

The conditional expectation is linear, monotone and satisfies the triangle and Jensen inequalities. The following properties can be verified immediately from the definition. Property (iii) is called the *law of total expectation* while (vi) is called the *pull out property*. If nothing else is said, then all relations concerning conditional expectations hold $\mathbb{P}$-a.s.

Theorem B.15 *Consider $\overline{\mathbb{R}}_+$-valued random variables X and Y and σ-fields $\mathcal{G}, \mathcal{G}_1, \mathcal{G}_2 \subset \mathcal{F}$. Then:*

(i) *If $\mathcal{G} = \{\emptyset, \Omega\}$, then $\mathbb{E}[X \mid \mathcal{G}] = \mathbb{E}[X]$.*
(ii) *If X is $\mathcal{G}$-measurable, then $\mathbb{E}[X \mid \mathcal{G}] = X$.*
(iii) $\mathbb{E}[\mathbb{E}[X \mid \mathcal{G}]] = \mathbb{E}[X]$.
(iv) *Suppose that $\mathcal{G}_1 \subset \mathcal{G}_2$ $\mathbb{P}$-a.s., i.e. suppose that for every $A \in \mathcal{G}_1$ there is a set $B \in \mathcal{G}_2$ with $\mathbb{P}((A\setminus B) \cup (B\setminus A)) = 0$. Then*

$$\mathbb{E}[\mathbb{E}[X \mid \mathcal{G}_2] \mid \mathcal{G}_1] = \mathbb{E}[X \mid \mathcal{G}_1].$$

(v) *Suppose that $\sigma(X)$ is independent of $\mathcal{G}$. Then $\mathbb{E}[X \mid \mathcal{G}] = \mathbb{E}[X]$.*
(vi) *Suppose that X is $\mathcal{G}$-measurable. Then $\mathbb{E}[XY \mid \mathcal{G}] = X\,\mathbb{E}[Y \mid \mathcal{G}]$.*

Let $(\mathbb{X}, \mathcal{X})$ be a measurable space and let $f \in \overline{\mathbb{R}}_+(\Omega \times \mathbb{X})$. Then, for any $x \in \mathbb{X}$, $f(x) := f(\cdot, x)$ (this is the mapping $\omega \mapsto f(\omega, x)$) is a random variable. Hence, if $\mathcal{G} \subset \mathcal{F}$ is a σ-field, we can form the conditional expectation $\mathbb{E}[f(x) \mid \mathcal{G}]$. A *measurable version* of this conditional expectation is a function $\tilde{f} \in \overline{\mathbb{R}}_+(\Omega\times\mathbb{X})$ such that $\tilde{f}(x) = \mathbb{E}[f(x) \mid \mathcal{G}]$ holds $\mathbb{P}$-a.s. for every $x \in \mathbb{X}$. Using the monotone class theorem the linearity of the conditional expectation can be extended as follows.

Lemma B.16 *Let $(\mathbb{X}, \mathcal{X}, \lambda)$ be an s-finite measure space and suppose that $f \in \overline{\mathbb{R}}_+(\Omega \times \mathbb{X})$ or $f \in L^1(\mathbb{P} \otimes \lambda)$. Let $\mathcal{G} \subset \mathcal{F}$ be a σ-field. Then there is a measurable version of $\mathbb{E}[f(x) \mid \mathcal{G}]$ satisfying*

$$\mathbb{E}\Big[\int f(x)\,\lambda(dx)\,\Big|\,\mathcal{G}\Big] = \int \mathbb{E}[f(x) \mid \mathcal{G}]\,\lambda(dx), \quad \mathbb{P}\text{-a.s.}$$

B.5 Gaussian Random Fields

Let $\mathbb{X}$ be a non-empty set, for instance a Borel subset of $\mathbb{R}^d$. A *random field* (on $\mathbb{X}$) is a family $Z = (Z(x))_{x\in\mathbb{X}}$ of real-valued random variables. Equivalently, Z is a random element of the space $\mathbb{R}^{\mathbb{X}}$ of all functions from

$\mathbb{X}$ to $\mathbb{R}$, equipped with the smallest σ-field making all projection mappings $f \mapsto f(t)$, $t \in \mathbb{X}$, measurable. It is customary to write $Z(\omega, x) := Z(\omega)(x)$ for $\omega \in \Omega$ and $x \in \mathbb{X}$. A random field $Z' = (Z'(x))_{x\in\mathbb{X}}$ (defined on the same probability space as the random field Z) is said to be a *version* of Z if $\mathbb{P}(Z(x) = Z'(x)) = 1$ for each $x \in \mathbb{X}$. In this case $Z \overset{d}{=} Z'$.

A random field $Z = (Z(x))_{x\in\mathbb{X}}$ is *square integrable* if $\mathbb{E}[Z(x)^2] < \infty$ for each $x \in \mathbb{X}$. In this case the *covariance function* K of Z is defined by

$$K(x, y) := \mathbb{E}[(Z(x) - \mathbb{E}[Z(x)])(Z(y) - \mathbb{E}[Z(y)])], \quad x, y \in \mathbb{X}.$$

This function is *non-negative definite*, that is

$$\sum_{i,j=1}^{m} c_i c_j K(x_i, x_j) \geq 0, \quad c_1, \dots, c_m \in \mathbb{R},\ x_1, \dots, x_m \in \mathbb{X},\ m \in \mathbb{N}. \tag{B.11}$$

A random field Z is said to be *Gaussian* if, for each $k \in \mathbb{N}$ and all $x_1, \dots, x_k \in \mathbb{X}$, the random vector $(Z(x_1), \dots, Z(x_k))$ has a multivariate normal distribution. Then the distribution of Z is determined by $\mathbb{E}[Z(x)]$, $x \in \mathbb{X}$, and the covariance function of Z. A random field $Z = (Z(x))_{x\in\mathbb{X}}$ is said to be *centred* if $\mathbb{E}[Z(x)] = 0$ for each $x \in \mathbb{X}$. The next theorem follows from Kolmogorov's existence theorem; see [63, Th. 6.16].

Theorem B.17 *Let $K: \mathbb{X} \times \mathbb{X} \to \mathbb{R}$ be symmetric and non-negative definite. Then there exists a centred Gaussian random field with covariance function K.*

The following result (see e.g. [59]) is an extension of the spectral theorem for symmetric non-negative matrices. It is helpful for the explicit construction of Gaussian random fields. Recall from Section A.2 that $\operatorname{supp} \nu$ denotes the support of a measure ν on a metric space.

Theorem B.18 (Mercer's theorem) *Suppose that $\mathbb{X}$ is a compact metric space. Let $K: \mathbb{X} \times \mathbb{X} \to \mathbb{R}$ be a symmetric, non-negative definite and continuous function. Let ν be a finite measure on $\mathbb{X}$. Then there exist $\gamma_j \geq 0$ and $v_j \in L^2(\nu)$, $j \in \mathbb{N}$, such that*

$$\int v_i(x) v_j(x)\, \nu(dx) = \mathbf{1}\{\gamma_i > 0\}\mathbf{1}\{i = j\}, \quad i, j \in \mathbb{N}, \tag{B.12}$$

and

$$K(x, y) = \sum_{j=1}^{\infty} \gamma_j v_j(x) v_j(y), \quad x, y \in \operatorname{supp} \nu, \tag{B.13}$$

where the convergence is absolute and uniform.

If in Mercer's theorem $j \in \mathbb{N}$ is such that $\gamma_j > 0$, then γ_j is an *eigenvalue* of K, that is $\int K(x, y)v_j(y)\, \nu(dy) = \gamma_j v_j(x)$, $x \in \operatorname{supp} \nu$. The *eigenfunction* v_j is then continuous on $\operatorname{supp} \nu$. There are no other positive eigenvalues.

A random field $(Z(x))_{x\in\mathbb{X}}$ is said to be *measurable* if $(\omega, x) \mapsto Z(\omega, x)$ is a measurable function.

Proposition B.19 *Let $\mathbb{X}$ be a locally compact separable metric space and let ν be a measure on $\mathbb{X}$ which is finite on compact sets. Let $\mathbb{X}^*$ denote the support of ν. Let $Z = (Z(x))_{x\in\mathbb{X}}$ be a centred Gaussian random field with a continuous covariance function. Then $(Z(x))_{x\in\mathbb{X}^*}$ has a measurable version.*

Proof In principle, the result can be derived from [29, Th. II.2.6]. We give here another argument based on the Gaussian nature of the random field.

The set $\mathbb{X}^*$ is closed and therefore a locally compact separable metric space in its own right; see Lemmas A.21 and A.22. Let ν^* be the measure on $\mathbb{X}^*$ defined as the restriction of ν to the measurable subsets of $\mathbb{X}^*$. It is easy to see that $\operatorname{supp} \nu^* = \mathbb{X}^*$. Therefore it is no restriction of generality to assume that $\mathbb{X} = \mathbb{X}^*$.

Let us first assume that $\mathbb{X}$ is compact. Then the assertion can be deduced from a more fundamental property of Z, namely the *Karhunen–Lòeve expansion*; see [2]. Let K be the covariance function of Z. With γ_j and v_j given as in Mercer's theorem (Theorem B.18), this expansion reads

$$Z(x) = \sum_{j=1}^{\infty} \sqrt{\gamma_j} Y_j v_j(x), \quad x \in \mathbb{X}, \tag{B.14}$$

where $Y_1, Y_2, \dots$ are independent and standard normal random variables and the convergence is in $L^2(\mathbb{P})$. Since (B.13) implies $\sum_{j=1}^{\infty} \gamma_j v_j(x)^2 < \infty$, Proposition B.7 shows that the series in (B.14) converges almost surely. Let $Z'(x)$ denote the right-hand side of (B.14), whenever the series converges. Otherwise set $Z'(x) := 0$. Then Z' is a measurable version of Z.

In the general case we find a monotone increasing sequence U_n, $n \geq 1$, of open sets with compact closures B_n and $\cup U_n = \mathbb{X}$. For $n \in \mathbb{N}$ let the measure ν_n on B_n be given as the restriction of ν to B_n and let $C_n \subset B_n$ be the support of ν_n. Let ν_n' be the measure on C_n given as the restriction of ν to C_n. Then it is easy to see that $\operatorname{supp} \nu_n' = C_n$. From the first part of the proof we know that there is a measurable version $(Z(x))_{x\in C_n}$. Since U_n is open it follows from the definition of the support of ν_n that $U_n \subset C_n$. Hence there is a measurable version of $(Z(x))_{x\in U_n}$. Since $\cup U_n = \mathbb{X}$ it is now clear how to construct a measurable version of Z. □

Appendix C

Historical Notes

1 Poisson and Other Discrete Distributions

The Poisson distribution was derived by Poisson [132] as the limit of binomial probabilities. Proposition 1.4 is a modern version of this limit theorem; see [63, Th. 5.7] for a complete statement. A certain Poisson approximation of binomial probabilities had already been used by de Moivre [111]. The early applications of the Poisson distribution were mostly directed to the "law of small numbers"; see von Bortkiewicz [17]. However, the fundamental work by de Finetti [39], Kolmogorov [77], Lévy [95] and Khinchin [71] clarified the role of the Poisson distribution as the basic building block of a pure jump type stochastic process with independent increments. Khinchin wrote in [70]: "... genau so, wie die Gauss–Laplacesche Verteilung die Struktur der stetigen stochastischen Prozesse beherrscht ..., erweist sich die Poissonsche Verteilung als elementarer Baustein des allgemeinen unstetigen (sprungweise erfolgenden) stochastischen Prozesses, was zweifellos den wahren Grund ihrer großen Anwendungsfähigkeit klarlegt." A possible English translation is: "... exactly as the Gauss–Laplace distribution governs the structure of continuous stochastic processes ..., it turns out that the Poisson distribution is the basic building block of the general discontinuous stochastic process (evolving by jumps), which undoubtedly reveals the true reason for its wide applicability."

2 Point Processes

The first systematic treatment of point processes (*discrete chaos*) on a general measurable phase space was given by Wiener and Wintner [160]. The modern approach via random counting measures (implicit in [160]) was first used by Moyal [116]. The results of this chapter along with historical comments can be found (in slightly less generality) in the monographs

[27, 62, 63, 65, 69, 103, 82, 134]. The idea of Proposition 2.7 can be traced back to Campbell [19].

3 Poisson Processes

The Poisson process on the non-negative half-line was discovered several times. In a remarkable paper Ellis [32] introduced renewal processes and derived the Gamma distribution for the special case of an exponentially distributed time between successive events. In his study of risk processes, Lundberg [97] introduced the compound Poisson process, using what is now called Kolmogorov's forward equation; see Cramér [26] for a review of Lundberg's work. A similar approach was taken by Bateman [9] to derive the Poisson distribution for the occurrence of α-particles. Erlang [33] introduced the Poisson process to model a stream of incoming telephone calls. He obtained the Poisson distribution by a limit argument. Bateman, Erlang and Lundberg all based their analysis on an (implicit) assumption of independent increments.

Newcomb [118] used a rudimentary version of a spatial Poisson process to model the locations of stars scattered at random. The great generality of Poisson processes had been anticipated by Abbe [1]; see [148] for a translation. The first rigorous derivation and definition of a spatial Poisson process (Poisson chaos) was given by Wiener [159]. A few years later Wiener and Wintner [160] introduced the Poisson process on a general phase space and called this the *completely independent* discrete chaos. The construction of Poisson processes as an infinite sum of independent mixed binomial processes (implicit in [159]), as well as Theorem 3.9, is due to Moyal [116]; see also [74] and [105]. The conditional binomial property in Proposition 3.8 (again implicit in [159]) was derived by Feller [36] in the case of a homogeneous Poisson process on the line; see also Ryll-Nardzewski [144]. Theorem 3.9 was proved by Ryll-Nardzewski [144] for homogeneous Poisson processes on the line and by Moyal [116] in the general case. Further comments on the history of the Poisson process can be found in [27, 44, 53, 63, 65, 103].

4 The Mecke Equation and Factorial Measures

Theorem 4.1 was proved by Mecke [105], who used a different (and very elegant) argument to prove that equation (4.2) implies the properties of a Poisson process. Wiener and Wintner [160] used factorial moment measures as the starting point for their theory of point processes. Proposition

4.3 is a slight generalisation of a result in [87]. Janossy measures and their relationship with moment measures were discussed in [160]. In the special case of a real phase space they were rediscovered by Bhabha [11]. The name was coined by Srinivasan [152], referring to Janossy [58]. Moment measures of random measures and factorial moment measures of simple point processes were thoroughly studied by Krickeberg [80] and Mecke [106]; see also [27] for an extensive discussion. Lemma 4.11 can be found in [116]. Proposition 4.12 can be derived from [162, Cor. 2.1].

5 Mappings, Markings and Thinnings

Mappings and markings are very special cases of so-called *cluster fields*, extensively studied by Kerstan, Matthes and Mecke in [69, 103]. The invariance of the Poisson process (on the line) under independent thinnings was observed by Rényi [136]. The general marking theorem (Theorem 5.6; see also Proposition 6.16) is due to Prékopa [133]. A special case was proved by Doob [29].

6 Characterisations of the Poisson Process

Proposition 6.7 was observed by Krickeberg [80]. A closely related result (for point processes) was derived in Wiener and Wintner [160, Sect. 12]. Theorem 6.10 was proved by Rényi [138], while the general point process version in Theorem 6.11 is due to Mönch [110]; see also Kallenberg [61]. Since a simple point process can be identified with its support, Theorem 6.10 is closely related to *Choquet capacities*; see [113, Th. 8.3] and [63, Th. 24.22]. A version of Rényi's theorem for more general phase spaces was proved by Kingman [76]. The fact that completely independent simple Poisson processes on the line are Poisson (Theorem 6.12) was noted by Erlang [33] and Bateman [9] and proved by Lévy [95] (in the homogeneous case) and by Copeland and Regan [24] (in the non-homogeneous case). In a more general Euclidean setting the result was proved in Doob [29] (in the homogeneous case) and Ryll-Nardzewski [143] (in the non-homogeneous case). For a general phase space the theorem was derived by Prékopa [133] and Moyal [116]. The general (and quite elegant) setting of a Borel state space was propagated in Kallenberg [63].

7 Poisson Processes on the Real Line

Some textbooks use the properties of Theorem 7.2 to define (homogeneous) Poisson processes on the real half-line. The theorem was proved by Doob [29], but might have been folklore ever since the Poisson process was introduced in [9, 32, 33, 97]. Feller [36] proved the conditional uniformity property of the points, a fact that is consistent with the interval theorem. Another short proof of the fact that a Poisson process has independent and exponentially distributed interarrival times can be based on the strong Markov property; see e.g. [63]. The argument given here might be new. Doob [29] discussed non-homogeneous Poisson processes in the more general context of stochastic processes with independent increments. More details on a dynamic (martingale) approach to marked point processes on the real line can be found in the monographs [18, 88]. The Poisson properties of the process of record levels (see Proposition 7.7) was observed by Dwass [31]. The result of Exercise 7.15 was derived by Rényi [137]. A nice introduction to extreme value theory is given in [139].

8 Stationary Point Processes

Stationary point processes were introduced by Wiener and Wintner [160] and are extensively studied in [27, 69, 103, 85, 157]. The pair correlation function (introduced in [160]) is a key tool of point process statistics; see e.g. [8, 23, 53]. Krickeberg [81] is a seminal book on point process statistics. Khinchin [72] attributes Proposition 8.11 to Korolyuk. Proposition 8.13 is a special case of [103, Prop. 6.3.7]. The ergodic theorem for spatial point processes was discussed in [160]. Theorem 8.14 also holds in an almost sure sense and was proved by Nguyen and Zessin [119] by using a general spatial ergodic theorem. More information on spatial ergodic theory can be found in Chap. 10 of [63].

9 The Palm Distribution

The idea of Palm distributions goes back to Palm [122]. For stationary point processes on the line the skew factorisation of Theorem 9.1 is due to Matthes [102]. His elegant approach was extended by Mecke [105] to accommodate point processes on a locally compact Abelian group. Theorem 9.4 was proved by Mecke [105]. A preliminary version for stationary Poisson processes on the line was obtained by Slivnyak [151]. The formulae of Exercise 9.4 are due to Palm [122] and Khinchin [72]. For stationary point processes on the line Proposition 9.5 was proved by Ryll-Nardzewski [145], while the general case is treated in [69]. Theorem 9.6 can be seen as a special case of the inversion formula proved in Mecke [105]. Volume biasing and debiasing is known as the *waiting time paradox*. It was studied independently by Nieuwenhuis [120] for stationary point processes on the line and by Thorisson [157], who also studied the spatial case; see the notes to Chaps. 8 and 9 in [157] for an extensive discussion and more references. Stationary Voronoi tessellations are studied in [23, 147]. Equation (9.22) is an example of a *harmonic mean formula*; see Aldous [3].

10 Extra Heads and Balanced Allocations

The extra head problem for a sequence of independent and identically distributed Bernoulli random variables was formulated and solved by Liggett [96]. This problem, as well as the point process version (10.1), are special cases of a *shift-coupling*. Thorisson [156] proved the existence of such couplings in a general group setting; see [157] for a discussion and more references. Stable allocations balancing Lebesgue measure and the stationary Poisson process were introduced and studied by Holroyd and Peres [49]. A discussion of balancing more general jointly stationary random measures can be found in [93] and [64]. Algorithm 10.6 (proposed in [48]) is a spatial version of an algorithm developed by Gale and Shapley [41] for the so-called stable marriage problem in a discrete non-spatial setting. Theorem 10.2 is taken from [84]. The modified Palm distribution was discussed in [69, Sec. 3.6] and [103, Sec. 9.1]. It was rediscovered in [120, 157].

11 Stable Allocations

The chapter is based on the article [48] by Hoffman, Holroyd and Peres. Algorithm 11.3 (taken from [49]) is another spatial version of the celebrated Gale–Shapley algorithm. In 2012 the Nobel Prize in Economics was awarded to Lloyd S. Shapley and Alvin E. Roth for the theory of stable allocations and the practice of market design. In the case of a finite measure $\mathbb{Q}$, Exercise 10.1 is a special case of [103, Prop. 6.3.7].

12 Poisson Integrals

Multiple stochastic integrals were introduced by Wiener [159] and Itô [55, 56]. The pathwise identity (12.9) was noted by Surgailis [154]. Multiple

Poisson integrals of more general integrands were studied by Kallenberg and Szulga [66]. Multiple point process integrals and an associated completeness property were discussed by Wiener and Wintner [160]. Krickeberg [80] proved a version of Prop. 12.6 for general point processes. Theorem 12.7 can be found in [92, 125, 154]. Theorems 12.14 and 12.16 as well as Corollary 12.18 are taken from [92]. The results of Exercises 12.7 and 12.8 are special cases of formulae for the product of stochastic integrals; see Kabanov [60] and Proposition 1.5.3 in [87].

13 Random Measures and Cox Processes

Doubly stochastic Poisson processes were introduced by Cox [25] and systematically studied in [43, 69, 103, 99]. Kallenberg [62, 65] gives an introduction to the general theory of random measures. Theorem 13.7 was proved by Krickeberg [79]. Theorem 13.11 was proved by Kallenberg [61]; see also Grandell [43]. The Poisson characterisation of Exercise 13.15 was proved by Fichtner [38]. The present short proof is taken from [117]. In the special case of random variables the result was found by Moran [115].

14 Permanental Processes

Permanental processes were introduced into the mathematics literature by Macchi [98, 99] as rigorous point process models for the description of bosons. In the case of a finite state space these processes were introduced and studied by Vere-Jones [158]. Theorem 14.6 is due to Macchi [98]; see also Shirai and Takahashi [150]. Theorem 14.8 was proved by Macchi [98] (in the case $\alpha = 1$) and Shirai and Takahashi [150]. Proposition 14.9 is from [150], although the present proof was inspired by [104]. Under a different assumption on the kernel it was proved in [150] that α-permanental processes exist for any $\alpha > 0$; see also [158]. A survey of the probabilistic properties of permanental and determinantal point processes can be found in [51]. Theorem 14.10 is taken from this source. The Wick formula of Lemma 14.5 can e.g. be found in [125].

We have assumed continuity of the covariance kernel to apply the classical Mercer theorem and to guarantee the existence of a measurable version of the associated Gaussian random field. However, it is enough to assume that the associated integral operator is locally of trace class; see [150].

15 Compound Poisson Processes

Proposition 15.4 can be found in Moyal [116]. Proposition 15.8 can be seen as a specific version of a general combinatorial relationship between the moments and cumulants of random variables; see e.g. [27, Chap. 5] or [125]. The explicit Lévy-Khinchin representation in Theorem 15.11 was derived by Kingman [74]. This representation also holds for Lévy processes (processes with homogeneous and independent increments), a generalisation of the subordinators discussed in Example 15.7. In this case the result was obtained in de Finetti [39], Kolmogorov [77], Lévy [95], Khinchin [71, 73] and Itô [54]. The reader might wish to read the textbook [63] for a modern derivation of this fundamental result. The present proof of Proposition 15.12 (a classical result) is taken from the monograph [65]. The shot noise Cox process from Example 15.14 was studied by Møller [109]. Exercises 15.13 and 15.15 indicate the close relationship between infinite divisibility and complete independence. Seminal contributions to the theory of infinitely divisible point processes were made by Kerstan and Matthes [68] and Lee [94]. We refer here to [62, 65, 69, 103] and to [63] for the case of random variables and Lévy processes. An early paper on the Dirichlet distribution is Ferguson [37].

16 The Boolean Model and the Gilbert Graph

The spherical Boolean model already has many features of the Boolean model with general grains (treated in Chapter 17) while avoiding the technicalities of working with probability measures on the space of compact (convex) sets. Theorem 16.4 on complete coverage can be found in Hall [45]. In the case of deterministic radii the Gilbert graph was introduced by Gilbert in [42] and was extensively studied in [126]. The process of isolated nodes is also known as the Matérn I process; see [23]. This dependent thinning procedure can be generalised in several ways; see e.g. [155].

17 The Boolean Model with General Grains

The first systematic treatment of the Boolean model was given by Matheron [101]. Theorem 17.10 is essentially from [101]. We refer to [23, 45, 147] for an extensive treatment of the Boolean model and to [101, 113, 147] for the theory of general random closed sets. Percolation properties of the Boolean model are studied in [42, 45, 107].

18 Fock Space and Chaos Expansion

The Fock space representation (Theorem 18.6) was proved in [90]. The chaos expansion of square integrable Poisson functionals (Theorem 18.10) as a series of orthogonal multiple Wiener–Itô integrals was proved by Itô [56]. The associated completeness property of multiple Poisson integrals was derived earlier in [160]; see also [159] for the Gaussian case. The present explicit version of the chaos expansion (based on the difference operators) was proved by Y. Ito [57] for homogeneous Poisson processes on the line, and in [90] for general Poisson processes. The Poincaré inequality of Theorem 18.7 was proved in Wu [161] using the *Clark–Ocone representation* of Poisson martingales. Chen [22] established this inequality for infinitely divisible random vectors with independent components.

19 Perturbation Analysis

In the context of a finite number of independent Bernoulli random variables Theorem 19.4 can be found in Esary and Proschan [34]. Later it was rediscovered by Margulis [100] and then again by Russo [142]. The Poisson version (19.3) is due to Zuyev [163] (for a bounded function f). In fact, this is nothing but Kolmogorov's forward equation for a pure birth process. Theorem 19.3 was proved (under stronger assumptions) by Molchanov and Zuyev [114]. For square integrable random variables it can be extended to certain (signed) σ-finite perturbations; see [86]. The present treatment of integrable random variables and finite signed perturbations might be new. A close relative of Theorem 19.4 for general point processes (based on a different difference operator) was derived in [14, 15]. Theorems 19.7 and 19.8 are classical results of stochastic geometry and were discovered by Miles [108] and Davy [28]. While the first result is easy to guess, Theorem 19.8 might come as a surprise. The result can be generalised to all intrinsic volumes of an isotropic Boolean model in $\mathbb{R}^d$; see [147]. The present approach via a perturbation formula is new and can be extended so as to cover the general case. The result of Exercise 19.8 is taken from [40]. Exercise 19.11 implies the classical derivative formula for independent Bernoulli random variables.

20 Covariance Identities

Mehler's formula from Lemma 20.1 was originally devised for Gaussian processes; see e.g. [121]. The present version for Poisson processes as well as Theorem 20.2 are taken from [89]. Other versions of the covariance identity of Theorem 20.2 were derived in [22, 50, 90, 129, 161]. Theorem 20.3 is closely related to the Clark–Ocone martingale representation; see [91]. The Harris–FKG inequality of Theorem 20.4 was proved by Roy [141] by reduction to the discrete version for Bernoulli random fields; see [46]. An elegant direct argument (close to the one presented here) was given by Wu [161].

21 Normal Approximation of Poisson Functionals

The fundamental Theorem 21.1 was proved by Stein [153]. Theorem 21.2 appears in the seminal paper by Peccati, Solé, Taqqu and Utzet [124] in a slightly different form. The *second order Poincaré inequality* of Theorem 21.3 was proved in [89] after Chatterjee [21] proved a corresponding result for Gaussian vectors. Abbe [1] derived a quantitative version of the normal approximation of the Poisson distribution; see Example 21.5. The normal approximation of higher order stochastic integrals and Poisson U-statistics was treated in [124] and in Reitzner and Schulte [135]. Many Poisson functionals arising in stochastic geometry have a property of *stabilisation* (local dependence); central limit and normal approximation theorems based on this property have been established in [128, 130, 131]. More examples for the application of Poisson process calculus to stochastic geometry can be found in [123].

22 Normal Approximation in the Boolean Model

Central limit theorems for intrinsic volumes and more general additive functions of the Boolean model (see Theorem 22.8) were proved in [52]. The volume was already studied in Baddeley [7]. The surface content was treated in Molchanov [112] before Heinrich and Molchanov [47] treated more general curvature-based non-negative functionals. A central limit theorem for the number of components in the Boolean model was established in [127]. Theorem 22.9 is new but closely related to a result from [52]. Using the geometric inequality [52, (3.19)] it is possible to prove that the condition from Exercise 22.4 is not only sufficient, but also necessary for the positivity (22.32) of the asymptotic variance.

References

[1] Abbe, E. (1895). Berechnung des wahrscheinlichen Fehlers bei der Bestimmung von Mittelwerthen durch Abzählen. In: Hensen, V., *Methodik der Untersuchungen bei der Plankton-Expedition der Humboldt-Stiftung.* Verlag von Lipsius & Tischer, Kiel, pp. 166–169.

[2] Adler, R. and Taylor, J.E. (2007). *Random Fields and Geometry.* Springer, New York.

[3] Aldous, D. (1988). *Probability Approximations via the Poisson Clumping Heuristic.* Springer, New York.

[4] Baccelli, F. and Błaszczyszyn, B. (2009). *Stochastic Geometry and Wireless Networks, Volume I - Theory.* NoW Publishers, Boston.

[5] Baccelli, F. and Błaszczyszyn, B. (2009). *Stochastic Geometry and Wireless Networks, Volume II - Applications.* NoW Publishers, Boston.

[6] Baccelli, F. and Brémaud, P. (2000). *Elements of Queueing Theory.* Springer, Berlin.

[7] Baddeley, A. (1980). A limit theorem for statistics of spatial data. *Adv. in Appl. Probab.* **12**, 447–461.

[8] Baddeley, A., Rubak, E. and Turner, R. (2015). *Spatial Point Patterns: Methodology and Applications with R.* Chapman & Hall and CRC Press, London.

[9] Bateman, H. (1910). Note on the probability distribution of α-particles. *Philos. Mag.* **20** (6), 704–707.

[10] Bertoin, J. (1996). *Lévy Processes.* Cambridge University Press, Cambridge.

[11] Bhabha, H.J. (1950). On the stochastic theory of continuous parametric systems and its application to electron cascades. *Proc. R. Soc. London Ser. A* **202**, 301–322.

[12] Billingsley, P. (1968). *Convergence of Probability Measures.* Wiley, New York.

[13] Billingsley, P. (1995). *Probability and Measure.* 3rd edn. Wiley, New York.

[14] Błaszczyszyn, B. (1995). Factorial moment expansion for stochastic systems. *Stoch. Proc. Appl.* **56**, 321–335.

[15] Błaszczyszyn, B., Merzbach, E. and Schmidt, V. (1997). A note on expansion for functionals of spatial marked point processes. *Statist. Probab. Lett.* **36**, 299–306.

[16] Bogachev, V.I. (2007). *Measure Theory.* Springer, Berlin.

[17] Bortkiewicz, L. von (1898). *Das Gesetz der kleinen Zahlen.* BG Teubner, Leipzig.

[18] Brémaud, P. (1981). *Point Processes and Queues.* Springer, New York.

[19] Campbell, N. (1909). The study of discontinuous phenomena. *Proc. Cambridge Philos. Soc.* **15**, 117–136.

[20] Chatterjee, S. (2008). A new method of normal approximation. *Ann. Probab.* **36**, 1584–1610.

[21] Chatterjee, S. (2009). Fluctuations of eigenvalues and second order Poincaré inequalities. *Probab. Theory Related Fields* **143**, 1–40.

[22] Chen, L. (1985). Poincaré-type inequalities via stochastic integrals. *Z. Wahrsch. verw. Gebiete* **69**, 251–277.

[23] Chiu, S.N., Stoyan, D., Kendall, W.S. and Mecke, J. (2013). *Stochastic Geometry and its Applications.* 3rd edn. Wiley, Chichester.

[24] Copeland, A.H. and Regan, F. (1936). A postulational treatment of the Poisson law. *Ann. of Math.* **37**, 357–362.

[25] Cox, D.R. (1955). Some statistical methods connected with series of events. *J. R. Statist. Soc. Ser. B* **17**, 129–164.

[26] Cramér, H. (1969). Historical review of Filip Lundberg's works on risk theory. *Scand. Actuar. J.* (suppl. 3), 6–12.

[27] Daley, D.J. and Vere-Jones, D. (2003/2008). *An Introduction to the Theory of Point Processes. Volume I: Elementary Theory and Methods, Volume II: General Theory and Structure.* 2nd edn. Springer, New York.

[28] Davy, P. (1976). Projected thick sections through multi-dimensional particle aggregates. *J. Appl. Probab.* **13**, 714–722. Correction: *J. Appl. Probab.* **15** (1978), 456.

[29] Doob, J.L. (1953). *Stochastic Processes.* Wiley, New York.

[30] Dudley, R.M. (2002). *Real Analysis and Probability.* Cambridge University Press, Cambridge.

[31] Dwass, M. (1964). Extremal processes. *Ann. Math. Statist.* **35**, 1718–1725.

[32] Ellis, R.L. (1844). On a question in the theory of probabilities. *Cambridge Math. J.* **4**, 127–133. [Reprinted in W. Walton (ed.) (1863). *The Mathematical and Other Writings of Robert Leslie Ellis.* Deighton Bell, Cambridge, pp. 173–179.]

[33] Erlang, A.K. (1909). The theory of probabilities and telephone conversations. *Nyt. Tidsskr. f. Mat. B* **20**, 33–39.

[34] Esary, J.D. and Proschan, F. (1963). Coherent structures of non-identical components. *Technometrics* **5**, 191–209.

[35] Federer, H. (1969). *Geometric Measure Theory.* Springer, New York.

[36] Feller, W. (1940). On the time distribution of so-called random events. *Phys. Rev.* **57**, 906–908.

[37] Ferguson, T.S. (1973). A Bayesian analysis of some nonparametric problems. *Ann. Statist.* **1**, 209–230.

[38] Fichtner, K.H. (1975). Charakterisierung Poissonscher zufälliger Punktfolgen und infinitesemale Verdünnungsschemata. *Math. Nachr.* **68**, 93–104.

[39] Finetti, B. de (1929). Sulle funzioni a incremento aleatorio. *Rend. Acc. Naz. Lincei* **10**, 163–168.

[40] Franceschetti, M., Penrose, M.D. and Rosoman, T. (2011). Strict inequalities of critical values in continuum percolation. *J. Stat. Phys.* **142**, 460–486.

[41] Gale, D. and Shapley, L.S. (1962). College admissions and the stability of marriage. *Amer. Math. Monthly* **69**, 9–14.

[42] Gilbert, E.N. (1961). Random plane networks. *J. Soc. Indust. Appl. Math.* **9**, 533–543.

[43] Grandell, J. (1976). *Doubly Stochastic Poisson Processes.* Lect. Notes in Math. **529**, Springer, Berlin.

[44] Guttorp, P. and Thorarinsdottir, T.L. (2012). What happened to discrete chaos, the Quenouille process, and the sharp Markov property? Some history of stochastic point processes. *Int. Stat. Rev.* **80**, 253–268.

[45] Hall, P. (1988). *Introduction to the Theory of Coverage Processes.* Wiley, New York.

[46] Harris, T.E. (1960). A lower bound for the critical probability in a certain percolation process. *Proc. Cambridge Philos. Soc.* **56**, 13–20.

[47] Heinrich, L. and Molchanov, I. (1999). Central limit theorem for a class of random measures associated with germ–grain models. *Adv. in Appl. Probab.* **31**, 283–314.

[48] Hoffman, C., Holroyd, A.E. and Peres, Y. (2006). A stable marriage of Poisson and Lebesgue. *Ann. Probab.* **34**, 1241–1272.

[49] Holroyd, A.E. and Peres, Y. (2005). Extra heads and invariant allocations. *Ann. Probab.* **33**, 31–52.

[50] Houdré, C. and Privault, N. (2002). Concentration and deviation inequalities in infinite dimensions via covariance representations. *Bernoulli* **8**, 697–720.

[51] Hough, J.B., Krishnapur, M., Peres, Y. and Virág, B. (2006). Determinantal processes and independence. *Probab. Surv.* **3**, 206–229.

[52] Hug, D., Last, G. and Schulte, M. (2016). Second order properties and central limit theorems for geometric functionals of Boolean models. *Ann. Appl. Probab.* **26**, 73–135.

[53] Illian, J., Penttinen, A., Stoyan, H. and Stoyan, D. (2008). *Statistical Analysis and Modelling of Spatial Point Patterns.* Wiley, Chichester.

[54] Itô, K. (1941). On stochastic processes (I). *Jpn. J. Math.* **18**, 261–301.

[55] Itô, K. (1951). Multiple Wiener integral. *J. Math. Soc. Japan* **3**, 157–169.

[56] Itô, K. (1956). Spectral type of the shift transformation of differential processes with stationary increments. *Trans. Amer. Math. Soc.* **81**, 253–263.

[57] Ito, Y. (1988). Generalized Poisson functionals. *Probab. Theory Related Fields* **77**, 1–28.

[58] Janossy, L. (1950). On the absorption of a nucleon cascade. *Proc. R. Irish Acad. Sci. Sec. A* **53**, 181–188.

[59] Jörgens, K. (1982). *Linear Integral Operators.* Pitman, Boston.

[60] Kabanov, Y.M. (1975). On extended stochastic integrals. *Theory Probab. Appl.* **20**, 710–722.

[61] Kallenberg, O. (1973). Characterization and convergence of random measures and point processes. *Z. Wahrsch. verw. Gebiete* **27**, 9–21.

[62] Kallenberg, O. (1986). *Random Measures.* 4th edn. Akademie-Verlag and Academic Press, Berlin and London.

[63] Kallenberg, O. (2002). *Foundations of Modern Probability.* 2nd edn. Springer, New York.

[64] Kallenberg, O. (2011). Invariant Palm and related disintegrations via skew factorization. *Probab. Theory Related Fields* **149**, 279–301.

[65] Kallenberg, O. (2017). *Random Measures, Theory and Applications.* Springer, Cham.

[66] Kallenberg, O. and Szulga, J. (1989). Multiple integration with respect to Poisson and Lévy processes. *Probab. Theory Related Fields* **83**, 101–134.

[67] Keane, M.S. (1991). Ergodic theory and subshifts of finite type. In: Bedford, T., Keane M. and Series, C. (eds.) *Ergodic Theory, Symbolic Dynamics and Hyperbolic Spaces.* Oxford University Press, Oxford.

[68] Kerstan, J., and Matthes, K. (1964). Stationäre zufällige Punktfolgen II. *Jahresber. Deutsch. Math. Ver.* **66**, 106–118.

[69] Kerstan, J., Matthes, K. and Mecke, J. (1974). *Unbegrenzt Teilbare Punktprozesse.* Akademie-Verlag, Berlin.

[70] Khinchin, A.Y. (1933). *Asymptotische Gesetze der Wahrscheinlichkeitsrechnung.* Springer, Berlin.

[71] Khinchin, A.Y. (1937). A new derivation of one formula by P. Lévy. *Bull. Moscow State Univ.* **1**, 1–5.

[72] Khinchin, A.Y. (1955). *Mathematical Methods in the Theory of Queuing* (in Russian). Trudy Mat. Inst. Steklov **49**. English transl. (1960): Griffin, London.

[73] Khinchin, A.Y. (1956). Sequences of chance events without after-effects. *Theory Probab. Appl.* **1**, 1–15.

[74] Kingman, J.F.C. (1967). Completely random measures. *Pacific J. Math.* **21**, 59–78.

[75] Kingman, J.F.C. (1993). *Poisson Processes.* Oxford University Press, Oxford.

[76] Kingman, J.F.C. (2006). Poisson processes revisited. *Probab. Math. Statist.* **26**, 77–95.

[77] Kolmogorov, A.N. (1932). Sulla forma generale di un processo stocastico omogeneo. *Atti Accad. Naz. Lincei* **15**, 805–808.

[78] Krantz, S. and Parks, H.R. (2002). *A Primer of Real Analytic Functions.* Birkhäuser, Boston.

[79] Krickeberg, K. (1972). The Cox process. *Sympos. Math.* **9**, 151–167.

[80] Krickeberg, K. (1974). Moments of point processes. In: Harding, E.F. and Kendall, D.G. (eds.) *Stochastic Geometry.* Wiley, London, pp. 89–113.

[81] Krickeberg, K. (1982). Processus ponctuels en statistique. In: Hennequin, P. (ed.) École d'été de probabilités de Saint-Flour X - 1980. *Lect. Notes in Math.* **929**, Springer, Berlin, pp. 205–313.

[82] Krickeberg, K. (2014). *Point Processes: A Random Radon Measure Approach.* Walter Warmuth Verlag, Berlin. (Augmented with several Scholia by Hans Zessin.)

[83] Kyprianou, A. (2006). *Introductory Lectures on Fluctuations of Lévy Processes with Applications.* Springer, Berlin.

[84] Last, G. (2006). Stationary partitions and Palm probabilities. *Adv. in Appl. Probab.* **37**, 603–620.

[85] Last, G. (2010). Modern random measures: Palm theory and related models. In: Kendall, W. and Molchanov, I. (eds.) *New Perspectives in Stochastic Geometry.* Oxford University Press, Oxford, pp. 77-110.

[86] Last, G. (2014). Perturbation analysis of Poisson processes. *Bernoulli* **20**, 486–513.

[87] Last, G. (2016). Stochastic analysis for Poisson processes. In: Peccati, G. and Reitzner, M. (eds.) *Stochastic Analysis for Poisson Point Processes.* Springer, Milan, pp.1–36.

[88] Last, G. and Brandt, A. (1995). *Marked Point Processes on the Real Line: The Dynamic Approach.* Springer, New York.

[89] Last, G., Peccati, G. and Schulte, M. (2016). Normal approximation on Poisson spaces: Mehler's formula, second order Poincaré inequalities and stabilization. *Probab. Theory Related Fields* **165**, 667–723.

[90] Last, G. and Penrose, M.D. (2011). Poisson process Fock space representation, chaos expansion and covariance inequalities. *Probab. Theory Related Fields* **150**, 663–690.

[91] Last, G. and Penrose, M.D. (2011). Martingale representation for Poisson processes with applications to minimal variance hedging. *Stoch. Proc. Appl.* **121**, 1588–1606.

[92] Last, G., Penrose, M.D., Schulte, M. and Thäle, C. (2014). Moments and central limit theorems for some multivariate Poisson functionals. *Adv. in Appl. Probab.* **46**, 348–364.

[93] Last, G. and Thorisson, H. (2009). Invariant transports of stationary random measures and mass-stationarity. *Ann. Probab.* **37**, 790–813.

[94] Lee, P.M. (1967). Infinitely divisible stochastic processes. *Z. Wahrsch. verw. Gebiete* **7**, 147–160.

[95] Lévy, P. (1934). Sur les intégrales dont les éléments sont des variables aléatoires indépendantes. *Ann. Scuola Norm. Sup. Pisa (Ser. II)* **3**, 337–366

[96] Liggett, T.M. (2002). Tagged particle distributions or how to choose a head at random. In: Sidoravicious, V. (ed.) *In and Out of Equlibrium.* Birkhäuser, Boston, pp. 133-162.

[97] Lundberg, F. (1903). *I. Approximerad Framställning av Sannolikhetsfunktionen. II. Återförsäkring av Kollektivrisker.* Akad. Afhandling, Almqvist & Wiksell, Uppsala.

[98] Macchi, O. (1971). Distribution statistique des instants d'émission des photoélectrons d'une lumière thermique. *C.R. Acad. Sci. Paris Ser. A* **272**, 437–440.

[99] Macchi, O. (1975). The coincidence approach to stochastic point processes. *Adv. in Appl. Probab.* **7**, 83–122.

[100] Margulis, G. (1974). Probabilistic characteristics of graphs with large connectivity. *Problemy Peredachi Informatsii* **10**, 101–108.

[101] Matheron, G. (1975). *Random Sets and Integral Geometry.* Wiley, New York.

[102] Matthes, K. (1964). Stationäre zufällige Punktfolgen I. *Jahresber. Dtsch. Math.-Ver.* **66**, 66–79.

[103] Matthes, K., Kerstan, J. and Mecke, J. (1978). *Infinitely Divisible Point Processes.* Wiley, Chichester (English edn. of [69]).

[104] McCullagh, P. and Møller, J. (2006). The permanental process. *Adv. in Appl. Probab.* **38**, 873–888.

[105] Mecke, J. (1967). Stationäre zufällige Maße auf lokalkompakten Abelschen Gruppen. *Z. Wahrsch. verw. Geb.* **9**, 36–58.

[106] Mecke, J. (2011). *Random Measures: Classical Lectures.* Walter Warmuth Verlag.

[107] Meester, R. and Roy, R. (1996). *Continuum Percolation*. Cambridge University Press, Cambridge.

[108] Miles, R.E. (1976). Estimating aggregate and overall characteristics from thick sections by transmission microscopy. *J. Microsc.* **107**, 227–233.

[109] Møller, J. (2003). Shot noise Cox processes. *Adv. in Appl. Probab.* **35**, 614–640.

[110] Mönch, G. (1971). Verallgemeinerung eines Satzes von A. Rényi. *Studia Sci. Math. Hung.* **6**, 81–90.

[111] Moivre, A. de (1711). On the measurement of chance, or, on the probability of events in games depending upon fortuitous chance. *Phil. Trans.* **329** (Jan.-Mar.) English transl. (1984): *Int. Stat. Rev.* **52**, 229–262.

[112] Molchanov, I. (1995). Statistics of the Boolean model: from the estimation of means to the estimation of distributions. *Adv. in Appl. Probab.* **27**, 63–86.

[113] Molchanov, I. (2005). *Theory of Random Sets*. Springer, London.

[114] Molchanov, I. and Zuyev, S. (2000). Variational analysis of functionals of Poisson processes. *Math. Operat. Res.* **25**, 485–508.

[115] Moran, P.A.P. (1952). A characteristic property of the Poisson distribution. *Proc. Cambridge Philos. Soc.* **48**, 206–207.

[116] Moyal, J.E. (1962). The general theory of stochastic population processes. *Acta Math.* **108**, 1–31.

[117] Nehring, B. (2014). A characterization of the Poisson process revisited. *Electron. Commun. Probab.* **19**, 1–5.

[118] Newcomb, S. (1860). Notes on the theory of probabilities. *The Mathematical Monthly* **2**, 134–140.

[119] Nguyen, X.X. and Zessin, H. (1979). Ergodic theorems for spatial processes. *Z. Wahrsch. verw. Geb.* **48**, 133–158.

[120] Nieuwenhuis, G. (1994). Bridging the gap between a stationary point process and its Palm distribution. *Stat. Neerl.* **48**, 37–62.

[121] Nourdin, I. and Peccati, G. (2012). *Normal Approximations with Malliavin Calculus: From Stein's Method to Universality*. Cambridge Tracts in Mathematics. Cambridge University Press, Cambridge.

[122] Palm, C. (1943). Intensity variations in telephone traffic. *Ericsson Technics* **44**, 1–189. English transl. (1988): North-Holland, Amsterdam.

[123] Peccati, G. and Reitzner, M. (eds.) (2016). *Stochastic Analysis for Poisson Point Processes: Malliavin Calculus, Wiener–Itô Chaos Expansions and Stochastic Geometry*. Bocconi & Springer Series 7. Springer.

[124] Peccati, G., Solé, J.L., Taqqu, M.S. and Utzet, F. (2010). Stein's method and normal approximation of Poisson functionals. *Ann. Probab.* **38**, 443–478.

[125] Peccati, G. and Taqqu, M. (2011). *Wiener Chaos: Moments, Cumulants and Diagrams: A Survey with Computer Implementation*. Springer, Milan.

[126] Penrose, M. (2003). *Random Geometric Graphs*. Oxford University Press, Oxford.

[127] Penrose, M.D. (2001). A central limit theorem with applications to percolation, epidemics and Boolean models. *Ann. Probab.* **29**, 1515–1546.

[128] Penrose, M.D. (2007). Gaussian limits for random geometric measures. *Electron. J. Probab.* **12** (35), 989–1035.

[129] Penrose, M.D. and Wade, A.R. (2008). Multivariate normal approximation in geometric probability. *J. Stat. Theory Pract.* **2**, 293–326.

[130] Penrose, M.D. and Yukich, J.E. (2001). Central limit theorems for some graphs in computational geometry. *Ann. Appl. Probab.* **11**, 1005–1041.

[131] Penrose, M.D. and Yukich, J.E. (2005). Normal approximation in geometric probability. In: Barbour, A.D. and Chen, L.H.Y. (eds.) *Stein's Method and Applications.* World Scientific, Singapore, pp. 37–58.

[132] Poisson, S.D. (1837). *Recherches sur la Probabilité des Judgements en Matière Criminelle et en Matière Civile, Précédées des Règles Générales du Calcul des Probabilités.* Bachelier, Paris.

[133] Prékopa, A. (1958). On secondary processes generated by a random point distribution of Poisson type. *Annales Univ. Sci. Budapest de Eotvos Nom. Sectio Math.* **1**, 153–170.

[134] Reiss, R.-D. (1993). *A Course on Point Processes.* Springer, New York.

[135] Reitzner, M. and Schulte, M. (2012). Central limit theorems for U-statistics of Poisson point processes. *Ann. Probab.* **41**, 3879–3909.

[136] Rényi, A. (1956). A characterization of Poisson processes. *Magyar Tud. Akad. Mat. Kutató Int. Közl* **1**, 519–527.

[137] Rényi, A. (1962). Théorie des éléments saillants d'une suite d'observations. *Annales scientifiques de l'Université de Clermont 2, tome 8, Mathématiques* **2**, 7–13.

[138] Rényi, A. (1967). Remarks on the Poisson process. *Studia Sci. Math. Hung.* **2**, 119–123.

[139] Resnick, S.I. (1987). *Extreme Values, Regular Variation and Point Processes.* Springer, New York.

[140] Revuz, D. and Yor, M. (1999). *Continuous Martingales and Brownian Motion.* Springer, Berlin.

[141] Roy, R. (1990). The Russo–Seymour–Welsh theorem and the equality of critical densities and the "dual" critical densities for continuum percolation on $\mathbb{R}^2$. *Ann. Probab.* **18**, 1563–1575.

[142] Russo, L. (1981). On the critical percolation probabilities. *Z. Wahrsch. verw. Geb.* **56**, 229–237.

[143] Ryll-Nardzewski, C. (1953). On the non-homogeneous Poisson process (I). *Studia Math.* **14**, 124–128.

[144] Ryll-Nardzewski, C. (1954). Remarks on the Poisson stochastic process (III). (On a property of the homogeneous Poisson process.) *Studia Math.* **14**, 314–318.

[145] Ryll-Nardzewski, C. (1961). Remarks on processes of calls. *Proc. 4th Berkeley Symp. on Math. Statist. Probab.* **2**, 455–465.

[146] Schneider, R. (2013). *Convex Bodies: The Brunn–Minkowski Theory.* 2nd (expanded) edn. Cambridge University Press, Cambridge.

[147] Schneider, R. and Weil, W. (2008). *Stochastic and Integral Geometry.* Springer, Berlin.

[148] Seneta, E. (1983). Modern probabilistic concepts in the work of E. Abbe and A. de Moivre. *Math. Sci.* **8**, 75–80.

[149] Serfozo, R. (1999). *Introduction to Stochastic Networks*. Springer, New York.

[150] Shirai, T. and Takahashi, Y. (2003). Random point fields associated with certain Fredholm determinants I: fermion, Poisson and boson point processes. *J. Funct. Anal.* **205**, 414–463.

[151] Slivnyak, I.M. (1962). Some properties of stationary flows of homogeneous random events. *Theory Probab. Appl.* **7**, 336–341.

[152] Srinivasan, S.K. (1969). *Stochastic Theory and Cascade Processes.* American Elsevier, New York.

[153] Stein, C. (1972). A bound for the error in the normal approximation to the distribution of a sum of dependent random variables. In: Le Cam, L., Neyman, J. and Scott, E.L. (eds.) *Proceedings of the Sixth Berkeley Symposium on Mathematical Statistics and Probability, Vol. 2: Probability Theory.* University of Berkeley Press, Berkeley, pp. 583–602.

[154] Surgailis, D. (1984). On multiple Poisson stochastic integrals and associated Markov semigroups. *Probab. Math. Statist.* **3**, 217–239.

[155] Teichmann, J., Ballani, F. and Boogaart, K.G. van den (2013). Generalizations of Matérn's hard-core point processes. *Spat. Stat.* **9**, 33–53.

[156] Thorisson, H. (1996). Transforming random elements and shifting random fields. *Ann. Probab.* **24**, 2057–2064.

[157] Thorisson, H. (2000). *Coupling, Stationarity, and Regeneration.* Springer, New York.

[158] Vere-Jones, D. (1997). Alpha permanents and their applications to multivariate Gamma, negative binomial and ordinary binomial distributions. *New Zealand J. Math.* **26**, 125–149.

[159] Wiener, N. (1938). The homogeneous chaos. *Amer. J. Math.* **60**, 897–936.

[160] Wiener, N. and Wintner A. (1943). The discrete chaos. *Amer. J. Math.* **65**, 279–298.

[161] Wu, L. (2000). A new modified logarithmic Sobolev inequality for Poisson point processes and several applications. *Probab. Theory Related Fields* **118**, 427–438.

[162] Zessin, H. (1983). The method of moments for random measures. *Z. Wahrsch. verw. Geb.* **83**, 395–409.

[163] Zuyev, S.A. (1992). Russo's formula for Poisson point fields and its applications. *Diskretnaya Matematika* **4**, 149–160 (in Russian). English transl. (1993): *Discrete Math. Appl.* **3**, 63–73.

Index